Preparation and Properties of Stereoregular Polymers

NATO ADVANCED STUDY INSTITUTES SERIES

*Proceedings of the Advanced Study Institute Programme, which aims
at the dissemination of advanced knowledge and
the formation of contacts among scientists from different countries*

The series is published by an international board of publishers in conjunction
with NATO Scientific Affairs Division

A	Life Sciences	Plenum Publishing Corporation
B	Physics	London and New York
C	Mathematical and Physical Sciences	D. Reidel Publishing Company Dordrecht, Boston and London
D	Behavioral and Social Sciences	Sijthoff International Publishing Company Leiden
E	Applied Sciences	Noordhoff International Publishing Leiden

Series C – Mathematical and Physical Sciences

Volume 51 – Preparation and Properties of Stereoregular Polymers

Preparation and Properties of Stereoregular Polymers

Based upon the Proceedings of the NATO Advanced Study Institute held at Tirrennia, Pisa, Italy, October 3–14, 1978

edited by

R O B E R T W. L E N Z
University of Massachusetts, Amherst, U.S.A.

and

F R A N C E S C O C I A R D E L L I
University of Pisa, Italy

D. Reidel Publishing Company

Dordrecht : Holland / Boston : U.S.A. / London : England

Published in cooperation with NATO Scientific Affairs Division

Library of Congress Cataloging in Publication Data

Nato Advanced Study Institute, Tirrenia, Italy, 1978.
    Preparation and properties of stereoregular polymers.

        (NATO advanced study institute series : Series C, Mathematical and physical
sciences ; v. 51)
        Includes index.
        1.   Polymers and polymerization–Congresses.   2.   Stereochemistry–Congresses.
I.   Lenz, Robert W.   II.   Ciardelli, Francesco.   III.   Title.   IV.   Series.
QD380.N35   1978   547′.84   79–24043

ISBN-13:978-94-011-7564-7   e-ISBN-13:978-94-011-7562-3
DOI: 10.1007/978-94-011-7562-3

Published by D. Reidel Publishing Company
P.O. Box 17, Dordrecht, Holland

Sold and distributed in the U.S.A., Canada, and Mexico
by D. Reidel Publishing Company, Inc.
Lincoln Building, 160 Old Derby Street, Hingham, Mass. 02043, U.S.A.

TABLE OF CONTENTS

FOREWORD

This book contains the texts of the main lectures presented
at the NATO Advanced Studies Institute on "Advances in Preparation
and Properties of Stereoregular Polymers" held at Tirrenia near
Pisa, Italy, from October 3 to 14, 1978. A few contributed papers
have also been included because they were concerned with topics
not included in the main lectures.

The primary objective of the Institute was to assist in the
further development of stereoregular polymers because of the
ever-increasing demand for new products with exceptional chemical
and physical properties. This need has reawakened interest in the
field. Indeed there is now a rapidly increasing activity in the
study of stereoregular polymerization and the preparation of
structurally-ordered polymers with the aim of achieving appreciable improvements in existing polymeric materials through new
developments in synthesis and properties as well as in discovering
new polymeric structures.

In order to achieve these objectives, a broad interdisciplinary cooperation among scientists involved in investigations on
the design, synthesis, characterization and application of stereoregular or structurally-ordered polymers will be necessary.
Accordingly, in this NATO Institute a selected group of polymer
chemists and physicists were brought together to open new avenues
of communication for the promotion of cooperative research. The
extent to which this goal was reached cannot be evaluated as yet,
but the large attendance of the participants during all of the
lectures throughout the duration of the Institute, and the fruit-
ful discussions which occurred between chemists and physicists,
may have already given us a positive answer about the success of
the program. Unfortunately the organization of the book cannot

reflect entirely the degree of integration achieved at the Institute, but it was necessary instead to organize the subjects treated in a more classical manner under the two general areas as follows: Section 1 - "Synthesis of Stereoregular Polymers and the Mechanism of Stereospecific Polymerization", and Section 2 - "Structure and Properties of Stereoregular Polymers".

The most difficult task in this connection is to maintain an integration of subject matter, among all aspects of the science of stereoregular polymers, in transferring the Institute program into printing. During the Institute itself this integration was achieved in treating synthetic procedures and mechanisms of stereospecific and stereoselective (or stereoelective) polymerization, with chemical transformations and the physical properties of stereoregular polymers (with particular emphasis on characterization, solution, and solid-state properties) and mechanical properties. By this approach, a continuous spectrum of research activities extending from synthesis to structure, structure to properties, and properties to application, was apparent. This spectrum is the characteristic which makes polymer science unique among the sciences, and it is of basic importance that all polymer scientists have an appreciation for their place and their interdependence within the spectrum.

We feel that the speakers have been able to transfer a feeling for this spectrum , and their place in it, to the printed page in this book. Hence it may not be too optimistic on our part to expect that the book will provide the reader, as the Institute did the participants, with the basic knowledge and the inspiration necessary for the continuing development of the science of stereoregular and structurally-ordered polymers, and we hope it will continue to promote cooperative research programs and interdisciplinary approaches to defined problems.

Certainly continued progress is needed in areas such as the improvement of existing polymerization processes, through the discovery of new catalytic systems with much higher activity and stereospecificity, and in the improvement of mechanical and stability properties of existing polymeric materials. Increasing attention is also being directed toward the custom synthesis of polymers with highly-defined molecular and morphological structure for use in a wide variety of new applications such as catalysis, drug-delivery systems, electronics, super-high tenacity fibers (based upon the newly-discovered concept of liquid-crystal order in polyesters and polyamides), processed foods, metal recovery, secondary recovery of oil, ecological and environmental control applications, and many others. Indeed, the field of customs designed macromolecules is destined to make an increasingly important contribution to mankind in the future.

We cannot conclude this short introduction without expressing our warm thanks to all speakers and participants who enthusiastically attended the twelve-day Institute and actively participated in the scientific program with both presentations and stimulating discussions.

Particular thanks are due to those who agreed to write the chapters for this book and who succeeded so well in balancing the basic information with the advanced aspects.

Our thanks are also due to the NATO Scientific Affairs Division for their generous support in covering the travel and living expenses of all speakers and of many student participants. We are also indebted to the National Research Council of Italy, the National Science Foundation of the United States, and to the companies in the United States and in Italy for their interest and their support towards the organizational expenses.

R.W.Lenz
Amherst, Mass., U.S.A.

F.Ciardelli
Pisa, Italy

PARTICIPANTS

Aklonis, J.   Dept. of Chemistry, University of Southern
                             California, Los Angeles, California, 90007,
                             U.S.A.
Alyürük, K.   Dept. of Chemistry, Middle East Technical
                             University, Ankara, Turkey
Ballard, D. S. H.   I.C.I. Corporate Laboratory, The Heath-
                             Runcorn Cheshire, United Kingdom
Benedetti, E.   Università Pisa, Italy
Bertucci, C.   CNR Centro Macromolecole Stereoordinate ed
                             Otticamente Attive Pisa, Italy
Blunt, H. W.   Hercules Research Center, Wilmington,
                             Delaware, U.S.A.
Bohn, W.   Farbwerke Hoechst, Frankfurt, West Germany
Brack, W.   Labofina, Bruxelles, Belgium
Brosse, J. C.   Laboratoire de Chimie Organique Macro-
                             molecules, Université du Maine, Route de
                             Laval, 7200 Le Mans, France
Butler, G. B.   Department of Chemistry, University of
                             Florida, Gainesville, Florida, 32611, U.S.A.
Bywater, S.   Chemistry Division, National Research
                             Council, Ottawa, Canada
Caunt, A. D.   I.C.I. Plastic Division, Hertz, United
                             Kingdom
Carlini, C.   Istituto di Chimica Organica Industriale,
                             Università Pisa, Italy
Chien, J. C.   Department of Chemistry, University of
                             Massachusetts, Amherst, Massachusetts 01003,
                             U.S.A.
Ciardelli, F.   Istituto di Chimica Organica Industriale,
                             Università, Pisa, Italy
Collette, J. W.   Centre Research and Development Dept., E.I.
                             du Pont de Nemours, Wilmington, Delaware,
                             U.S.A.
Conciatori, A. B.   Celanese Research Corporation, Summit, New
                             Jersey, U.S.A.
Corradini, P.   Istituto Chimico, Università, Napoli, Italy

Dandge, D. K.   Institute for Kemiindustry, Technical
                           University of Denmark, Lyngby, Denmark
Da Silva, M. A.   Departamento de Quimica, Universidade,
                           Coimbra, Portugal
Deslandes, Y.   Department of Chemistry, Universiti de
                           Montreal, C.P. 6210 Succ. A, Montreal,
                           Canada
Eisenbach, C.   Institut für Makromolekulare Chemie, Univ.
                           of Freiburg, Stefan-Meier-Str. 3, D7800
                           Freiburg i. Br., West Germany
Farina, M.   Istituto di Chimica Industriale, Università,
                           Milano, Italy
Fatti, G.   Università, Pisa, Italy
Figueiredo, J.L.C.C.   Centro de Engenharia Quimica, Faculdade de
                           Engenharia, Rua das Bragas, Universidade
                           Oporto, Porto, Portugal
Filippone, M.   Solvay & Cie. S.A., Rosignano Solvay, Italy
Fink, G.   Institut für Technische Chemie, Technische
                           Universität, Lichtenbergstr. 4, D8046
                           Garching bei München, West Germany
Fles, D.   INA Industrija Nafte, Zagreb, Yugoslavia
Frias, J.   OGFE (JNICT), Universidade, Lisboa,
                           Portugal
Frias, M.   OGFE (JNICT), Universidade, Lisboa,
                           Portugal
Gauthier, J. M.   Laboratoire de Chimie Organique Macro-
                           molecules, Université du Maine, Route de
                           Laval, 72000 Le Mans, France
Giacomelli, G.   Università, Pisa, Italy
Goethals, E. J.   Laboratorium voor Organische Chemie, Rijks-
                           universiteit, Krygslaan 271-S4, B-9000
                           Gent, Belgium
Guerra, G.   Istituto Chimico, Università, Napoli, Italy
Harwood, H. J.   Institute of Polymer Science, University of
                           Akron, Akron, Ohio, 44325, U.S.A.
Hatada, K.   Dept. of Chemistry, University, Osaka, Japan
Heggs, T. G.   I.C.I. Plastic Division, Hertz, United
                           Kingdom
Höcker, H.   Institut für Organische Chemie, Universität
                           München, Karlstrasse 23, D-8000 München 2,
                           West Germany
Hogen Esch, T.   Department of Chemistry, University of
                           Florida, Gainesville, Florida, 32611, U.S.A.
Hvilsted, S.   Chemical Engineering Department, University
                           of Massachusetts, Amherst, Massachusetts,
                           01003, U.S.A.
Herrington, D.R.   Standard Oil Ohio, Cleveland, Ohio, U.S.A.
Johnson, B. H.   Exxon Chemical Company, Plastics Technology
                           Division, Baytown, Texas, U.S.A.

Julemont, M.	Laboratoire de Chimie Macromoléculaire et Catalyse Organique, Université, Liège, Belgium
Karasz, F. C.	Polymer Science and Engineering Department, University of Massachusetts, Amherst, Massachusetts, 01003, U.S.A.
Keii, T.	Department of Chemical Engineering, Tokyo Institute of Technology, Ookayama, Japan
Klesper, E.	Institut für Makromolekulare Chemie, Univ. of Freiburg, Stefan-Meier-Str. 3, D7800 Freiburg i. Br., West Germany
Koide, N.	Department of Chemistry, Science University of Tokyo, Tokyo, Japan
Kops, J.	Instituttet for Kemiindustri, Dt H Building 227, Technical University of Denmark, DK-2800 Lyngby, Denmark
Kresge, E.	Exxon Chemical Company, Linden, New Jersey, 07036, U.S.A.
Kuiper, J.	Koninklijke/Shell-Laboratorium, Amsterdam Netherlands
Lardicci, L.	Istituto di Chimica Organica, Università, Pisa, Italy
Leborgne, A.	Laboratoire de Chimie Macromoléculaire, Université Pierre et Marie Curie, 75230 Paris Cedex 05, France
Lenz, R. W.	Chemical Engineering Department, Goessmann Lab, University of Massachusetts, Amherst, Massachusetts, 01003, U.S.A.
Longo, F.	ANIC S.p.A., S. Donato, Milanese, Italy
Lora, S.	CNR, Laboratoire FRAE, Legnaro, Italy
Lorenzi, G. P.	Technisch-Chemisches Laboratorium, ETH, Zürich, Switzerland
Lotz, B.	Centre de Recherches sur les Macromolécules, 6 rue Boussingault, 67083 Strasbourg, France
Loureiro, J. M.	Departmento de Engenharia Quimica, Faculdade de Engenharia, Rua dos Bragas, Universidade Oporto, Porto, Portugal
Macknight, W. J.	Polymer Science and Engineering Department, University of Massachusetts, Amherst, Massachusetts, 01003, U.S.A.
Mah, T.	Department of Chemistry, King's College, Strand, London WC 2, England
Majumdar, R. N.	Pisa, Italy
Marchessault, R. H.	Xerox Research Centre of Canada Limited, Mississagua, Canada
Mark, J.	Department of Chemistry, University of Cincinnati, Cincinnati, Ohio, 45221, U.S.A.
Mathias, L. J.	Chemistry Dept., Auburn University, Auburn, Alabama, 36830, U.S.A.

Menicagli, R.   Università, Pisa, Italy
Millich, F.   Chemistry Department, University of Missouri
                        5100 Rockhill Road, Kansas City, Missouri,
                        64110, U.S.A.
Montaudo, G.   Istituto Chimico, Università, Cantania,
                        Italy
Mülkapt, R.   University of Braunschweig, West Germany
Müller, A. H. E.   Institut für Physikalische Chemie,
                        Universität Mainz, D-6500, Mainz, West
                        Germany
Napolitano, R.   Istituto Chimico, Università, Napoli, Italy
Ng, H.   Department of Chemistry, University of
                        Toronto, Toronto, Ontario, M5S 1A4, Canada
Noristi, L.   Montedison Plastic Division, Ferrara, Italy
Pedemonte, E.   Istituto di Chimica Industriale, Università
                        Genova, Italy
Pelosi, L. F.   Elastomer Chemistry Dept., E. I. du Pont de
                        Nemours, Wilmington, Delaware, U.S.A.
Perez, S.   C.E.R.M.A.V. (CNRS) Cedex 53, Grenoble
                        38000, France
Pezzin, G.   CNR, Centro Fisica delle Macromolecole,
                        Bologna, Italy
Piccolo, O.   Technisch-Chemisches Laboratorium, ETH,
                        Zürich, Switzerland
Pino, P.   Technisch-Chemisches Laboratorium, ETH,
                        Universität Str. 6, 8092 Zürich, Switzer-
                        land
Pirozzi, B.   Istituto Chimico, Università, Napoli, Italy
Pizzoli, M.   Istituto Ciamician, Università, Bologna,
                        Italy
Prud'homme, R. E.   Chemistry Department, Laval University,
                        Quebec 10, Quebec, G1K 7P4, Canada
Rodrigues, A.   Departmento de Engenharia Quimica, Faculdade
                        de Engenharia, Rua dos Bragas, University
                        Oporto, Porto, Portugal
Rudin, A.   Chemistry Department, University of Water-
                        loo, Waterloo, Ontario, N2L 3G1, Canada
Russo, S.   CNR, Centro Chimica-Fisica di Macromolecole
                        Sintetiche e Naturali, Genova, Italy
Rytter, E.   Institute of Inorganic Chemistry, University
                        of Trondheim, N-7034, Trandheim-WTH, Norway
Salvadori, P.   CNR, Centro Macromolecole Stereoordinate ed
                        Otticamente Attive, Pisa, Italy
Sawan, S. P.   University of Akron, Akron, Ohio, 44325,
                        U.S.A.
Sepulchre, M.   Laboratoire de Chimie Macromolécularie,
                        Université Pierre et Marie Curie, 75-230
                        Paris Cedex 05, France
Sharifi-Sandjani, N.   Faculty of Science, University, Tehran, Iran

Solaro, R.   Istituto di Chimica Organica, Università,
                      Pisa, Italy

Spassky, N.   Laboratoire de Chimie Macromoléculaire,
                      Université Pierre et Marie Curie, 4 Place
                      Jussieu, 75230 Paris Cedex 05, France

Stacht, W.   Institut für Physikalische Chemie,
                      Universität Koln, Koln, West Germany

Standt, U. D.   Institute für Chemische Technologie, Hans-
                      Sommer Str., 10, D-3300 Braunschweig,
                      West Germany

Stigliani, G.   S.I.R. S.p.A., Solbiate Olona, Italy

Tait, P. J. T.   Department of Chemistry, U.M.I.S.T., P.O.
                      Box 88, Manchester M60 1QD, England

Teyssié, P.   Laboratoire de Chimie Macromoléculaire et
                      Catalyse Organique, Université Liège,
                      Sart Tilman, 4000 Liège, Belgium

Tuzi, A.   Istituto Chimico, Università, Napoli, Italy

Watterson, A.   Department of Chemistry, University of
                      Lowell, Lowell, Massachusetts, 01854,
                      U.S.A.

Wegner, G.   Institute für Makromolekulare Chemie,
                      Univ. of Freiburg, Stefan-Meier-Str., D7800
                      Freiburg i. Br., West Germany

Zambelli, A.   Istituto di Chimica delle Macromolecole
                      del CNR, Milano, Italy

Zerbi, G.   Istituto di Chimica, Università Trieste,
                      Prazzale Europa I, 34100 Trieste, Italy

STEREOREGULAR POLYMERS: STEREOCHEMICAL ASPECTS AND SYNTHETIC APPROACHES

Piero Pino and Gian Paolo Lorenzi

Swiss Federal Institute of Technology
Department of Industrial and Engineering Chemistry
Universitätstrasse 6, 8092 Zurich, Switzerland

Table of Contents

1

R. W. Lenz and F. Ciardelli (eds.), Preparation and Properties of Stereoregular Polymers, 1-71.

"Die Tatsachen zwingen dazu, die Verschie-
denheit isomerer Moleküle von gleicher
Strukturformel durch verschiedene Lagerung
der Atome im Raum zu erklären"(J.Wislicenus[1])

1.   INTRODUCTION

     Stereoregular polymers constitute a large class of macro-
molecular compounds that are endowed with many interesting and
valuable properties. The natural ones are of vital importance.
The properties that these polymers possess are determined in
large part by their steric regularity. The essential role of
stereochemistry in determining the physical and chemical properties of chemical compounds was envisaged by Wislicenus (1) more
than one hundred years ago. That this applies to high as well as
to low molecular weight compounds was recognized early during
the development of macromolecular science. In the late twenties
Meyer and Mark (2,3) interpreted the differences between natural
rubber and gutta-percha by suggesting that the former represents
the cis- and the latter the trans-form of 1,4-polyisoprene, and
Staudinger proposed stereoirregularity as the cause of the non-
crystallizability of polyindene (4,5), polystyrene (5) and poly-
vinylacetate (5). At that time researchers were gradually beginning to realize that all organic polymers found in nature and
containing sites of stereoisomerism were stereoregular. The development in the understanding of the structure of various natu-

ral polymers has been described by Meyer (6). Thus it must have
become obvious that stereoregularity is an essential condition
in order for these natural polymers to carry out their specific
biologic functions. However, fundamental studies on the influence
of stereochemistry on the physical and chemical properties of
polymers started only in the late forties, when the first stereoregular polymers were synthesized from monomers non containing
stereoisomerism sites. Great impetus to developments in this area
was given by the first synthesis of a highly stereoregular polypropylene by Natta and his group in 1955 (7). This originated a
large interest in the synthesis and the study of other stereoregular polymers, thus forming the basis for the remarkable
progresses achieved in this area in the last 20 years. From this
development it clearly emerged that stereochemistry strongly in-
fluences the reactivity, the conformational equilibria and many
physical properties of polymers in dilute solution and in the
melt. For chemically regular polymers, stereochemistry determines
the type of crystal structure and the degree of crystallinity,
and thus determines the properties of these polymers in the solid
state. Therefore, it is not surprising that, in some cases (e.g.
polypropylene) stereochemistry may even determine the success of
the industrial application of a polymeric product. The relevance
of synthetic stereoregular polymers is by no means limited to the
study and applications of the relationships between polymer
stereochemistry and properties. Progress in many other areas of
polymer science has been made possible or has been accelerated
by studies on stereoregular synthetic polymers. In particular,
significant advances in the understanding of the mechanism of
various polymerization processes have been achieved through the
efforts made to elucidate the factors that determine the formation of stereoregular polymers. Furthermore, some factors influ-
encing the mode of crystallization of the polymer chains and the
morphology of the polymers have been clarified by the investigation of the large number of stereoregular polymers which have
been synthesized.

In this review we shall discuss first the concept of stereoregularity in polymers. Then we shall consider the various possible ways of synthesizing stereoregular polymers, distinguishing
between the formation of:

a) Stereoregular polymers from monomers not containing stereoisomerism sites;
b) Stereoregular polymers from monomers containing stereoisomerism
   sites by processes involving neither modification of the sites
   already existing nor formation of new ones;
c) Stereoregular polymers from monomers containing stereoisomerism
   sites by processes involving formation of new stereoisomerism

sites, but not modification of those already present in the
monomers;

d) Stereoregular polymers from monomers containing stereoisomerism
sites by processes affecting at least one of these sites.

In compiling this review we have tried to consider all existing polymerization methods for synthesizing stereoregular
polymers. However, only a limited number of references has been
given for each method. The literature is so extensive that a
complete listing would have been prohibitive. Our choice of these
references may have been sometimes arbitrary. Also the different
emphasis given to different methods may reflect more a personal
preference than an actual interest of these methods.

2. THE CONCEPT OF STEREOREGULARITY IN LINEAR POLYMERS

2.1. Definition of Stereoregular Polymers

The IUPAC commission on macromolecular nomenclature defines
(8) a stereoregular polymer as "a regular polymer whose molecules
can be described by only one species of stereorepeating unit in
a single sequential arrangement". A stereorepeating unit is
further defined as "a configurational repeating unit having defined configuration at all sites of stereoisomerism in the main
chain of a polymer molecule". Tactic * polymers are considered
to be those which exhibit a regular pattern of configurations for

---

* From the words isotactic and syndiotactic proposed by Natta (9)
  to designate stereoregular polymers of mono- and 1,1-disub-
  stituted ethylenes, the IUPAC commission has proposed the word
  tacticity to indicate steric order in the main chain. Tacticity,
  meaning only order, might be used also to indicate structural
  order in polymers (e.g. absence of head-to-head, tail-to-tail
  arrangements of the monomeric units), and therefore stereotacticity would be a more appropriate word. However, if consistently used only to indicate steric order in polymers
  (tacticity = stereotacticity), the word tacticity can be useful
  for classifying polymers containing more than one site of
  steric isomerism per monomeric unit (ditactic and tritactic
  polymers).

at least one type of stereoisomerism sites in the main chain. According to these nomenclature proposals, one can have tactic stereoregular polymers and tactic non-stereoregular polymers, according to whether every type or only one type of stereoisomerism site in the main chain is in a configurationally ordered sequence. Cases in which steric order exists only for stereoisomerism sites in the side chains, the main chain being nontactic (not containing stereoisomerism sites) or atactic are not considered.

The definition of stereoregular polymers as given by the IUPAC commission refers to ideal structures, but it is conceded that it may be applied also to practical cases where deviations from the ideality are not too large. Despite this, the IUPAC definition, is very restrictive. We shall consider as stereoregular polymers not only those which fall under this definition, but also any tactic polymer and any copolymer that contains in the main chain stereoisomerism sites of the same type in a configurationally regular sequence. As stereoregular will be considered also those polymers and copolymers which contain stereoisomerism sites in the side chains, and which exhibit a regular pattern of configurations for these sites, even if no stereoregularity exists in the main chains. We shall call stereoirregular those polymers that contain stereoisomerism sites, but do not exhibit any recognizable regular sequence of configurations, and non-stereoregular those polymers that do not contain stereoisomerism sites. Examples of these types of polymers are given in Scheme 1.

2.2. Sites of Steric Isomerism

As sites of stereoisomerism atoms or groups of atoms are considered which can give rise to enantiomerism (e.g. asymmetric carbon atoms) or diastereoisomerism (e.g. 1,2-disubstituted ethylene units). Only the cases will be considered, in which the steric structure of an isolated macromolecule does not change substantially with time at room temperature, that is, the case in which the activation energy for changing the steric structure of the steric isomerism sites in dilute solution is larger than 20 kcal/mole. This condition is satisfied by cis- and trans-substituted ethylene units, by asymmetric carbon atoms, by certain single-bonded C-C units with hindered rotation (atropoisomers) and by certain chiral structures such as, for instance, the helicenes (Scheme 2).

We shall not consider cis-trans isomerism occurring in polyamides or polyurethanes due to the existence of cis or trans amide or urethane groups, since isomerization in these cases

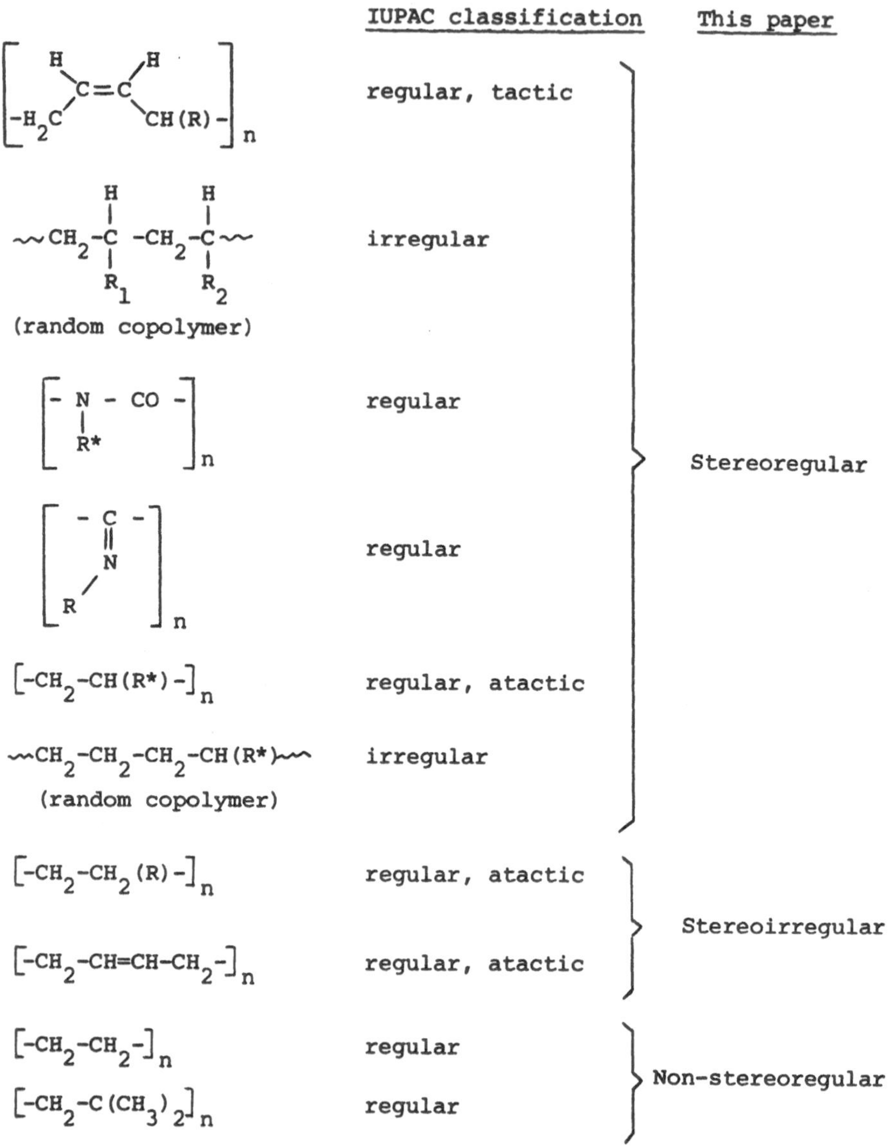

	IUPAC classification	This paper

$$\left[\begin{array}{c} \underset{-H_2C}{\overset{H}{C}} = \underset{CH(R)-}{\overset{H}{C}} \end{array} \right]_n$$ regular, tactic

$$\sim CH_2-\underset{R_1}{\overset{H}{\underset{|}{C}}} -CH_2-\underset{R_2}{\overset{H}{\underset{|}{C}}} \sim$$ irregular

(random copolymer)

$$\left[\begin{array}{c} - N - CO - \\ \underset{R*}{\overset{|}{}} \end{array} \right]_n$$ regular

$$\left[\begin{array}{c} - \underset{\underset{N}{\overset{\|}{}}}{C} - \\ R^{\diagup} \end{array} \right]_n$$ regular

$$\left[-CH_2-CH(R*)- \right]_n$$ regular, atactic

$$\sim CH_2-CH_2-CH_2-CH(R*) \sim$$ irregular

(random copolymer)

$$\left[-CH_2-CH_2(R)- \right]_n$$ regular, atactic

$$\left[-CH_2-CH=CH-CH_2- \right]_n$$ regular, atactic

$$\left[-CH_2-CH_2- \right]_n$$ regular

$$\left[-CH_2-C(CH_3)_2 \right]_n$$ regular

Stereoregular

Stereoirregular

Non-stereoregular

Scheme 1. Types of polymers considered as stereoregular in this paper, but not conforming to the IUPAC definition of stereoregular polymers, and types of stereoirregular and non-stereoregular polymers (The R*s are substituents containing one site of stereoisomerism, and the polymers which contain them are considered to exhibit a regular pattern of configurations for these sites).

(R_1, R_2 : Cl , NO_2 , COOH)

Scheme 2. Typical examples of stereoisomerism sites in which
           change of configuration does not occur with
           measurable rate at room temperature ($\Delta E > 20$ kcal/mole)

requires an activation energy generally smaller than 20 kcal/mole
(10,11). For the same reason, we shall not consider the conformational isomerism phenomena arising from the existence of helical
conformations, which are thermodynamically stable, but in which
the shift from one helical sense to the other requires activation energies which are well below 10 kcal/mole (Scheme 3). However, we shall consider as stereoregular polymer poly-<u>tert</u>.butyl-
isocyanide, where right-handed or left-handed helical conformations can be isolated and are stable at room temperature (12).

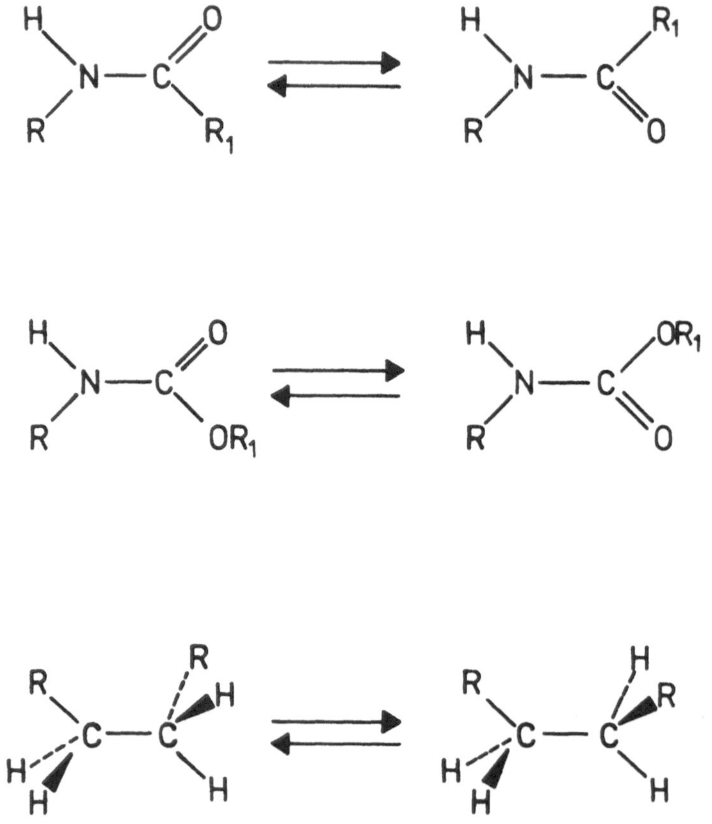

Scheme 3.   Typical examples of stereoisomerism sites in which the
            equilibrium between the steric isomers is rapidly
            reached at room temperature (ΔE<20 kcal/mole)

2.3. "Partially Stereoregular" Polymers

The meaning of stereoregularity becomes ambiguous, when a polymer showing a significant content of steric irregularities is considered. For this "partially stereoregular" polymer two extreme possibilities can be envisaged: i) the polymer is a mixture of stereoregular and stereoirregular (atactic) macromolecules; ii) the polymer consists of macromolecules all having a comparable degree of order of the same type (e.g., all the macromolecules contain isotactic sections having the same average length). These extreme cases are illustrated in Scheme 4 *. These possibilities were not clear to most of the polymer chemists until the fifties, after a large number of "partially stereoregular" polymers had been prepared, and suitable methods for the separation of stereoregular and stereoirregular fractions were found. Then it was noted that for vinyl polymers prepared by ionic or coordination polymerization the most common situation is a polymer which is a mixture of substantially stereoregular macromolecules (containing less than 5% of steric irregularities), less ordered macromolecules with the steric irregularities either concentrated in blocks (stereoblock-macromolecules) or randomly distributed (Scheme 5), and substantially stereoirregular macromolecules.

A meaningful investigation of "partially stereoregular" polymers is only possible if a fractionation according to stereoregularity can be achieved to a reasonable extent. As it was shown for the first time by Natta and his school (7), in the case of "partially stereoregular" vinyl polymers, solvent extraction, which is known to be a rather inefficient method for separating polymers according to molecular weight, is an excellent method for separating macromolecules having different degrees of stereoregularity. This method takes advantage of the fact that, in general, stereoregular polymers crystallize and stereoirregular polymers do not. Thus, by extraction with boiling solvents of increasing dissolving powers and boiling points, polymer fractions of increasing stereoregularity degree can be separated. More sophisticated methods can be used if necessary, and in some cases chromatographic separations have been very successful (13,14). However, solvent extraction, because of its simplicity, remains by far the most used method for fractionating

---

* An appropriate nomenclature to distinguish between these cases does not exist, and we shall use the imprecise expressions "partially stereoregular" polymer and "partial stereoregularity", when no data exist about the diastereomeric composition of the polymer.

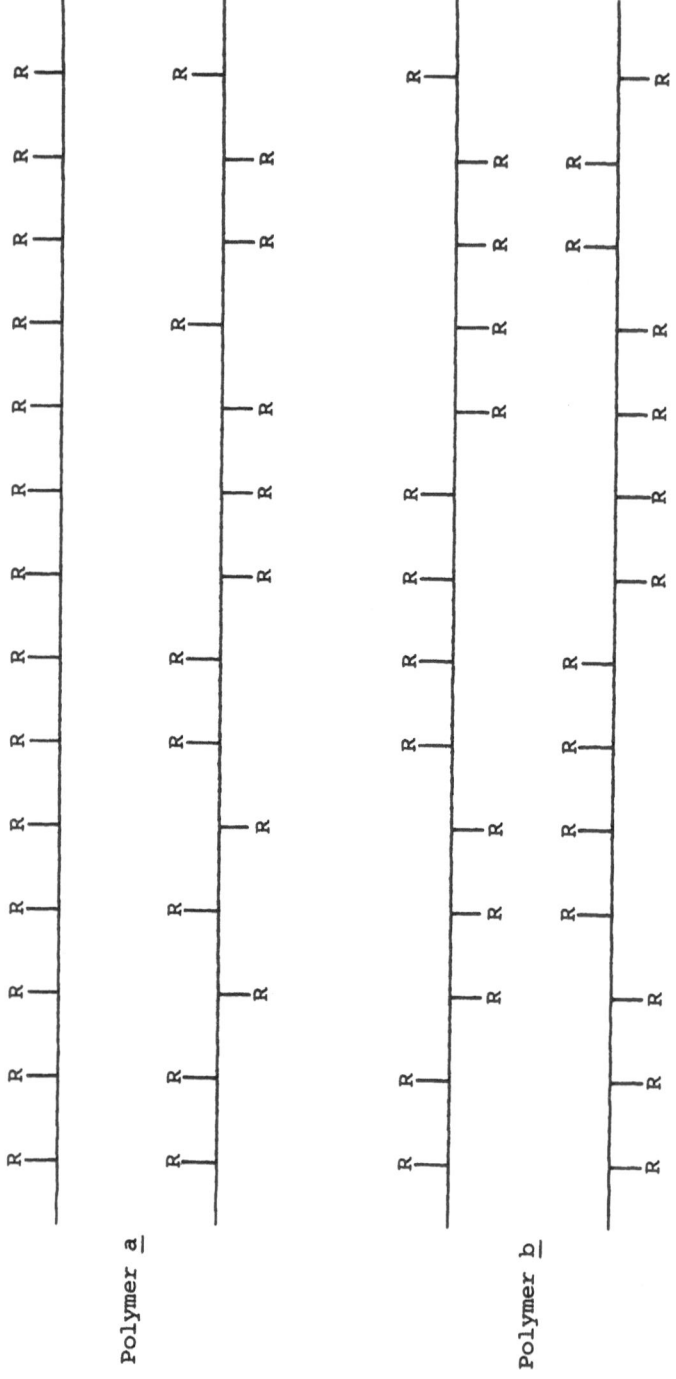

Polymer a

Polymer b

Scheme 4. Two extreme cases of "partially stereoregular" polymers. Polymer a is a mixture of stereoregular (isotactic) and stereoirregular (atactic) macromolecules; polymer b is a mixture of macromolecules having comparable degree of order of the same type.

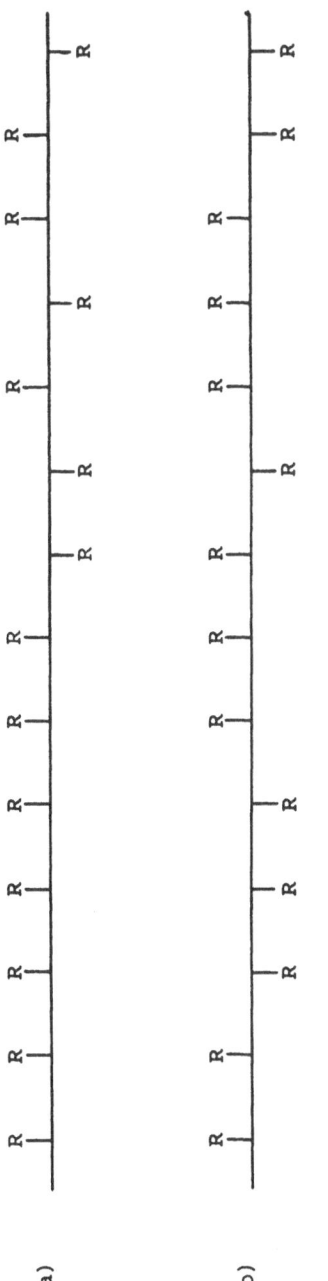

Scheme 5. Two types of macromolecules commonly present in "partially stereoregular" polymers: a) stereoblock macromolecule; b) macromolecule with randomly distributed steric irregularities.

polymers according to stereoregularity.

A further difficulty in investigating stereoregularity in
polymers is to find suitable analytical tools to determine quantitatively the degree of stereoregularity and the distribution
of the steric irregularities possibly present in the polymers. As
an absolute analytical method ^{1}H- or ^{13}C-NMR spectroscopy is
commonly used. In particular cases, where the stereoisomerism
site is already present in the monomer, as for instance in the
case of natural polypeptides and polysaccharides, chemical or
biochemical degradation can be used. The most important relative
methods are the measurement of the relative intensity of IR
stereoregularity or crystallinity bands, the estimation of crystallinity from X-ray diffraction or from calorimetric or density
measurements, and the measurement of chiroptical properties. Largely used semi-quantitative methods are based on measurements of
solubility in suitable solvents. NMR is the only method which
yields at least some quantitative information about the type and
distribution of steric irregularities in polymers.

2.4. Types Of Stereoregular Polymers So Far Recognized In Nature
     Or Synthesized

Among the stereoregular polymers known, with only one stereoisomerism site per monomeric unit in the main chain, the case in
which this site (e.g. a double bond or an asymmetric carbon atom)
is repeated along the polymer main chain with the same configuration is much more frequent both in natural (natural rubber, natural polypeptides, poly-β-hydroxybutyrate (15)) and synthetic
polymers (isotactic vinyl polymers) than the case in which stereoregularity arises from ordered successions of different configurations. Indeed, up to now, only the ordered alternation of the two
possible configurations of asymmetric carbon atoms (syndiotactic
vinyl polymers and alternating copolymers of D- and L-amino acids)
has been recognized. A cis, trans-alternating configuration of
the double bond has been proposed for a polybutadiene obtained
under specific conditions (16). More complicated repeating sequences of asymmetric carbon atoms having opposite configurations
have been proposed in some cases, but have not been confirmed. An
example is the structure I, which has been proposed (17,18) for
certain samples of polymethylmethacrylate.

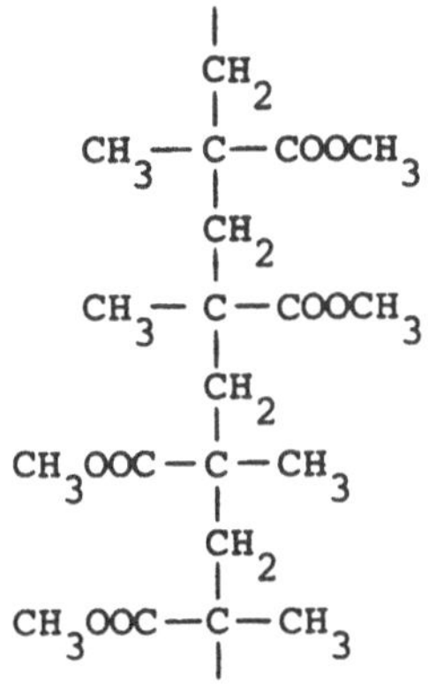

$$
\begin{array}{c}
| \\
CH_2 \\
| \\
CH_3-C-COOCH_3 \\
| \\
CH_2 \\
| \\
CH_3-C-COOCH_3 \\
| \\
CH_2 \\
| \\
CH_3OOC-C-CH_3 \\
| \\
CH_2 \\
| \\
CH_3OOC-C-CH_3 \\
|
\end{array}
$$

I

Stereoregular polymers with more than one stereoisomerism site
per monomeric unit in the main chain have been identified in
nature (polysaccharides and nucleic acids), and have been synthesized (e.g. erythro-diisotactic, trans-polymethylsorbate (19)).

3. SYNTHETIC APPROACHES

    The synthesis of stereoregular polymers that are monodisperse
(i.e. that consist of macromolecules that are of the same length)
cannot be accomplished by any of the known polymerization methods.
Stepwise syntheses of identical, stereoregular chains can be accomplished in certain cases by using monomers already containing
stereoisomerism sites and conditions not involving changes in
the configuration of these sites (20,21), but are impractical for
producing large macromolecules.

    Polydisperse stereoregular polymers can be synthesized
either from monomers having no stereoisomerism sites, the stereoisomerism sites being formed during the polymerization, or from
monomers containing sites of steric isomerism. In the latter case,
the synthesis of stereoregular polymers may or may not involve
formation of new stereoisomerism sites, without affecting those
already present in the monomers, or else it may affect these sites.

3.1. Stereoregular Polymers From Monomers Not Containing Stereo-
     isomerism Sites

    For producing stereoregular polymers from monomers not con-
taining stereoisomerism sites, stereospecific processes are needed.
In the past 40 years stereospecific polymerization processes have

become available for a large number of these monomers. The monomers are generally unsaturated compounds containing double or
triple bonds either between carbon atoms (mono- and 1,1-disub-
stituted ethylenes, dienes, and acetylenes), or carbon and oxygen
(aldehydes), or carbon and nitrogen (cyanides). In these cases,
at least two atoms connected with a multiple bond of the monomer
remain incorporated in the polymer chains. Other unsaturated monomers, for which this does not occur, are isocyanides and diazoalkanes (22). In principle, also saturated compounds not containing stereoisomerism sites (e.g. $C_6H_5SiCl_3$ (23)) can be used to
produce stereoregular polymers.

　　3.1.1. <u>Stereospecific polymerization of mono- or 1,1-disub-
stituted ethylenes. Historical developments</u>. In a polymerization
of mono- or 1,1-disubstituted ethylenes, including the 1,2-poly-
merization of butadiene, macromolecular chains are generated that
have a backbone consisting exclusively of carbon atoms, and, in
general, every second carbon atom is asymmetric *.

　　The possibility of the existence of stereoisomerism in polymers derived from vinyl monomers not containing stereoisomerism
sites was foreseen by Staudinger (5) in the course of his pioneering work. This possibility is immediately apparent when one applies
the concepts of organic stereochemistry developed by Le Bel and
Vant'Hoff at the end of the nineteenth century. Based on the
difficulty of carrying out the asymmetric synthesis of low molecular weight compounds, the synthetic organic chemists of
Staudinger's time should have considered the possibility to obtain
stereoregular polymers to be very remote. Indeed, the synthesis
e.g. of an isotactic macromolecule with a stereoregularity of
about 99%, which is rather common in the propylene stereospecific
polymerization, corresponds to a 1% error in the process of enantioface discrimination on which the above synthesis is based (see
below). A similar specificity in the enantioface discrimination
would allow one to synthesize a compound containing one chirality
center, with an optical purity of 99%. Even today such a demand-
ing task can be performed by non-enzymatic reactions only in a
very few cases. And yet, in the following years, some work provided hints that some stereoregulation during the polymerization
was possible. In 1936 Natta and Rigamonti (25) used electron
diffraction to examine some polystyrenes having a very high mole-

---

* A special case is the hydrogen-transfer polymerization of acryl-
　amide or methacrylamide leading to a polyamide. The synthesis of
　an optically active, "partially stereoregular" poly-methyl-β-
　alanine by polymerization of methacrylamide with an optically
　active basic catalyst has been reported (24).

cular weight and found some crystallinity. Some years later
Alfrey, Bartovics and Mark (26) observed differences in behavior
in solution for polystyrenes that had been produced at different
temperatures. They suggested that different amounts of branching
of the polymer chains in the different samples might account for
the observed results. Huggins (27), however, attributed these
results to differences in the steric structure of the samples. An
attempt to accomplish an asymmetric radical polymerization of
vinyl monomers was carried out by Marvel, Frank and Prill (28) by
using an optically active acyl peroxide, but no difference was
found between the polymers thus produced and those synthesized
by using optically inactive initiators. In 1948, on carrying out
a cationic polymerization of vinyl isobutyl ether with $BF_3.(C_2H_5)_2O$
in a polyphase system, Schildknecht and coworkers (29) obtained
a polymer which showed X-ray crystallinity. They concluded that
this polymer was "partially stereoregular". However, despite the
fact that, in a theoretical paper in 1942 Bunn (30) had already
proposed a helical conformation for stereoregular vinyl polymers
having sterically equivalent monomeric units, they considered
(31) only a planar zig-zag conformation, and, citing steric
reasons, they proposed that the structure was one in which asymmetric carbon atoms of opposite configuration alternate in the
main chain. Some years later, Natta et al. (32) showed that the
stereoregularity of Schildknecht's crystalline polyvinyl isobutyl
ether was of the isotactic type, the conformation of the main
chain in the crystalline state being helical, as foreseen by Bunn.
In 1950 styrene was polymerized (33) by means of Alfin catalysts
(34) (a slurry of NaCl, $(CH_3)_2CHONa$ and $CH_2=CH-CH_2Na$);the polymer
obtained was amorphous and was not thoroughly investigated. That
this polymer was stereoregular was established only some years
later, after the accomplishment of the stereospecific polymerization of styrene by the Natta's group (35) using Ziegler-type catalysts. By suitable annealing, the amorphous Alfin-polymerized
polystyrene could be crystallized to varying degrees (36). The
X-ray diagram of the crystallized polymer reproduced that reported by Natta for isotactic polystyrene.

The true breakthrough in the production of stereoregular
polymers occurred in the mid-fifties, when at least three different catalytic systems were found to produce stereoregular polymers
of propylene and other α-olefins (Scheme 6). It was the Natta
group that, having prepared a partially stereoregular polypropylene using a Ziegler-type catalytic system, conceived for the
first time that this polymer could consist of stereoregular,
crystallizable and stereoirregular, non-crystallizable macromolecules, and they separated a highly crystalline fraction by solvent
extraction, and established by X-ray the type of stereoregularity

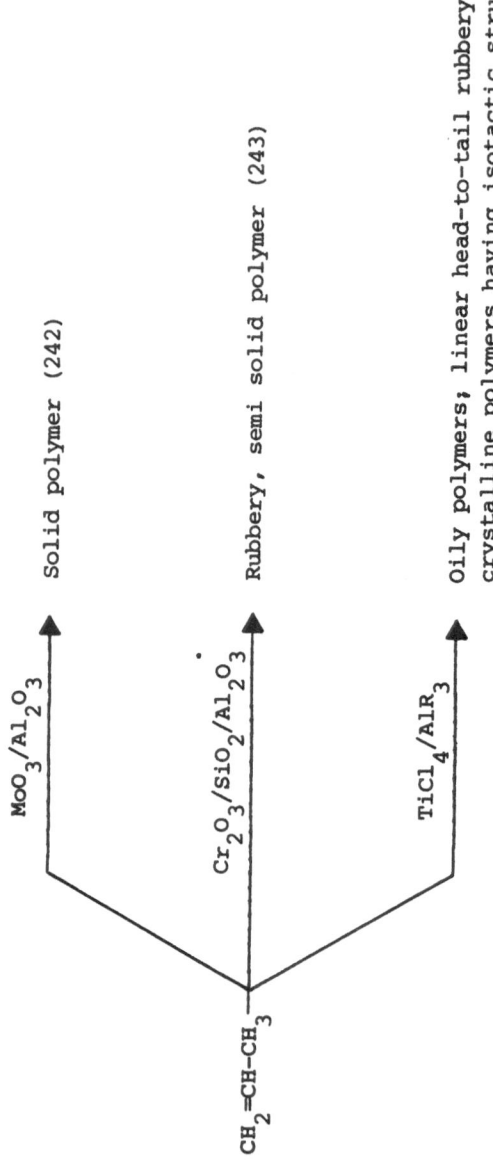

Scheme 6. Some types of catalytic systems for the polymerization of propylene

existing in the crystalline fraction. We now know from NMR studies,
that the stereoregularity degree of this highly crystalline fraction can be 98% or more (37). In the following years, many stereoregular polymers were obtained from various vinyl monomers. A
comprehensive review covering these polymers in detail has been
published by Pasquon (38).

3.1.2. Stereospecific polymerization of mono- or 1,1-disub-
stituted ethylenes. Stereoregulating factors. The investigation
of epimerization equilibria of low molecular weight models has
indicated that stereoregular polymers are energetically less
favored when compared to the atactic polymers (39). Therefore
stereospecific polymerization must be kinetically controlled. In
radical polymerizations the stereospecificity can be originated
by the difference between the free activation energies of the
reactions between the non-prochiral carbon atom of the monomer
and one face or the other of the essentially planar, prochiral
terminal carbon atom of the growing chain (Scheme 7). In an anionic

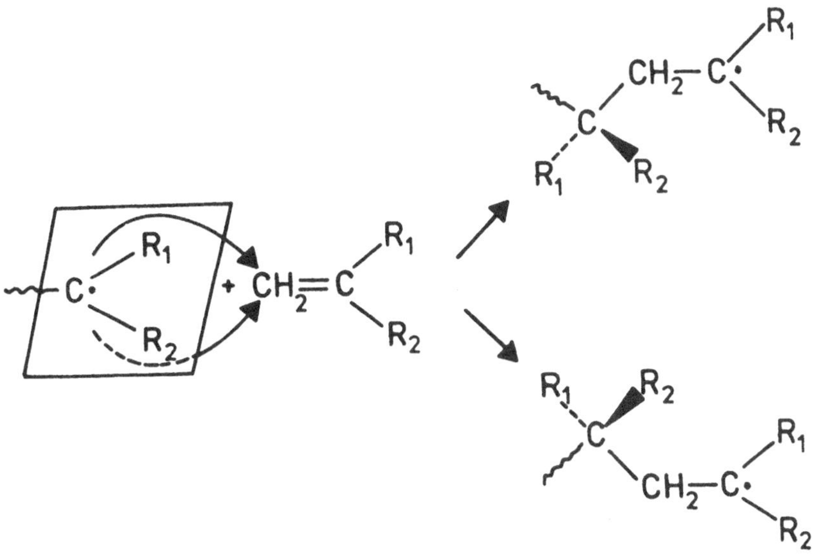

Scheme 7.   Possibilities for the addition of a monomer molecule
            to the growing chain end in a radical polymerization.

polymerization of the type shown in Scheme 8, the stereospecificity can be originated by the difference between the free activation energies of the reactions of the non-prochiral terminal
atom of the growing chain and one prochiral face or the other of
the monomer.

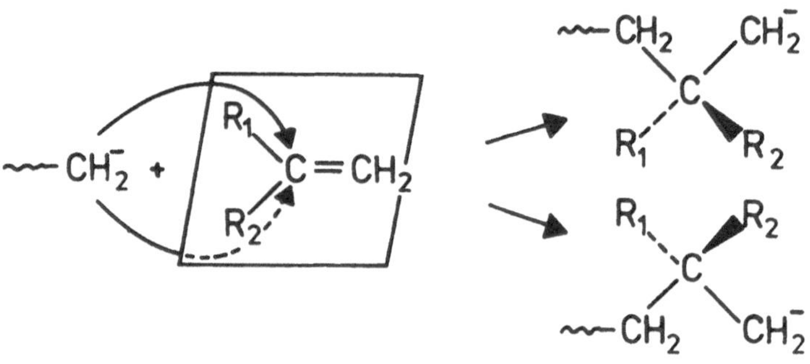

Scheme 8.   Possibilities for the addition of a monomer molecule
            to the growing chain in an anionic polymerization

In radical or ionic polymerizations, the step controlling the
stereoregularity is in general also the step controlling the rate
of chain growth. If the growth reaction is a multistep process,
as in general it is postulated for coordination polymerization,
the step controlling the rate of chain growth does not need to
be also the step controlling the stereoregularity.

    In radical polymerization, the degree of stereoregulation
achieved up to now is very low in general, the largest average
length for a stereoregular section of the chain corresponding to

about three monomeric units (40). At the present, the only way
to achieve stereoregulation in radical polymerization seems to
be the polymerization of monomers which have been previously
ordered in a crystalline matrix (41,42). In homogeneous ionic
polymerizations, both the growing chain end and the counterion
can participate in the stereoregulation. In this case the chiral
component necessary for the enantioface discrimination is probably the growing chain end, although the presence of chiral
counterions including one or more monomer molecules in many cases
cannot be excluded. The stereoregulating capability increases on
going from highly solvated free ions to tight ion pairs. Tentative models for diastereomeric transition states involving cyclic
structures for the growing chain ends have been proposed for
cationic polymerizations of vinyl ethers (43) and for anionic
polymerizations of methacrylates (44,45).

The most efficient enantioface discriminating agents seem
to be transition metal complexes covalently bound to the growing
chain end, which are also able to achieve a very high regioselectivity in the attack to the double bond. Unfortunately, the
type of monomers which are polymerized stereospecifically with
this type of catalysts are mainly unsaturated hydrocarbons.
Propylene (14) and butadiene (46) can be polymerized by the
above catalysts both to isotactic and syndiotactic polymers.

The catalysts that polymerize α-olefins to isotactic polymers are highly regioselective, the bond between metal and
growing chain being exclusively of the type $Me-CH_2-CH(R)-$ (47).
In the case of the stereospecific polymerization of propylene to
syndiotactic polymer, the reaction is also regioselective, even
if to a smaller extent. However, in this case bonds of the type
$Me-CH(CH_3)-CH_2-$ are prevailingly formed, at least with the
soluble catalytic system $VOCl_3/AlR_2Cl/anisole$ (48,49). In the
case of the polymerization of α-olefins to isotactic polymers
involving $Me-CH_2-CH(R)-$groups, it has been shown that the main
stereoregulating agent is the metallic complex, which therefore
must be chiral (40). On the contrary, in syndiotactic polymerization it seems that, at least in the cases investigated which
involve $Me-CH(CH_3)-CH_2-$ groups, the main stereoregulating
factor is the growing chain end (40) (Scheme 9). Since the
details of the polymerization mechanism are not known, it is not
known whether regioselectivity and enantioface discrimination
are regulated in the same reaction step. It appears, however,
that in the synthesis of isotactic polymers, the difference in
activation energies for the primary and secondary insertions and
for the attack to one of the other enantioface must be, at least
for some catalytic centers, higher than 4 kcal/mole, the process

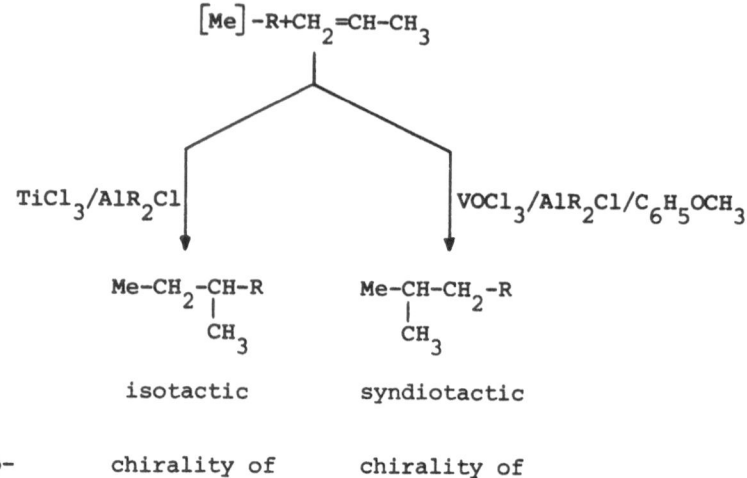

Scheme 9.   Stereospecific polymerization of
            propylene to isotactic or syndiotactic polymer.

being highly regio- and stereospecifc even at 180°C (50). In the
case of polymerization to syndiotactic polymers, the above differences in activation energies are smaller but probably still
in the range of 2 kcal/mole.

     In order to identify the factors causing such large activa-
tion energy differences, many models have been proposed for the
catalytic centers (Scheme 10). Accepting the hypothesis that the
polymerization occurs in two steps, it is possible to admit that
both regioselectivity and enantioface discrimination occur during
the first step corresponding to a π-complex formation, the second
step, the insertion reaction, involving substantially similar
activation energies for the four diastereoisomers of the complex.
It is, however, not possible to exclude that the opposite is true,
that is, that the four diastereoisomers have essentially the
same stability, but the activation energies for the insertions
are substantially different (Scheme 11).

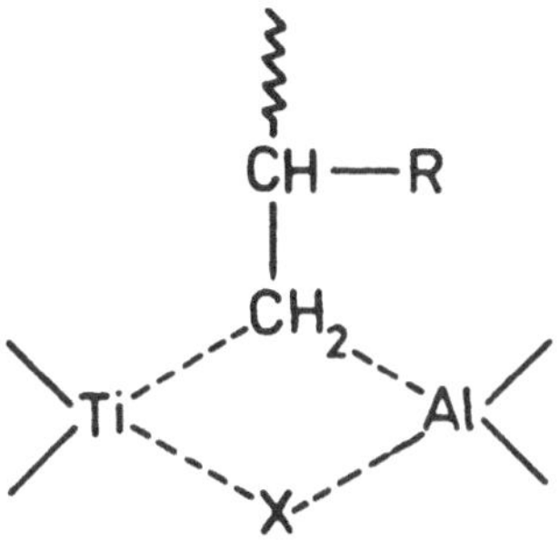

F. Patat and H. Sinn (51)

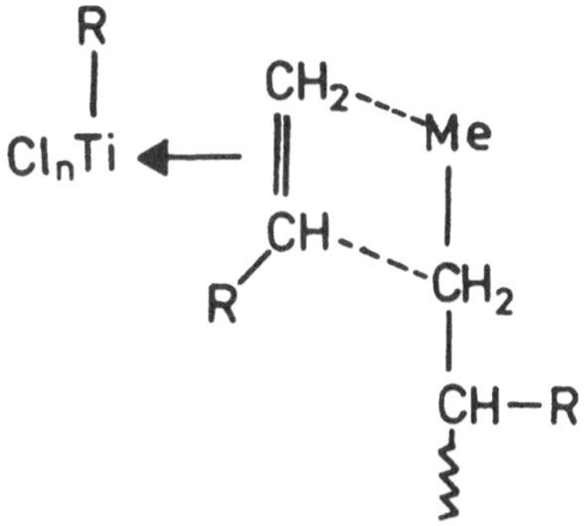

P. Pino (52)

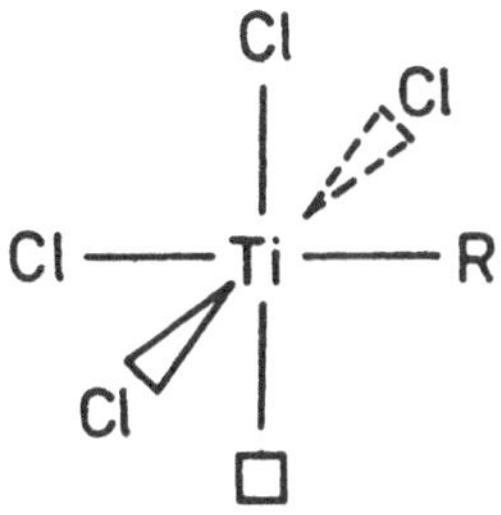

P. Cossee (53)

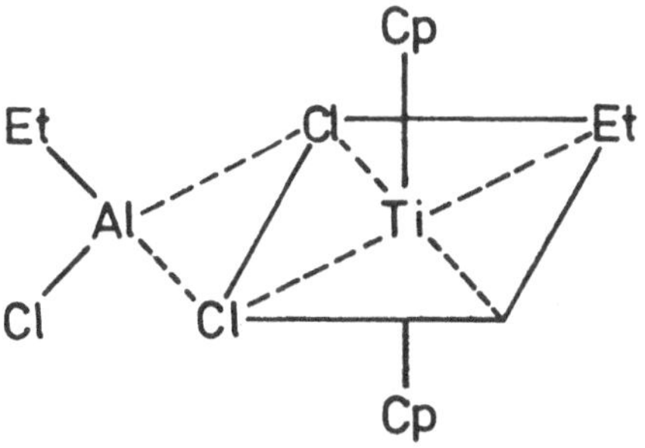

G. Henrici-Olivé and S. Olivé (54)

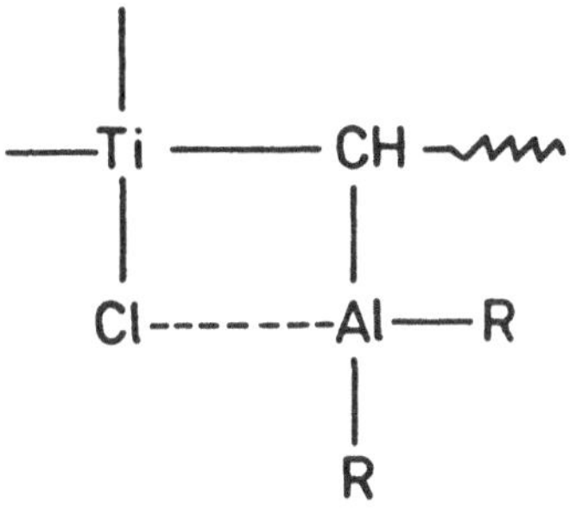

P. Pino, G. Consiglio and
H. Ringger (55)

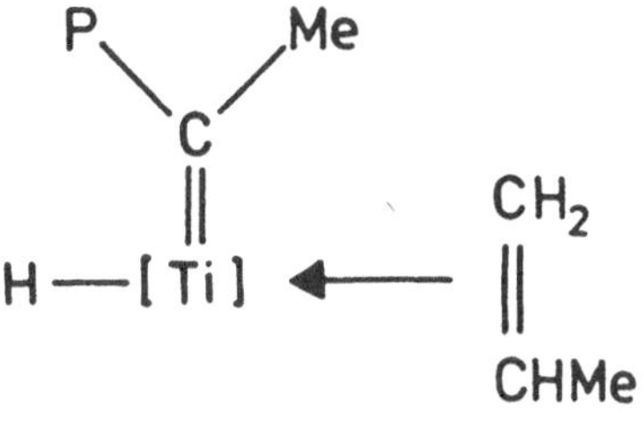

K.J. Ivin et al. (56)

Scheme 10. Some types of catalytic centers proposed for the
stereospecific polymerization of ethylene and
α-olefins

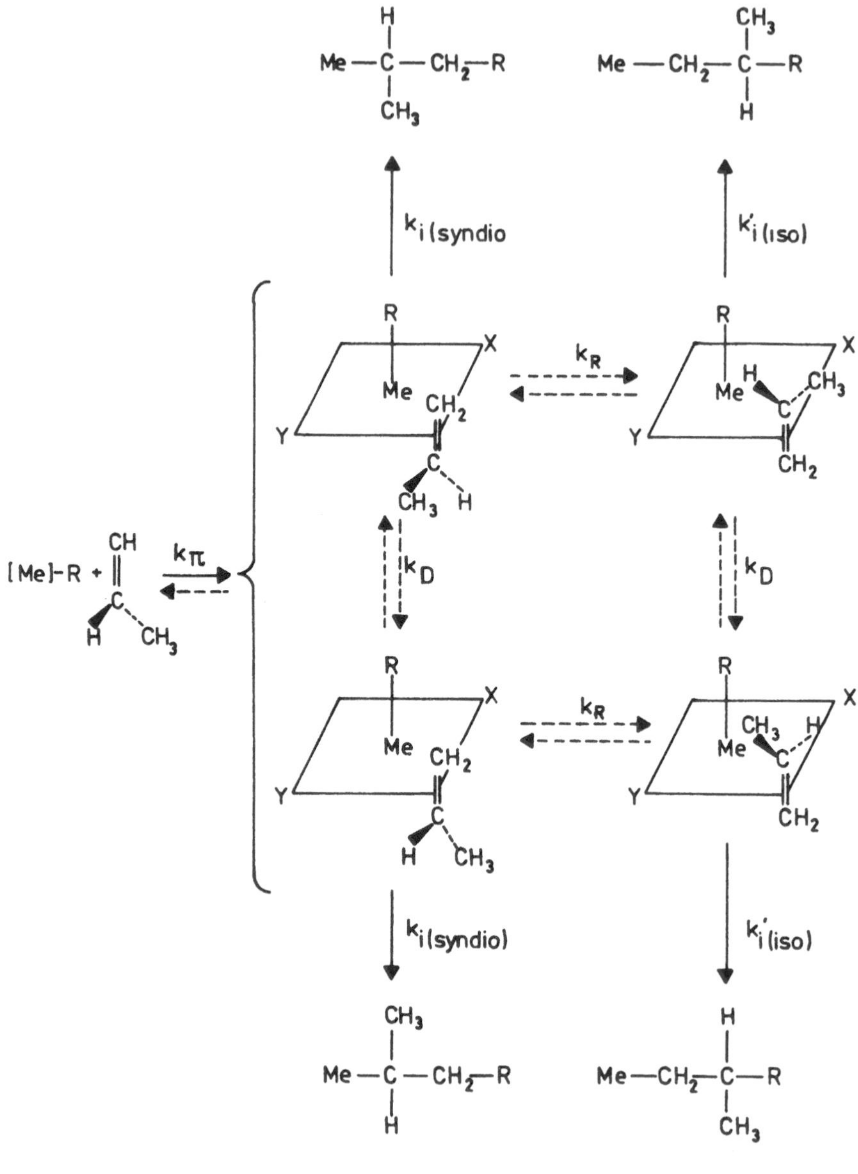

Scheme 11.   Possible diastereomeric π-complexes and insertion
             reactions leading to isotactic or syndiotactic
             polypropylene.

3.1.3. Stereospecific polymerization of conjugated dienes to 1,4-polymers. Synthetic 1,4-polydienes with very high degrees of stereoregularity have been obtained by various methods. By radical polymerization, stereoregular 1,4-trans polymers have been obtained when polymerizing the monomers as inclusion complexes (Table 1). Polymers with either 1,4-trans or 1,4-cis structure can be produced by coordination polymerization in the presence of various organometallic derivatives of alkali metals or of transition metals.

The first stereoregular 1,4-polybutadiene was obtained by Morton and his associates by using an Alfin catalyst in 1947 (34) and had a predominantly trans structure. The structure of this polymer was determined by careful X-ray and IR-examination (61). A few years later the first synthesis of 1,4-cis-polyisoprene * was accomplished (62) by using a Li metal dispersion. In the same period, Ziegler-type catalysts were reported to polymerize stereospecifically to 1,4-trans- or 1,4-cis-polymers a number of conjugated dienes. More recently, new types of catalysts for the stereospecific polymerization of dienes were prepared starting with π-allyl derivatives of transition metals. Some catalysts for the stereospecific 1,4-polymerization of butadiene are indicated in Scheme 12.

The factors controlling the structural (1,4-units versus 1,2- or 3,4-units) and geometrical (1,4-cis versus 1,4-trans) isomerism in the polymerization of conjugated dienes are not completely clear. Concerning the geometrical isomerism it is worth noting that completely stereospecific 1,4-additions are known in synthetic organic chemistry (see e.g. the chlorination of butadiene). In the Diels-Alder reaction only cis double bonds are necessarily formed, being part of the six membered ring structure of the product. For the conjugated dienes' polymerization catalyzed by organometallic derivatives, high stereospecificity of cis- or trans-type can easily be achieved depending on the catalyst used, which evidently not only causes the polymerization, but also controls the geometrical isomerism. Based on the isolation of a catalytic intermediate in the butadiene

---

* 1,4-polyisoprenes contain trisubstituted double bonds. In agreement with the organic compounds nomenclature, the 1,4-cis-polyisoprene should be named 1,4-Z-polyisoprene. However, in the polymer chemistry the cis-trans nomenclature is consistently used also for chains containing trisubstituted double bonds, making reference to the stereochemistry of the double bond with respect to the main chain and disregarding the nature of the lateral chains.

Table 1. Inclusion polymerization of some conjugated dienes

Monomer	Host	Type of stereoregularity	Ref.
$CH_2=CH-CH=CH_2$	Urea PHTP a)	trans-1,4	41 57
$CH_2=C-C=CH_2$ CH_3 CH_3	Thiourea PHTP a)	trans-1,4	58 59
$CH_2=C-C=CH_2$ Cl Cl	Thiourea	trans-1,4	58,60

a) PHTP = trans, anti, trans, anti, trans-perhydrotriphenylene

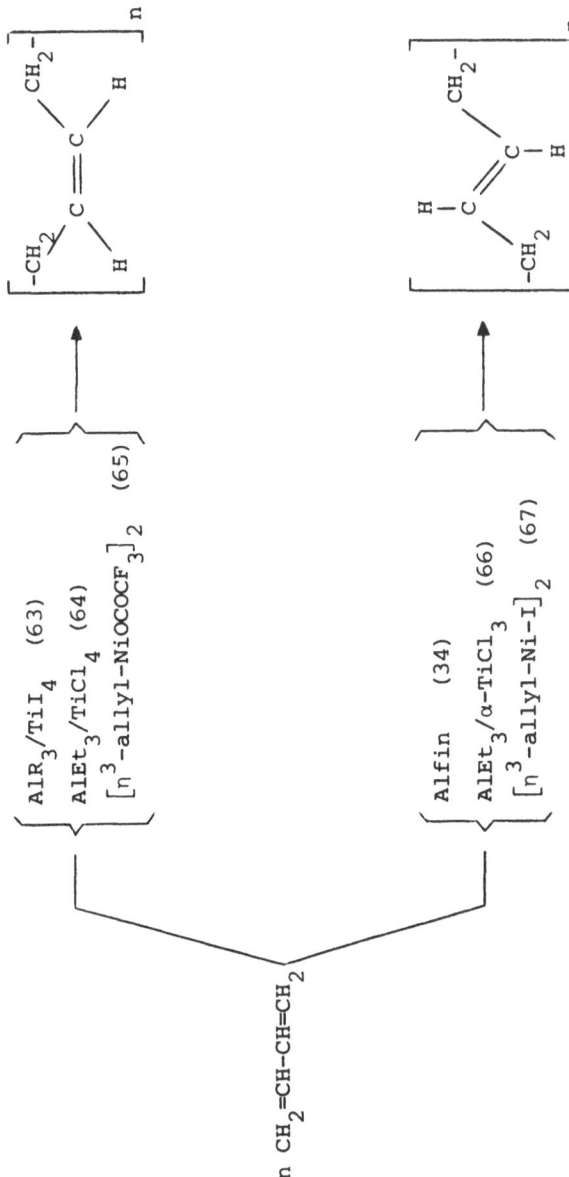

Scheme 12. Some catalytic systems for the stereospecific polymerization of butadiene to 1,4-polymers.

dimerization (68), the structure of which was established by
X-ray examination, there is no doubt that allyl complexes play
an overwhelmingly important role in stereoregulation. According
to some authors (69,71), for the case of butadiene polymerization, the syn-anti isomerism of the η^3-allyl-bonded chain
(Scheme 13) is responsible for the formation of a cis or trans
monomeric unit in the growing chain.

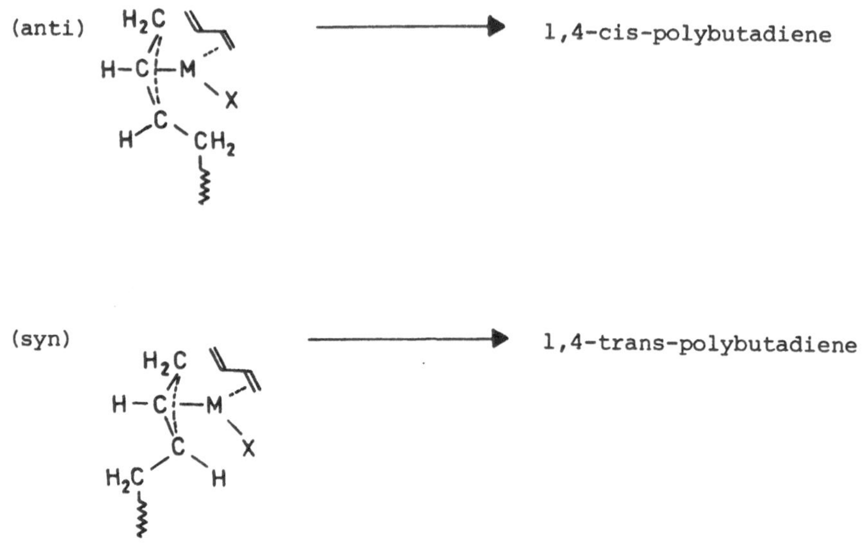

(anti)   → 1,4-cis-polybutadiene

(syn)   → 1,4-trans-polybutadiene

Scheme 13.   Possible control of the cis-trans isomerism in 1,4-
             polybutadiene by the configuration of the η^3-allyl-
             complex between metal and growing chain.

However, the possibility that the type of conformation of the
monomer which can be coordinated on the catalytic complex in a
cisoid or transoid conformation plays a role, cannot be excluded.
This role is decisive in Cossee's mechanism (70), according to
which a monodentate-transoid or bidentate-cisoid coordination of
the diolefin is responsible for the formation of either trans or
cis 1,4-units respectively in the polymer (Scheme 14). As for the
mode of the addition of metal and growing chain to the entering
unit, the results obtained by Porri and Aglietto (71) in the
study of the stereospecific polymerization of cis,cis-1,4-di-

deuterio-1,3-butadiene indicate that it is of the cis type.

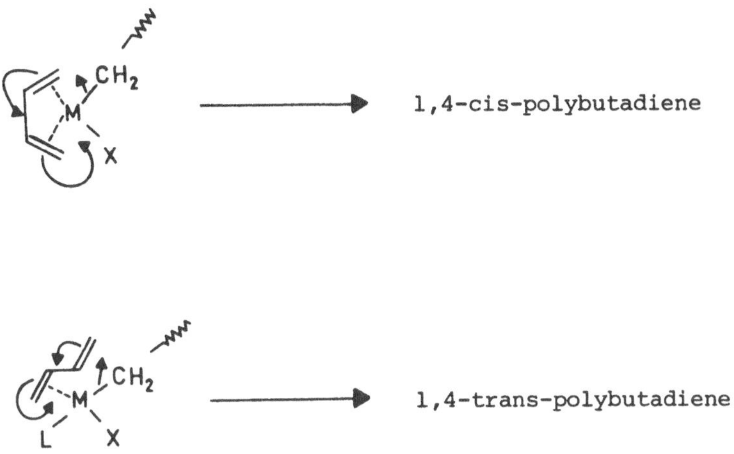

$\longrightarrow$ 1,4-cis-polybutadiene

$\longrightarrow$ 1,4-trans-polybutadiene

Scheme 14.   Possible control of the cis-trans isomerism in 1,4-
             polybutadiene by the conformation of the coordinated
             monomers.

The factors responsible for the stereospecific polymerization of isoprene (38,72) have been less thoroughly investigated.

It is worth mentioning that, under certain conditions,speci-
fic catalytic systems can lead to the formation of butadiene or
isoprene polymers containing equal amounts of cis and trans
units (73). For these polymers the term "equibinary polydienes"
has been proposed. It now appears that depending on the nature
of the solvent and on the temperature of polymerization, the
distribution of the two isomeric units can range from purely
random, to highly alternate or even to sequential (16). These
results hint at an interesting new type of catalytic control.

3.1.4.Polymerization of monosubstituted allenes. Under
suitable conditions the polymerization of allenic hydrocarbons
of the type $CH_2=C=CHR$ can proceed mainly through opening of the
less-substituted double bond. Thus polymers having monomeric
units of the type II result. In this case, as it was pointed out
by Otsuka, Mori, Suminoe and Imaizumi (74), stereoisomerism is
expected, due to the possibility that the pendant double
bonds may assume different, relative configurations. Of course,

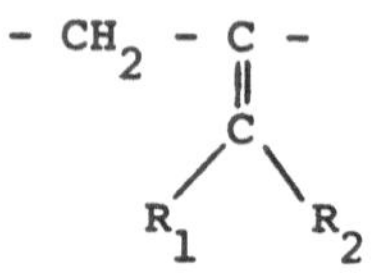

$$- CH_2 - \underset{\underset{\underset{R_1 \quad R_2}{\diagdown}}{C}}{\overset{\|}{C}} -$$

II

the absolute configuration (E or Z) of each double bond can be
established only by knowing the groups attached to the main chain
carbon atom. The only case in which this type of stereoisomerism
in polyallenes seems to have been investigated is that of poly-
methoxyallene. Ghalamkar-Moazzam and Jacobs (75) have studied
a polymer obtained by means of π-allylnickel halides and they
found three NMR-peaks for the allylic methylenes X,Y and Z of
the polymer III. The ratios of peak areas X:Y:Z were 3:1:1, so
that the structure has about 60% of type X, 20% type Y, and 20%
type Z.

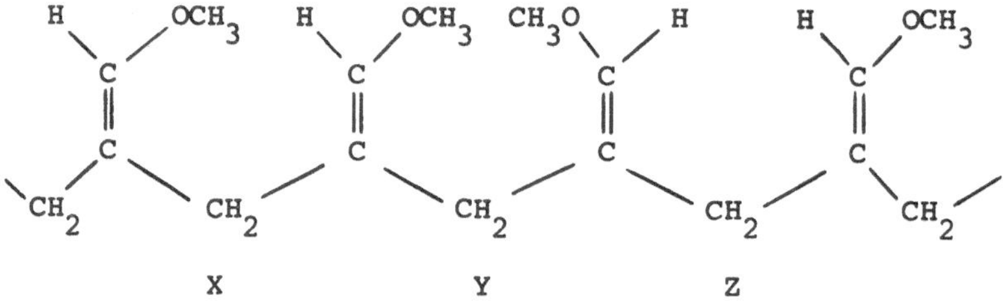

III

3.1.5. Stereospecific polymerization of acetylenes. Stereo-
regular polymers derived from acetylene or substituted acetylenes
represent a very interesting class of compounds which has become
increasingly more significant to polymer scientists in recent
years. Characteristic of these polymers is the presence of
conjugated, cis or trans double bonds *. In these polyconjugated
compounds, the π-electrons of the double bonds tend to be delocalized along the polymer chains. This implies that the polymer molecules may prevailingly assume cisoid or transoid conformations. As a consequence, these compounds display a number of
unusual properties (76).

---

* See footnote of section 3.1.3.

Polyacetylene, the simplest example of this class of polyconjugated systems was synthesized in 1958 by Natta et al.(77)
as the trans form by means of typical Ziegler catalysts. Cis and
trans configurations of polyacetylene were reported first by
Watson, McMordie and Lands (78) in 1961. Shirakawa and Ikeda
(79-81) synthesized an all-cis and an all-transpolyacetylene
and pointed out the cis-transoid(IV)and trans-transoid(V)struc-
ture of these polymers.

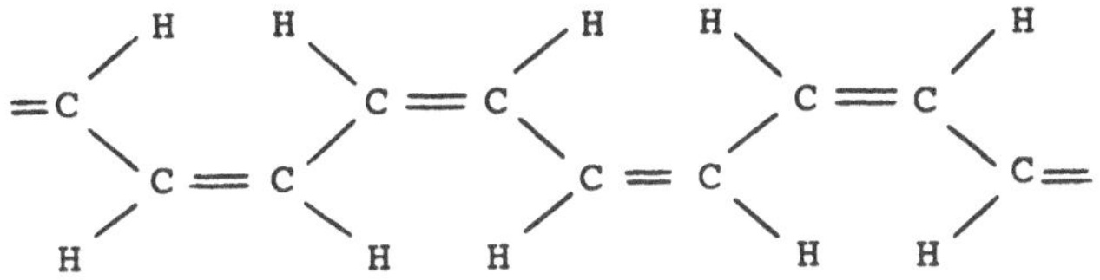

IV

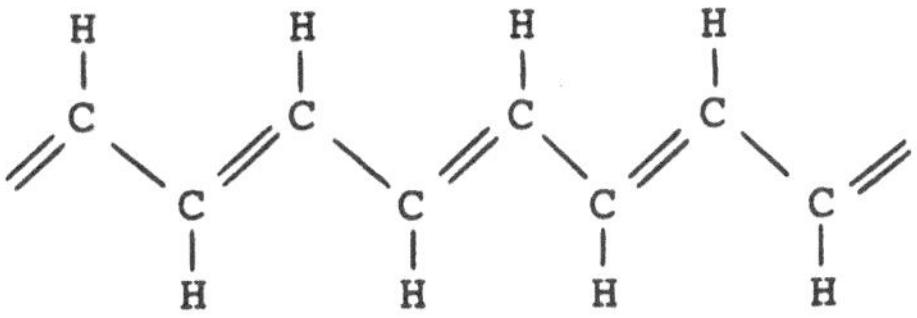

V

Since polyacetylene is insoluble in all solvents tested, identification of the isomers has been made by vibrational spectroscopic studies on thin films or by NMR spectra on the solid polymers
(82). The structural aspects of polymers of acetylene or acetylene
derivatives has recently been discussed by Simionescu et al. (83).
The mechanism of stereoregulation in the stereospecific polymerization of acetylene or substituted acetylenes is still unclear.

Stereoregular polymers with conjugated double and triple
bonds in the main chain have been obtained by solid state polymerization from a number of oligoacetylenes with conjugated
triple bonds (84). The polymerization of two such monomers is
illustrated in Scheme 15. The configuration of the double bonds
connecting the monomeric units in these polymers is always trans.
The crystalline monomer phase functions as a three-dimensional

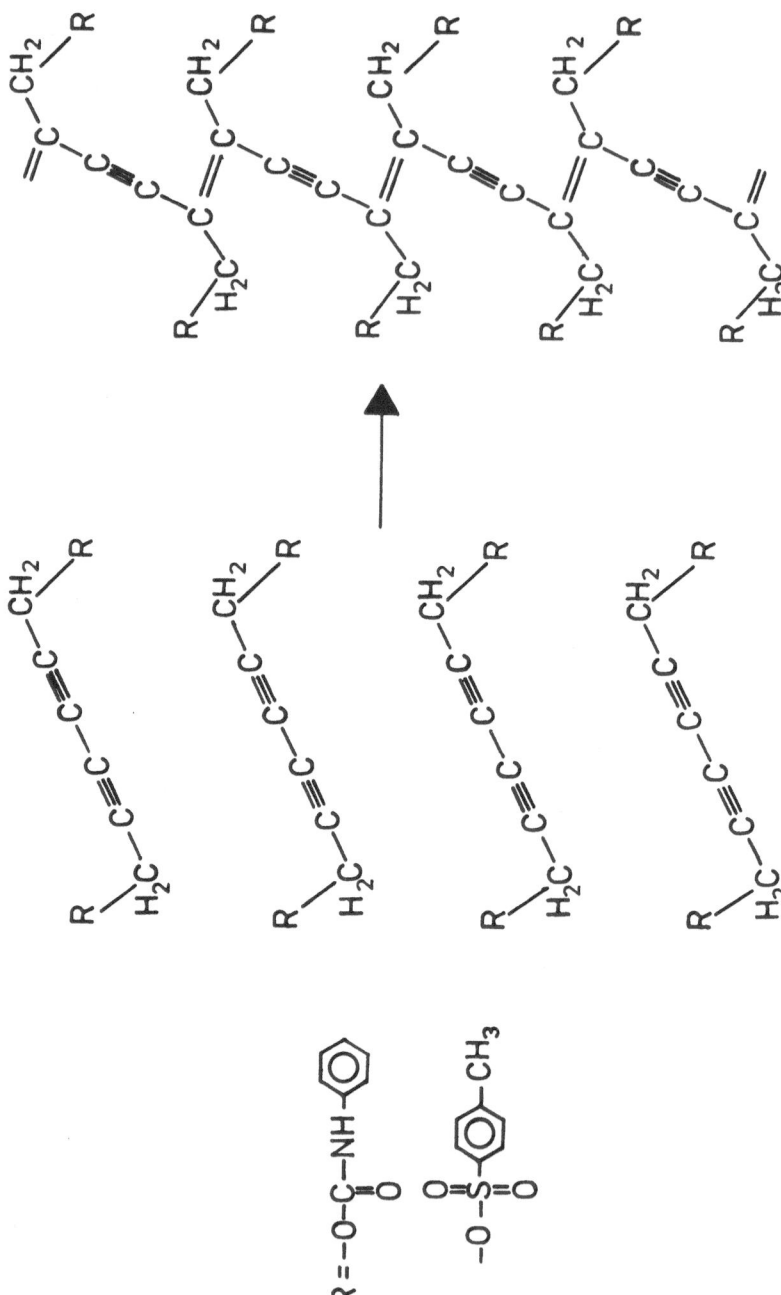

Scheme 15.   Topochemical polymerization of monomers with conjugated triple bonds (85)

template. The structural order of the crystalline monomer is preserved during the solid state reaction, and the resulting polymer is obtained in the form of a single crystal. These polymeric crystals are of considerable importance for investigations of the properties of the nearly defect-free polymeric state.

3.1.6. <u>Stereospecific polymerization of aldehydes</u>. A number of aliphatic aldehydes and haloaldehydes has been polymerized to stereoregular, isotactic polymers often using simple initiators such as organoaluminum or organozinc compounds (86). Especially in the case of aldehydes with long side chains or bulky side groups isotactic polymers are readily formed. According to Yasuda and Tani (87), in the stereospecific polymerization of an aliphatic aldehyde, the bulkiness of the alkyl group is the most important factor in the stereoregulation of the polymerization. The catalyst enhances the degree of stereoregulation by controlling the mode of approach of the incoming monomer sterically, through the coordination.

3.1.7. <u>Stereospecific polymerization of cyanides and iso-cyanides</u>. Geometrical isomerism around a C=N-double bond (syn and anti forms) is possible in principle. Polymers with C=N-double bonds in the main chain have been obtained by the polymerization of cyanides and by the ring-opening polymerization of pyridine (88). The polymer obtained from pyridine has monomeric units of the type VI. The configuration of the double bonds in

$$- CH = CH - CH = CH - CH = N -$$

VI

these polymers, however, does not seem to have been thoroughly investigated. Polymers (VII) with C=N-double bonds pendant on each backbone carbon atom have been obtained from isocyanides(89).

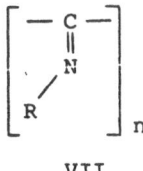

VII

Various kinds of experimental data (solubility, viscosity X-ray-diagrams) suggest that polyisocyanides with bulky substituents exist in the form of rod-like helical macromolecules, and contain long sequences of monomeric units of the same configuration (either syn or anti). In the case of poly-<u>tert</u>-butyl-isocyanide,

the helical conformation is stable in solution even at room
temperature, as shown by the separation of right-handed and left-
handed helices by elution chromatography using poly[(+)-sec-butyl-
isocyanide] as support (12) (Scheme 16). This polymer represents
a rare case of atropoisomerism in aliphatic compounds. In the
stereospecific polymerization, by which this polymer is formed,
two different types of stereoisomerism sites are created from
a monomer not containing stereoisomerism sites.

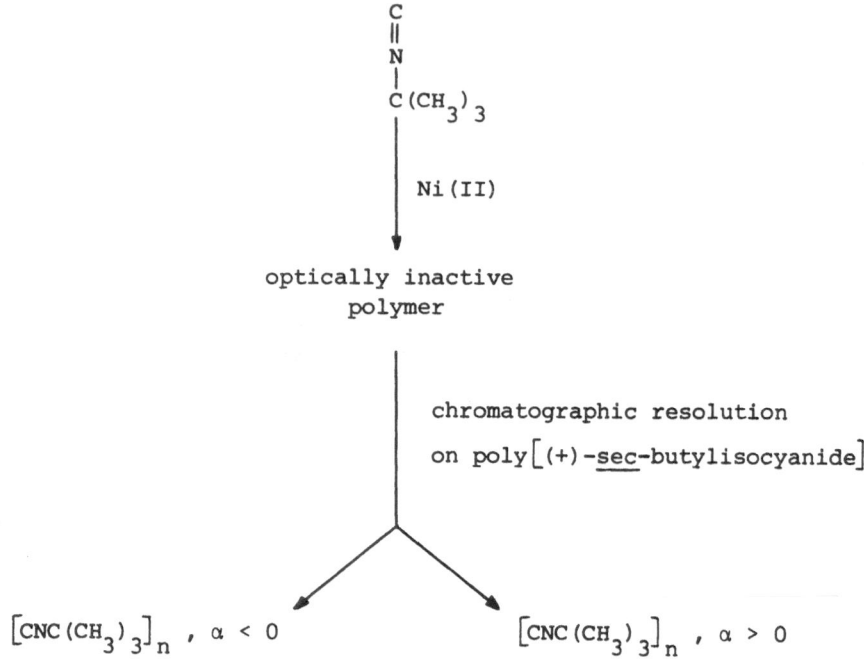

Scheme 16.   Chromatographic separation of poly-tert-butyl-
             isocyanide

3.2.   Stereoregular Polymers From Monomers Containing Stereo-
       isomerism Sites By Processes Involving Neither Modification
       Of The Sites Already Existing In The Monomers Nor Formation
       Of New Ones

       A very obvious way of producing stereoregular polymers
using this synthetic approach is to start from monomers of high
steric purity. Since the polymerization processes considered

involve neither modification of the stereoisomerism sites already existing nor formation of new sites, highly stereoregular polymers must result in this case. This approach has often been used, the most serious drawback being that monomers of high steric purity are generally difficult to prepare.

A second, more interesting method for producing stereoregular polymers by this approach is to polymerize sterically inhomogeneous monomers (e.g. racemic monomers or mixture of diastereomers) with polymerization catalysts which are able to discriminate between the enantiomeric or diastereomeric forms of the monomers. As a limiting case, with these catalysts macromolecular chains form, each of which derives from one steric isomer only, and hence is stereoregular. Only in a very few cases have highly stereoregular polymers been obtained in this way, and in all those cases mixtures of enantiomeric isomers have been used as the starting monomer. Of these enantiomer discriminating syntheses, two versions are known. In one version, chiral racemic catalysts are used, whereby the two enantiomorphic forms of these catalysts polymerize selectively the one or the other enantiomer of the monomer. Thus, in the ideal case, the polymer produced from a racemic monomer consists of a mixture of macromolecules, each derived exclusively from the one or the other enantiomer of the monomer (Scheme 17). This process has been

$$
n(R) + n(S) \xrightarrow{\text{racemic catalyst}} \left\{ \begin{array}{l} \sim\!\!\!\sim (R)\,(R)\,(R)\,(R)\,(R)\,(R)\,(R) \sim\!\!\!\sim \\[2ex] \sim\!\!\!\sim (S)\,(S)\,(S)\,(S)\,(S)\,(S)\,(S) \sim\!\!\!\sim \end{array} \right.
$$

Scheme 17.   Ideal stereoselective polymerization of a racemic monomer.

called "stereoselective" (90). In the second version, an optically active catalyst, that is a catalyst containing an excess of one of the two enantiomorphic catalytic sites, is used, so that one of the enantiomers of the monomer is preferentially polymerized. In the ideal case, with an optically pure catalyst, only one of the enantiomeric monomers should polymerize (Scheme 18).

$$
n(R) + n(S) \xrightarrow{\text{optically pure catalyst}} \sim\!\!\!\sim (R)\,(R)\,(R)\,(R)\,(R)\,(R)\,(R) \sim\!\!\!\sim + n(S)
$$

Scheme 18.   Ideal stereoelective polymerization of the enantiomer (R) of a racemic monomer.

This kind of polymerization, which has been called "stereoelective" (91) or "asymmetric selective" (92), corresponds to
a kinetic resolution of a racemic monomer.

It is clear that stereoselectivity implies stereoelectivity,
and that a relationship must exist between the actual stereoselectivity and stereoelectivity observed when a racemic monomer
is polymerized using a given chiral catalyst. If the relative
overall rate constants of polymerization of the two enantiomers
on a single chiral catalytic center are known (e.g. from a
stereoelective polymerization experiment), the average relative
amounts of the two enantiomeric monomeric units in a polymer
chain formed in the stereoselective polymerization can, with
certain assumptions, be evaluated.

A special type of stereoelectivity can be foreseen in a
copolymerization of a racemic monomer with an optically active
comonomer. Under the influence of the latter, one   enantiomer
of the racemic monomer can be preferentially incorporated in the
polymer chains. Only a few systems of this type have been investigated. In the following sections we shall consider separately
the synthesis of stereoregular polymers from monomers of high
steric purity, and from racemic monomers.

3.2.1. <u>Polymerization of monomers having high steric purity</u>.
Among the polymers containing cis or trans carbon-carbon double
bonds as sites of stereoisomerism, polyesters obtained from unsaturated dicarboxylic acids or their derivatives predominate.
Polyesters from maleic anhydride or fumaric acid were prepared
as long ago as 1929 by Carothers and Arvin (93). A number of
polyamides have also been prepared from derivatives of unsaturated acids (94,95). It is worth noting that the steric purity
of the monomers does not guarantee the stereoregularity of the
polymers. Isomerization of the double bonds may take place
during the polymerization, especially when acids are present (96).

Stereoregular polymers containing asymmetric carbon atoms
in the main chain as sites of stereoisomerism have been obtained
from a large number of optically active monomers. Many stereoregular polymers of this kind have been obtained by polycondensation. In the case of some chiral monomers having two functional
groups of the same type, the polycondensation may lead to
different, structurally isomeric polymers (97). This is illustrated by the polycondensation of hexamethylenediamine with β-methyl
adipic acid (Scheme 19). In this case (98), and in similar cases,
steric control is easily achieved by using optically pure monomers.
The control of structural isomerism is more difficult, and up to
now it has not been thoroughly investigated (99).

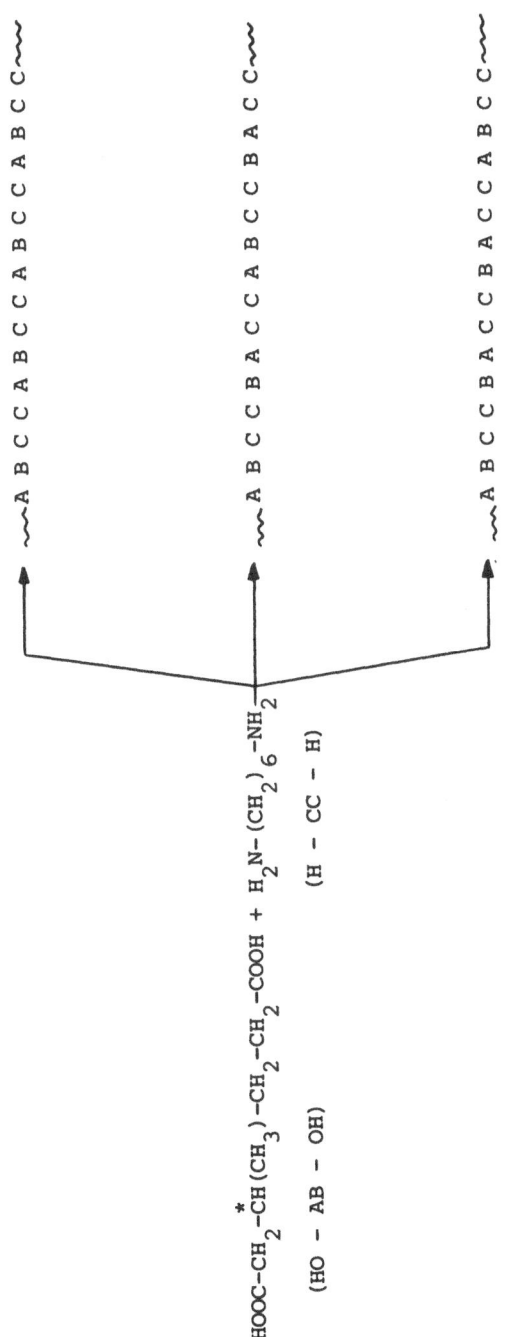

Scheme 19. Structurally isomeric polymer chains that may form in the polycondensation of β-methyladipic acid with hexamethylene diamine.

Other stereoregular polymers with asymmetric carbon atoms in the main chain as stereoisomerism sites have been obtained by ring-opening polymerization of optically active cyclic monomers. These monomers include epoxides (100-104), episulfides (105,106), aziridines (107-109), lactides(110), lactones (111-116), thiolactones (117), lactames (118-120), cyclic acetals (121), and N-carboxy anhydrides (122). Some stereoregular polymers containing atropoisomeric units as stereoisomerism sites (VIII (123) and IX (124)) have also been synthesized.

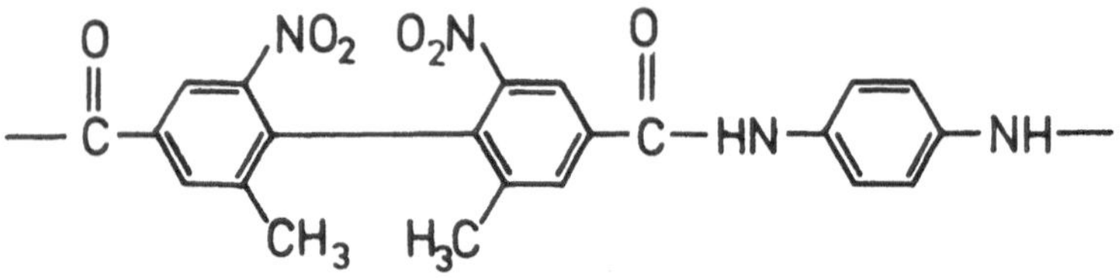

VIII

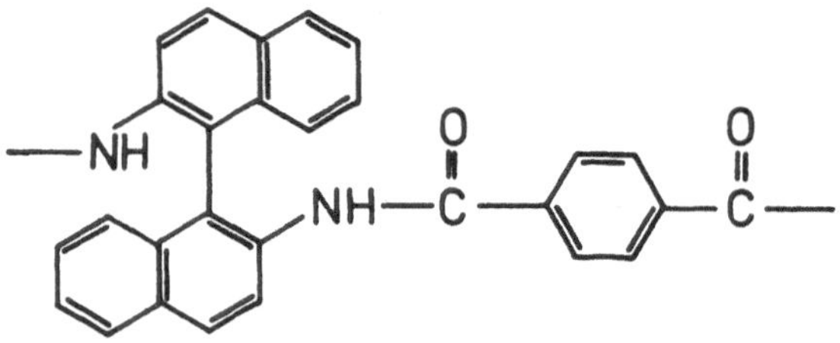

IX

Finally, stereoregular polymers with asymmetric carbon atoms in the side chains as stereoisomerism sites have been produced by polycondensation (125) and by polymerization of optically active isocyanates (X (126)). As stressed in section 2.2. we do not consider the amide groups in the main chain of polyiso-cyanates as sites of steric isomerism.

3.2.2. Stereoselective or stereoelective polymerization of racemic monomers. Stereoregular polymers have been obtained by stereoselective polymerization from many cyclic, racemic monomers (100,121,127-140). Some of these monomers are indicated in Table 2. The identification of the stereoselective character of a polymerization process, which leads to isotactic optically inactive polymers, requires a very accurate characterization of the polymer obtained. For the polymers reported in Table 2, elution chromatography on an optically active support, spectroscopic (IR, or NMR) measurements, comparison of the X-ray pattern with that of the corresponding optically active polymer, and enzymic degradation have been used. In many cases the stereoselectivity of the polymerization process has been confirmed by the stereoelectivity observed when using the catalyst in an optically active form.

One case of stereospecific, but not stereoselective polymerization of a racemic cyclic monomer is known. The polymerization of racemic-tert-butylethyleneoxide using tert-BuOK as the catalyst has yielded a crystalline polymer, for which an heterotactic structure has been proposed (141).

No example of stereoregular polymer prepared by stereoselective polycondensation has been reported up to now.

Optically active, stereoregular polymers have been obtained by stereoelective polymerization with optically active catalysts from racemic epoxides (104,142-145), episulfides (146-148), aziridines (146), lactones (149), and N-carboxy anhydrides (150-152).

Stereoselectivity and stereoelectivity in the ring-opening polymerization of racemic epoxides, as well of other cyclic monomers, is very likely determined by a coordination step which

Table 2. Some examples of stereoselective polymerization

Racemic monomer	Catalytic system	Method used for studying the stereoselectivity	% isotactic dyads	Ref.
$CH_3-CH\overset{O}{\underset{}{\diagdown}}CH_2$	$ZnEt_2/CH_3OH$	chromatographic separation		129
		NMR	93	134
$CH_3-CH\overset{S}{\underset{}{\diagdown}}CH_2$	$ZnEt_2/H_2O$	X-ray pattern		130
		NMR	79	131
CH_3-CH-O / CH_2-CO	$\left[Et_2Al-O-C(Ph)=NPh\right]_2$	X-ray pattern and IR	not determined	135
$CH_3-CH-CO$ / $NH-CO$	$rac.Ni\left[C_2H_5(CH_3)CHCOO\right]_2/$ $n.Bu_3P$	IR	not determined	136
$CH_3-CH-CH_3$ / $CH_2-CH-CO$ / $NH-CO$	$R-NH_2$	enzymic degradation	~95	139

precedes the addition of the monomer to the growing chain. As
shown in Scheme 20 for the case of propylene oxide, due to the
presence of the substituent, the coordination of the oxygenation
can be more or less hindered depending on the side of approach
of the coordinating agent. If the coordinating agent is chiral,

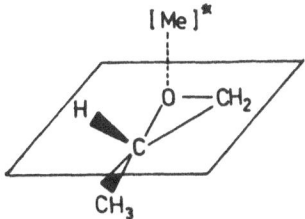

Scheme 20. Type of coordination complex which may be preferen-
            tially formed between a chiral catalytic center and
            (S)-propylene oxide.

one antipode of the monomer will be more readily coordinated
than the other. In stereoselective polymerization, an equal
number of enantiomorphic catalytic sites should exist, the
catalyst being optically inactive, and therefore the polymerization rate of the two antipodes of the monomer must be the same.
In stereoelective polymerization the number of catalytic sites
having opposite type of chirality is different, and therefore
the rate of polymerization of the two antipodes of the monomer
is different.

   In the case of the synthesis of poly-α-amino acids from
N-carboxy anhydrides it is possible that the conformational
rigidity of the growing chain plays an important role in determining the stereoselective or stereoelective character of the polymerization.

   In general, the overall polymerization rate constants of the
two enantiomers are not very different, the ratio r * between
these rate constants being in most cases lower than three (Table 3).

---

* Provided certain kinetic conditions are satisfied, r can be
  calculated by the formula (153)

$$(1-x)^{r-1} = \frac{1 + \alpha/\alpha_0}{(1 - \alpha/\alpha_0)^r}$$

where x is the conversion, α_0 the optical rotation of the pure
enantiomer, and α the optical rotation of the non-polymerized
monomer.

Table 3. Some examples of stereoelective polymerization

Racemic monomer	Catalytic system	Degree of electivity r [a]	Ref.
$CH_3-CH-CH_2$ (O)	$ZnEt_2-(+)borneol$	1.5–1.6	143,154
$CH_3-CH-CH_2$ (S)	$CdMe_2-(-)t.BuCHOHCH_2OH$	1.9	155
$C(CH_3)_3-C-CH_2$ (S)	$ZnEt_2-(-)t.BuCHOHCH_2OH$	2.8	148
$CH_3-CHCH_3-CH_2-CH-CO-O-NH-CO$	L-proline methylester	1.3	152

a) Overall rate constant of polymerization of faster polymerizing enantiomer/overall rate constant of polymerization of slower polymerizing enantiomer.

The low stereoelectivity can be due to an intrinsic property of the catalytic center, and/or to the fact that the optical purity of the catalyst is small (that is, the prevalence of catalytic centers with one configuration is small).

Examples of stereoelective copolymerization have been reported. Matsuura, Tsuruta, Terada, and Inoue (156) prepared a polyester from 3-phenyl-Δ^4-tetrahydrophthalic acid anhydride and propylene oxide using diethyl zinc-(+)borneol as an optically active catalyst and obtained an optically active polyester. The authors reported that the stereoelective copolymerization of 3-phenyl-tetrahydrophthalic acid anhydride had occurred, but that of propylene oxide was not observed. Kumata, Furukawa, and Saegusa (157) copolymerized racemic propylene oxide with ethylene oxide using the $ZnEt_2/(+)$-borneol system, and found that stereoelectivity of propylene oxide was not hindered by the occurrence of an ethylene oxide unit at the end of the growing chain.

A special case of stereoelectivity has been encountered in polymerizing a racemic monomer with an optically active one. Minoura et al. (158) carried out the ring-opening copolymerization of propylene oxide with d-camphoric acid anhydride using diethyl zinc and triethyl amine as catalyst. It was found that the products were alternating copolymers which were optically active. The propylene oxide recovered from the copolymerization system was also optically active. An attempt to incorporate preferentially in a polyamide chain one of the optical antipodes of trans-cyclopropane-1,2-dicarbonyl chloride, by polycondensing this racemic monomer with optically active 1,2-propylene diamine failed (159).

3.3. Stereoregular Polymers From Monomers Containing Stereo-isomerism Sites By Processes Involving Formation Of New Stereoisomerism Sites, But Without Modification Of Those Already Present In The Monomers

As far as we know, there is only one example of a stereoregular polymer belonging to this class that has been synthesized from a monomer containing a double bond as a site of stereoisomerism. This is the case of the 1,2-syndiotactic polymer derived from 1,3-trans pentadiene (XI) (160). This polymer has side chains with the trans double bonds already present in the monomer. In all other cases of this class, the stereoregular polymers have been obtained from unsaturated monomers containing an asymmetric carbon atom. The asymmetric carbon atom remains in

the lateral chains of the polymers and constitutes a source of
steric isomerism. Additional stereoisomerism sites, either

$$
\left[-CH_2 - \underset{\underset{\underset{CH_3}{|}}{\underset{\|}{\underset{C-H}{H-C}}}}{\overset{\overset{H}{|}}{\underset{|}{C}}} - CH_2 - \underset{\underset{H}{|}}{\overset{\overset{\overset{\overset{CH_3}{|}}{H-C}}{\underset{\|}{\underset{C-H}{C-H}}}}{C}} - \right]_n
$$

XI

asymmetric carbon atoms or double bonds, form during the polymerization and become part of the main chain.

The synthesis of these polymers may be carried out starting
from either optically active monomers of very high optical purity
or racemic monomers. In the first case polymers having steric
order in the side chains are obtained irrespective of whether
the polymerization process is stereospecific or not. In the
second case the polymers exhibit steric order in the side chains
only if the process is such that each macromolecule predominantly
derives from one of the antipodes of the monomer (stereoselective
or stereoelective process).

3.3.1. Polymerization of monomers having high optical purity.
The polymers belonging to this class are optically active, and
many of them have been synthesized in order to use optical activity as a probe for the study of the conformational equilibria
of polymer chains in solution. The investigations carried out
by Walden (161) in 1896 on a polymer derived from the isoamyl
ester of itaconic acid may be considered the very first approach
of this kind.

The polymers most investigated up to now are those derived
from optically active α-olefins, vinyl ethers, vinyl ketones,
acrylic or methacrylic esters, acrylamides, and aldehydes. As
indicated in Table 4, isotactic and, in certain cases, isotactic
and syndiotactic polymers have been obtained by using stereospecific processes. The stereospecific processes used with these
chiral monomers are the same as those employed for producing
stereoregular polymers from the corresponding achiral monomers.

Table 4. Classes of optically active, unsaturated monomers,
that have been polymerized to stereoregular polymers

| Monomers | Polymers obtained | | Ref. |
	Repeating unit	Main chain stereoregularity	
α-olefins	$-CH_2-CH-$ $|$ R^*	isotactic	162–166
Vinyl ethers	$-CH_2-CH-$ $|$ O $|$ R^*	isotactic	167–169
Vinyl ketones	$-CH_2-CH-$ $|$ CO $|$ R^*	isotactic	170–171
Acrylic esters	$-CH_2-CH-$ $|$ CO $|$ O $|$ R^*	not determined	172
Methacrylic esters	CH_3 $|$ $-CH_2-C-$ $|$ CO $|$ O $|$ R^*	isotactic or syndiotactic	173–176
Acrylamides	$-CH_2-CH-$ $|$ CO $|$ $N-R$ $|$ R^*	not determined	177
Aldehydes	$-CH-O-$ $|$ R^*	isotactic	178

It is worth noting that the two faces of these chiral monomers are diastereofaces, and therefore they should have different reactivities. This has been experimentally shown in the
case of some chiral α-olefins by synthesizing the corresponding cis- and trans-dichloro, amino platinum complexes (179). It
was found that, when the olefin is (S) and the asymmetric carbon
atom of the olefin is in the α- or β-position with respect to
the double bond, the face prevailingly bound to Pt is the si-si.
The prevalence depends on the structure of the olefin and is
larger the nearer the asymmetric carbon atom is to the double
bond. In certain cases, this prevalence may reach 80% or more.
Because of the different reactivities of the two diastereofaces,
a small degree of stereoregularity should be expected, even in
polymers produced by processes that are non-stereospecific for
achiral monomers. Although conclusive experimental confirmation
is still lacking, some hints in this direction have appeared in
the literature. For instance, Wulff (180) has copolymerized a
p-vinyl phenylboronic ester of D-mannitol with methyl methacrylate by radical initiation. After complete removal of the mannitol
groups by hydrolysis, the copolymer was still optically active,
indicating some prevalent configuration of the asymmetric carbon
atoms of the main chain (Scheme 21). Probably the monomers should

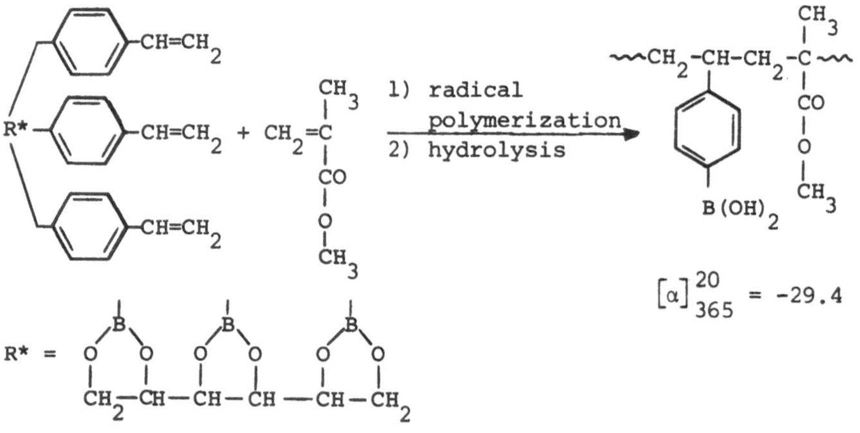

Scheme 21.   Synthesis of a "partially stereoregular" copolymer
             by radical copolymerization of methyl methacrylate
             with an optically active comonomer.

be more carefully designed in order to use NMR to make a quantitative evaluation of the small differences in degrees of stereoregularity.

Other examples of stereoregular polymers obtained by chiral
monomers having high optical purity are chiral poly-1-alkynes
(XII) (181) and polyisocyanides (XIII (182), XIV (183)), although
the configuration of the double bonds present in the macromolecules has not been thoroughly investigated up to now.

$$\begin{bmatrix} - \text{ C} = \text{CH} - \\ | \\ \text{CH}_2 \\ | \\ *\text{CH-CH}_3 \\ | \\ \text{C}_2\text{H}_5 \end{bmatrix}_n \qquad\qquad \begin{bmatrix} - \text{ C} - \\ || \\ \text{N} \\ | \\ *\text{CH-CH}_3 \\ | \\ \text{C}_6\text{H}_5 \end{bmatrix}_n$$

XII   XIII

(poly-(S)-4-methyl-1-hexyne)   (poly-(S)-α-phenyl ethyl iso-
                                     cyanide or
                                     poly-(R)-α-phenyl ethyl iso-
                                     cyanide)

$$\begin{bmatrix} - \text{ C} - \\ || \\ \text{N} \\ | \\ *\text{CH-CH}_3 \\ | \\ \text{C}_2\text{H}_5 \end{bmatrix}_n$$

XIV

(poly-(S)-sec.butyl isocyanide)

3.3.2. Stereoselective or stereoelective polymerization of
racemic monomers. The stereoselective polymerization of racemic
monomers was first investigated in the case of racemic α-olefins
(184). When the asymmetric carbon atom was in the α-position with
respect to the double bond, essentially stereoregular polymers
were obtained, the single macromolecules having isotactic main
chain and carbon atoms with mostly the same configuration in the
lateral chains. The stereoselectivity decreases when the asymmetric carbon atom of the monomer is in the β-position with

respect to the double bond, and disappears when the asymmetric
carbon atom is in the γ-position (185) (Table 5).

Table 5.   Stereoselectivity in the stereospecific polymerization
            of racemic α-olefins of the type $CH_2=CH-(CH_2)_n-CH-C_2-H_5$
            (catalytic system: $TiCl_3/AlR_3$) (185) $|$
                                                                      CH_3

n	Isotactic fraction considered	Stereoselectivity
0	ethylacetate ins., diethylether sol.	$\sim$90%
1	Diisopropylether ins., isooctane sol.	$\sim$20%
2	methanol ins., diethylether sol.	$\sim$ 0%

In the latter case, macromolecules have been obtained with isotactic main chains, but without order in the distribution of the
configurations of the asymmetric carbon atoms of the lateral
chains. Similar results were obtained in the stereospecific
polymerization of vinyl ethers (186), where, however, the asymmetric carbon atom nearest to the main chain can only be in the
β-position with respect to the main chain, and therefore stereoselectivity is low.

   Stereoselectivity has also been observed in the copolymerization of racemic α-olefins with ethylene (187,188) and in the
copolymerization of racemic α-olefins with an optically active
α-olefin (189,190). In the latter case a copolymer of the
optically active α-olefin and of the antipode with the same
chirality of the racemic monomer, together with a homopolymer of
the remaining antipode, was obtained (Scheme 22).

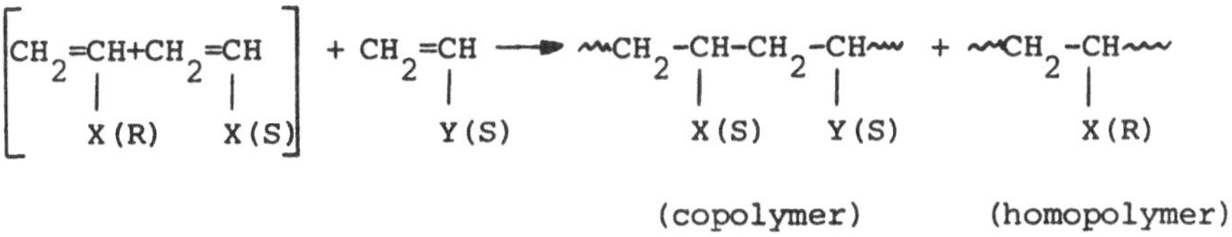

$$\left[\begin{array}{cc} CH_2=CH+CH_2=CH \\ | \quad\quad\quad | \\ X\,(R) \quad\quad X\,(S) \end{array}\right] + \begin{array}{c} CH_2=CH \\ | \\ Y\,(S) \end{array} \longrightarrow \begin{array}{cc} \sim\!\!CH_2-CH-CH_2-CH\!\!\sim \\ | \quad\quad\quad\quad | \\ X\,(S) \quad\quad Y\,(S) \end{array} + \begin{array}{c} \sim\!\!CH_2-CH\!\!\sim \\ | \\ X\,(R) \end{array}$$

$$\qquad\qquad\qquad\qquad\qquad\qquad\qquad (\text{copolymer}) \qquad\qquad (\text{homopolymer})$$

Scheme 22.   Example of a stereoselective copolymerization of a racemic α-olefin with an optically active α-olefin. The configuration of the asymmetric carbon atoms in the lateral chain is indicated in parentheses.

Stereoelectivity was observed in the polymerization of racemic α-olefins containing an asymmetric carbon atom in the α- or β-position with respect to the double bond by using Ziegler-Natta catalysts obtained from a transition metal halide and an optically active metal alkyl (185). However, the degree of electivity reached was very small, being in general less than 1.5. In the case of racemic α-olefins having the asymmetric carbon atom in the α-position with respect to the double bond, the fact that a low stereoelectivity is combined with a high stereoselectivity (see Table 5), shows that the reason for the low stereoelectivity is not the lack of capacity of the single centers to discriminate between the antipodes, but rather the small optical purity of the catalyst. Recently, Ciardelli and co-workers have reported (191) that stereoelective polymerization of racemic α-olefins can also be achieved with catalytic systems prepared from $TiCl_4$ or $TiCl_3$ and $Al(i-C_4H_9)_3$ which have been modified by addition of an optically active third component, such as (-)-α-pinene.

The stereoelective polymerization of a racemic vinyl ether in the presence of an optically active catalyst has been attempted (192), but not achieved. On the other hand, the cationic co-polymerization of racemic l-methylpropyl vinyl ether with (S)-l-phenylethyl vinyl ether or (R)-l-phenylethyl vinyl ether was shown (193) to be stereoelective in the presence of the heterogeneous catalyst $Al(O-i.C_3H_7)_3/H_2SO_4$.

A remarkably high stereoelectivity has recently been reported in the polymerization of α-methylbenzyl methacrylate (194,195). For the anionic polymerization of this monomer by a Grignard reagent/ (-) sparteine system in toluene at -78°C, the optical purity of the unreacted monomer reached nearly 100% at about 65% conversion. Additionally, the polymer obtained was shown by NMR to have a very high content of isotactic triads.

Stereoselectivity and stereoelectivity in the polymerization of racemic α-olefins have probably the same origin as in
the polymerization of epoxides (Section 3.2.2.), and are determined by the chiral character of the catalysts used. Chiral
catalytic centers of a given configuration attack prevailingly
one face of the double bond of the monomer and centers of the
opposite configuration attack prevailingly the other face. In the
case of chiral olefins, the two diastereofaces of a monomer
molecule have different reactivities, and whether the re-re face
or the si-si face of this molecule is more reactive depends on
the type of chirality of the asymmetric carbon atom. For instance,
on the basis of the investigation on Pt-complexes (179), the
si-si face is the more reactive face in an (S)-olefin. Thus
chiral catalytic centers which attack the si-si face may prefer
the (S)-antipode, whereas those which attack the re-re face may
prefer the (R)-antipode (Scheme 23).

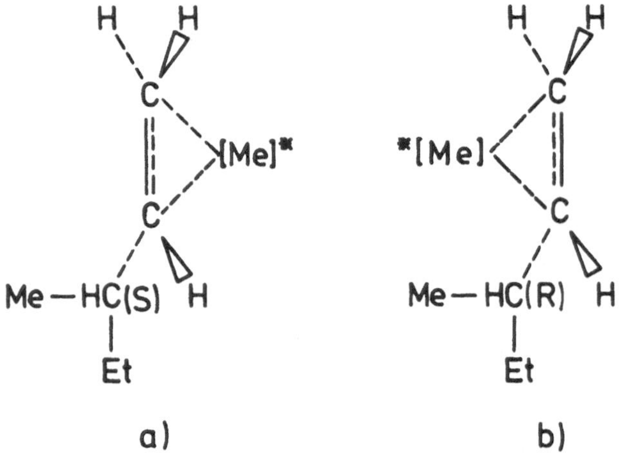

Scheme 23.   Type of coordination complexes which may be prefer-
             entially formed between chiral catalytic centers of
             opposite configuration and the antipodes of a chiral
             α-olefin.
             a) (S)-3-methyl-1-pentene; b) (R)-3-methyl-1-pentene.

The factors regulating stereoselectivity and stereoelectivity in the polymerization of vinyl ethers and of methacrylic esters have not been sufficiently investigated up to now.

3.4. Stereoregular Polymers From Monomers Containing Stereoisomerism Sites By Processes Affecting At Least One Of These Sites

Numerous stereospecific polymerization processes for monomers containing stereoisomerism sites, one or more of which are affected by the polymerization, are known. The sites may be either internal double bonds or asymmetric carbon atoms. The first case is encountered with the stereospecific ring-opening polymerization of cycloolefins by metathesis catalysts, and with the stereospecific polymerization of cycloolefins and other unsaturated monomers by double bond opening, and of substituted, conjugated dienes. The second case can be exemplified by the stereospecific ring-opening polymerization of epoxides or episulfides with two asymmetric carbon atoms in the ring. The polymerization (196) of α-methyl benzylchloride to a crystalline poly-α-methylbenzyl containing a significant proportion of isotactic sequences is an additional example of this type of stereospecific synthesis.

3.4.1. Stereospecific ring-opening polymerization of cyclic olefins by metathesis catalysts. Stereoregular polymers belonging to this class are obtained by the ring opening polymerization of cycloolefins. This type of polymerization, which was first reported by Natta and co-workers (197-199), is catalyzed by compounds of several transition metals, especially those of tungsten, which are known catalysts for olefin metathesis (200). In 1971 Dall'Asta and Motroni (201) gave evidence that the cleavage of the double bond was taking place during the polymerization of cycloolefins, and a metathesis-type mechanism was proposed. This demonstrated that the stereoisomerism site of the monomers is involved in the polymerization. In this connection it is relevant that the stereoregularity of the resulting polymer depends on the stereospecificity of the polymerization process and not on the stereochemistry of the double bond existing in the monomer. As Table 6 shows, cyclopentene, which contains necessarily a cis double bond, yields, depending on the catalytic system used, either a cis- or trans-polymer, and cis-cyclooctene yields a trans-polymer. Alkyl substituted cycloolefins have also been polymerized by metathesis catalysts. From 3-methyl-cyclooctene, a polymer containing ca. 65% of the double bonds in the trans configuration has been obtained (203).

Table 6. Examples of cycloolefin polymerization by metathesis

Monomer	Catalytic System	Type of Polymer	Ref.
Cyclobutene	$RuCl_3-C_2H_5OH$		198
Cyclopentene	$WCl_6-Al(C_2H_5)_3$		199
Cyclopentene	$MoCl_5-Al(C_2H_5)_3$		199
Cycloheptene	$WCl_6-Al(C_2H_5)_2Cl$		202
Cis-cyclooctene	$WCl_6-Al(C_2H_5)_3$		202

3.4.2. <u>Stereospecific polymerization or copolymerization of</u> <u>unsaturated monomers by opening of an internal double bond</u>. Linear internal olefins could not be homopolymerized to stereoregular polymers, but they have been copolymerized with ethylene to stereoregular polymers by stereospecific catalysts (204). For instance, a stereoregular crystalline alternating copolymer was obtained from cis-2-butene and ethylene. This copolymer was found by X-ray analysis to have an erythro diisotactic structure (205) (Scheme 24). In contrast to linear internal olefins, β-substituted vinyl ethers could be homopolymerized to stereoregular diisotactic polymers by polymerization at low temperatures in the presence of alkyl aluminum halides. For example, threo-di-isotactic polymers were obtained from trans-1-methyl-2-alkoxy-ethylenes (206) and from trans-1-chloro-2-alkoxy-ethylenes(207), and erythro-di-isotactic polymers were obtained from cis-chloro-2-alkoxy-ethylenes (207) (Table 7). Racemic cis- and trans-1-methylpropyl propenyl ether have been polymerized by asymmetric alkoxyaluminum dichlorides, and in the case of the cis monomer stereoelectivity has been observed (208).

Stereoregular polymers and copolymers have been obtained from a number of unsaturated cyclic compounds. Crystalline cyclobutene polymers having cyclobutene units have been obtained using catalysts, both homogeneous and heterogeneous, prepared from vanadium salts and organoaluminum compounds (198). The X-ray and IR-spectra of the polymers obtained are different, depending on which type of catalysts was used. A di-isotactic and disyndiotactic structure respectively has been attributed to the polymers (198) (Table 7). In the case of cycloolefins with more than 4 carbon atoms in the ring, no homopolymerization in the presence of co-ordinated anionic catalysts has been observed *. On the other hand, the copolymerization of these cycloolefins with ethylene is possible with these catalysts. In the case of cyclopentene (210) and cycloheptene (211), crystalline alternating copolymers with stereoregularity of the erythro-di-isotactic type have been obtained using catalysts from vanadium salts.

Diisotactic polymers were prepared from cyclic vinyl ethers, such as 2,3-dihydropyran (XV) (212), and from benzofuran (XVI) α-naphthofuran (XVII) and β-naphthofuran (XVIII). From the last three monomers predominantly isotactic, optically active polymers

---

* Acenaphthylene has been homopolymerized by radical emulsion polymerization to a product having predominantly a threo disyndiotactic structure (209).

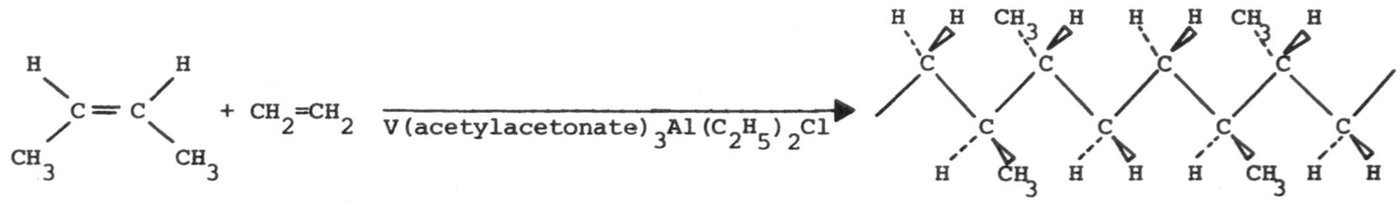

Scheme 24. Synthesis of an alternating copolymer with erythro-di-isotactic structure from cis-2-butene and ethylene.

Table 7. Some diisotactic polymers obtained from monomers with an internal double bond

Monomer	Catalyst	Type of Polymer	Ref.
	AlR_2Cl		207
	AlR_2Cl		207
	VCl_3/AlR_3 $V(acac)_3/Al(C_2H_5)_2Cl$		198 198

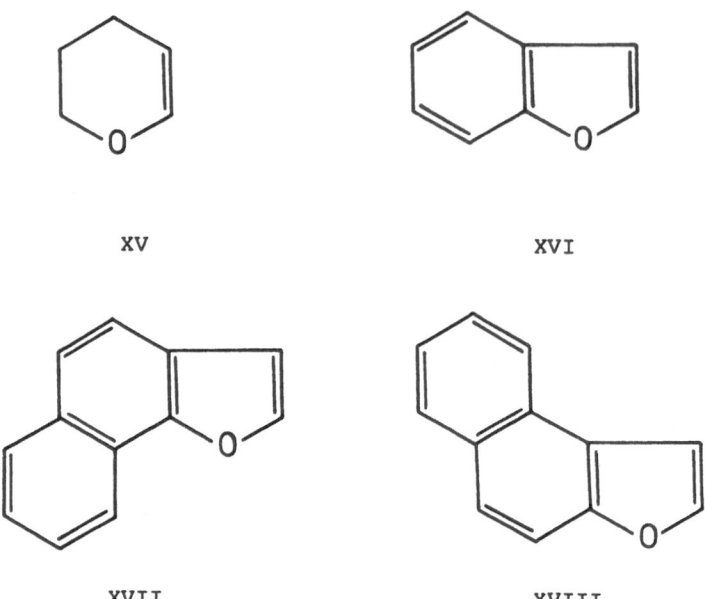

XV   XVI

XVII   XVIII

were obtained by polymerizing with a system based on an alkyl
aluminum halide and an optically active Lewis base (213,214).
Earlier attempts to produce stereoregular polymers from cyclic
monomers by an asymmetric polymer synthesis had been carried out
by Schmitt and Schuerch (215), but in no case did the polymers
obtained show a detectable optical activity.

Partially stereoregular copolymers have been obtained by
asymmetric polymer synthesis using maleic anhydride as one comonomer and optically active α-methylbenzyl methacrylate (216)
or α-methylbenzyl vinyl ether (217) as the other comonomer. These
copolymers were optically active even after removal of their
α-methylbenzyl groups (Scheme 25). Analogous results have been
obtained by Minoura's group (218, 219) on copolymerizing optically active α,β-disubstituted olefins with achiral vinyl monomers.

Finally, in a series of recent papers, Minoura and coworkers
have reported the formation of optically active copolymers from
styrene/maleic acid(220), styrene/maleic anhydride (221),
indene/maleic anhydride (222), and indene/acrylic acid (223) by

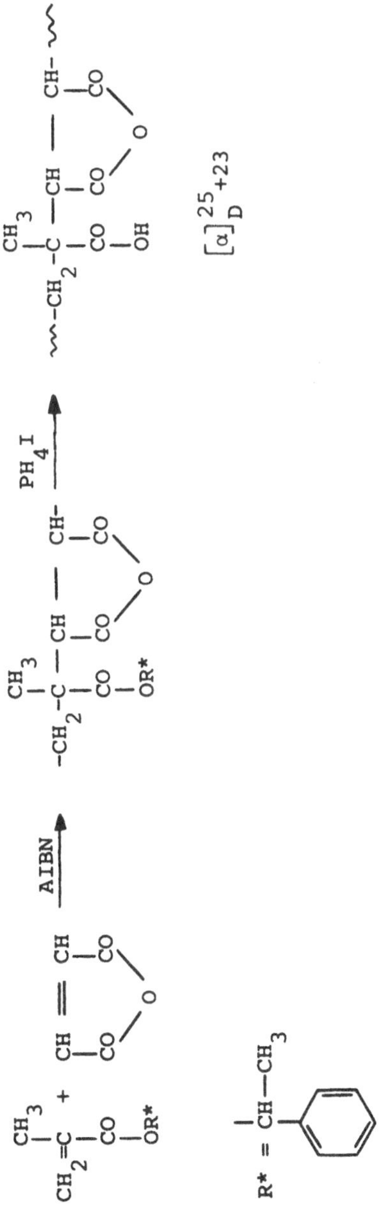

Scheme 25. Example of an asymmetric copolymer synthesis.

radical initiation in the presence of lecithin. The observed
optical rotation indicates that an excess of one configuration
is present in the chain backbone of these copolymers.

    3.4.3. <u>Stereospecific polymerization of substituted,conju-
gated dienes</u>. Stereoregular polytactic polymers have been
obtained from a number of substituted dienes, including one
optically active 1,3-substituted propadiene, and various 1- or
1,4-substituted butadienes. (R)-penta-2,3-diene has been polymerized by means of π-allyl-Ni-iodide to an optically active
polymer, to which an interesting stereoregular structure has been
attributed (Scheme 26) (224). Some of the stereoregular polymers

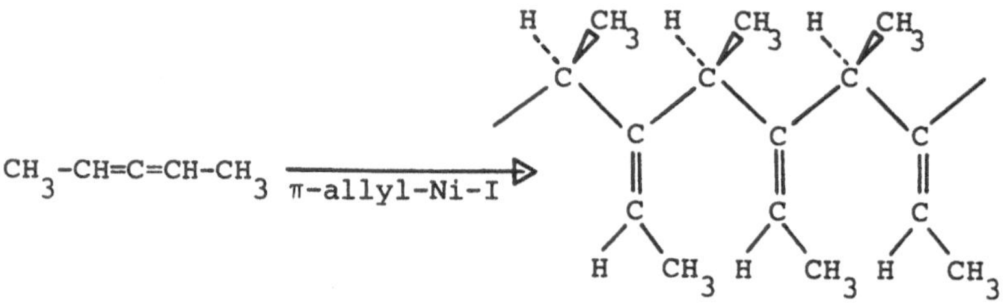

$$CH_3-CH=C=CH-CH_3 \xrightarrow{\pi-allyl-Ni-I}$$

Scheme 26. Stereospecific polymerization of (R)-penta-2,3-diene.

obtained from substituted butadienes are indicated in Table 8.
Racemic 5-methyl-1,3-pentadiene has also been polymerized to
trans-1,4-isotactic polymer by the $Al(i-C_4H_9)_3/TiCl_3$ system and
some evidence has been provided that shows that the polymerization proceeds stereoselectively, to some degree (227).

    Starting from these substituted dienes, asymmetric syntheses
of optically active polymers are possible, since the chirality
of the asymmetric carbon atoms of the main chain is determined
by the local environment. Indeed these syntheses have been
successfully carried out with many of them. The synthesis of
optically active polysorbates by Natta et al. dates back to
1960 (228), and is the first example of an asymmetric synthesis
of homopolymers. A conclusive proof of the asymmetric induction
was obtained by oxidative degradation of the polymers to succinic
acid derivatives (229). This synthesis, as well as those performed with trans 1,3-pentadiene (230), 1-phenyl-butadiene (231),
and 1-phenyl-4-methyl-butadiene (146) have been carried out using
optically active initiators. A new kind of asymmetric polymerization was obtained by Farina et al (232) by γ-irradiation of
trans-1,3-pentadiene included in (-)perhydrotriphenylene (XIX).

Table 8. Stereospecific polymerization of some 1- or 1,4-substituted butadienes

Monomer	Catalytic system	Type of polymer	Tacticity	Ref.
cis-or trans-1,3-pentadiene	$AlEt_3/VCl_3$	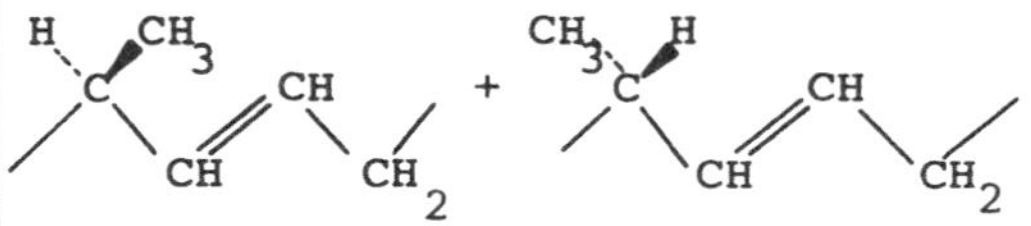 [structure: CH_3, CH, CH_2, CH, C, H + CH_3, H, C, CH, CH, CH_2]	isotactic trans-1,4	225
trans-1,3-pentadiene	$AlEt_3/Ti(O\text{-}n.Bu)_4$	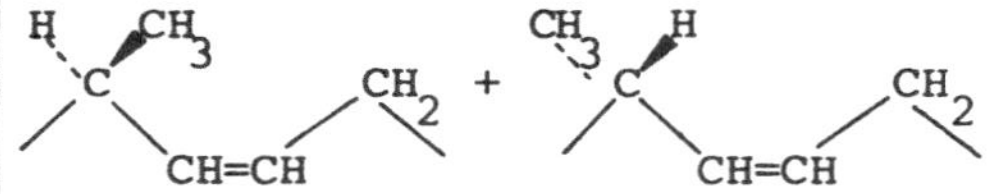 [structure: CH_3, CH, CH_2, CH=CH + CH_3, H, C, CH=CH, CH_2]	isotactic cis-1,4	226
trans-trans methylsorbate	LiBu	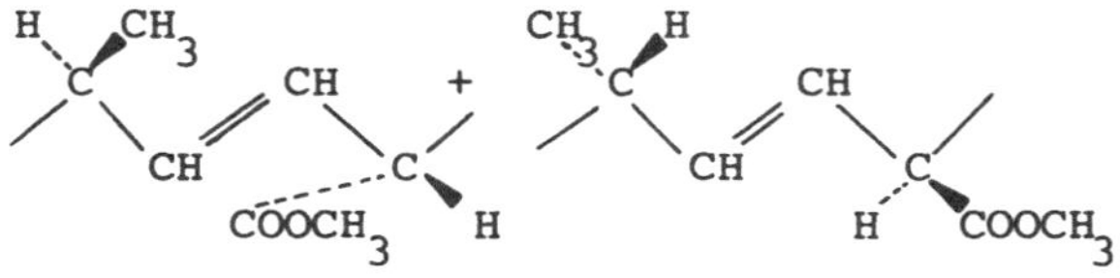 [structure: CH_3, CH, CH, $COOCH_3$, H + CH_3, H, C, CH, CH, C, H, $COOCH_3$]	erythro diiso-tactic, trans	19
trans-(S)-5-methyl-1,3-heptadiene	$Al(i.C_4H_9)_3/TiCl_3$	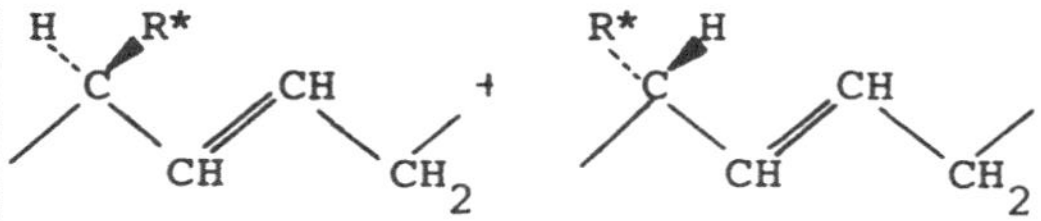 [structure: R^*, CH, CH_2, CH, H, R^* + CH, CH_2, CH] a)	isotactic trans-1,4	227

$* = -CH\text{-}C_2H_5$
$\quad\ \ |$
$\quad\ \ CH_3$

More recently, asymmetric polymerizations have been reported for
cis and trans 1,3-pentadiene (233) and for trans-2-methyl-1,3-
pentadiene (234) included in deoxycholic acid (XX). The optically active polymers obtained from these monomers possess an
essentially trans-isotactic structure.

XIX   XX

3.4.4. <u>Stereospecific ring-opening polymerization of epoxides or episulfides with two asymmetric carbon atoms in the ring.</u>
Highly stereoregular ditactic polymers have been obtained from
2,3-epoxybutanes (235), 1,4-dichloro-2,3-epoxybutanes (236), cis-
2-butene-episulfide (237,238) and cis-cyclohexene episulfide
(238). By examining the products obtained upon cleaving the
polyethers obtained from the optically active, or racemic trans-
2,3-epoxybutane, and from the meso cis-2,3-epoxybutane, Vandenberg (239) could conclusively establish that the polymerization
of these monomers proceeds always with complete inversion of
configuration of the ring-opening carbon atom. The stereochemical
relationships between 2,3-epoxybutane monomers and polymers are
illustrated in Table 9. The mechanism with inversion of configuration at the ring-opening carbon atom proposed by Vandenberg
has been confirmed by Price and Spector (240) for the case of
deuterated ethylene oxides. In keeping with these results is the
asymmetric synthesis of an optically active polymer that was
accomplished on polymerizing cis-2,3-epoxybutane with an (1)-men-
thol modified catalyst (241). Asymmetric polymerization experiments of a similar kind have been carried out recently by Spassky,
Sigwalt and coworkers (238). These authors obtained optically
active polymers upon polymerizing cis-2-butene episulfide and cis-
cyclohexene episulfide with diethyl zinc /(R) 3,3-dimethyl-1,2-
butandiol as catalyst (Scheme 27).

Table 9. Stereochemical relationships in 2,3-epoxybutane polymerization

Monomer	Polymerization type	Stereochemistry of a dimer unit
(S)(R) structure: CH_3, H, CH_3, H around C–C–O ring	cationic or coordination	~ RS – RS ~ (meso-diisotactic)
(R)(S) structure: CH_3, CH_3, H, H around C–C–O ring	cationic	~~ RR – SS ~~ (disyndiotactic)
	coordination	~ RR – RR ~ (racemic diisotactic)
		~ SS – SS ~

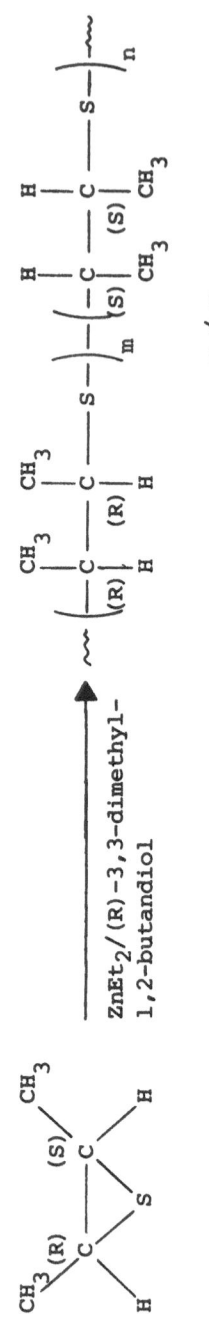

Scheme 27. Asymmetric polymerization of cis-2-butene episulfide

4. FINAL REMARKS

Beside the practical importance of stereoregular polymers, the investigation of the synthesis and properties of the stereo-regular polymers has strongly influenced progress in our under-standing of the polymerization mechanism and of the relationships between molecular structure and physical properties of the polymeric materials. Concerning the progresses in the understanding of reaction mechanism, the factors affecting the steric regularity of the polymers are at the present fairly well understood,even if the structure of the catalytic centers in most of the cases is still not completely clear.

The existence of enantiomorphic active centers in hetero-geneous catalysis and the possibility of modifying their structures and their reactivities has been shown for the first time in polymerization catalysts, and this concept will probably find interesting applications in synthetic organic chemistry. The stereospecificity of some of the catalysts for the polymerization of unsaturated monomers is exceptional and comparable to that of enzymes. Indeed, new analytical tools are necessary for detecting the extremely low percentage of structural and steric irregularities in some stereoregular macromolecules. The relationships between structure and conformation of the isolated stereoregular macromolecules,at least in some classes of polymers, are now well understood, and the general knowledge of the relationships between structure and crystallinity in polymers has improved. The possibility of using stereoregular polymers with a well defined structure opens new possibilities for the investigation of the chemical and physical properties of linear polymers in the solid state, in the melt and in solution.

The main trends of research in the field of stereoregular polymers include the investigation of the structure of the catalytic centers responsible for the stereoregulation and of the details of the mechanism of stereospecific polymerization, with the hope of achieving a better control of the percent and distribution of steric irregularities in the polymer chains. Only in this way can one hope to obtain the synthesis of a complete series of polymers with properties ranging gradually between that of an amorphous material and that of a highly crystalline material starting from a given monomer.

Apart from the future developments, the present achievements in this field already show that the synthesis of stereoregular polymers can be considered one of the major developments of polymer science in the second half of this century.

REFERENCES

(1) Quoted by F. Ebel, in "Stereochemie", K. Freudenberg, ed., F. Deuticke, Leipzig and Vienna, 1933, p. 534.

(2) K.H. Meyer and H. Mark, Ber., 61, 1939 (1928).

(3) K.H. Meyer and H. Mark, "Der Aufbau der hochpolymeren organischen Naturstoffe", Akadem. Verlagsgesellschaft,Leipzig, 1930, p. 205.

(4) H. Staudinger, A.A. Ashdown, M. Brunner, H.A. Bruson and S. Wehrli, Helv. Chim. Acta, 12, 934 (1929).

(5) H. Staudinger, "Die hochmolekularen organischen Verbindungen", J. Springer, Berlin, 1932, pp. 113-114.

(6) K.H. Meyer, "Natural and Synthetic High Polymers", Interscience Publishers, Inc., New York, 1942.

(7) G. Natta, P. Pino, P. Corradini, F. Danusso, E. Mantica, G. Mazzanti and G. Moraglio, J. Am. Chem. Soc., 77, 1708 (1955).

(8) Pure Appl. Chem., 40, 479 (1974).

(9) G. Natta, Chim. Ind. (Milan), 37, 888 (1955).

(10) M.B. Robin, F.A. Bovey and H. Bash,"Molecular and Electronic Structure of the Amide Group", in "The Chemistry of the Amides", J. Zabicky, ed., Wiley, New York, 1970.

(11) M. Branik and H. Kessler, Tetrahedron, 30, 781 (1974).

(12) R.J.M. Nolte, A.J.M. Van Beijnen and W. Drenth, J. Am. Chem. Soc., 96, 5932 (1974).

(13) G. Natta, M. Pegoraro and M. Peraldo, Ric. Sci, Suppl., 28, 1473 (1958).

(14) G. Natta, I. Pasquon, P. Corradini, M. Peraldo, M. Pegoraro and A. Zambelli, Atti Accad.Nazl. Lincei, Rend., Classe Sci. Fis. Mat. Nat., (8) 28, 539 (1960).

(15) J.N. Baptist, "Simple Stereoregular Polymers in Biological Systems", in "The Stereochemistry of Macromolecules", Vol. 2, A.D. Ketley, ed., M. Dekker, Inc., New York, 1967.

(16) Ph.Teyssié, M. Julémont, J.M. Thomassin, E. Walckiers and R. Warin, in "Coordination Polymerization", J.C.W. Chien, ed., Academic Press, Inc., New York, 1975, p. 327.

(17) A. Nishioka, H. Watanabe, K. Abe and Y. Sono, J. Polym. Sci., 48, 241 (1960).

(18) J.C.H. Hwa, J. Polym. Sci., 60, S 12 (1962).

(19) G. Natta, M. Farina, P. Corradini, M. Peraldo, M. Donati and P. Ganis, Chim. Ind. (Milan), 42, 1361 (1960).

(20) R.B. Merrifield, J. Am. Chem. Soc., 86, 304 (1964).

(21) G. Khorana, Pure Appl. Chem., 17, 349 (1968).

(22) A.G. Nasini and L. Trossarelli, J. Polym. Sci., Part C, 4, 167 (1964).

(23) J.F. Brown, Jr., L.H. Voght, Jr., A. Katchman, J.W. Eustance,

K.M. Kiser and K.W. Krantz, J. Am. Chem. Soc., 82, 6194 (1960).

(24) K. Yamaguchi and Y. Minoura, J. Polym. Sci., Part A-1, 10, 1217 (1972).

(25) G. Natta and R. Rigamonti, Atti Accad.Sci. Lincei, 24, 381 (1936).

(26) T. Alfrey, A. Bartovics and H. Mark, J. Am. Chem. Soc., 65, 2319 (1943).

(27) M.L. Huggins, J. Am. Chem. Soc., 66, 1991 (1944).

(28) C.S. Marvel, R.L. Frank and E. Prill, J. Am. Chem. Soc., 65, 1647 (1943).

(29) C.E. Schildknecht, S.T. Gross, H.R. Davidson, T.M. Lambert and A.O. Zoss, Ind. Eng. Chem., 41, 2104 (1948).

(30) C.W. Bunn, Proc. Roy. Soc., 180, 67 (1942).

(31) C.E. Schildknecht, S.T. Gross and A.O. Zoss, Ind. Eng. Chem., 41, 1998 (1949).

(32) G. Natta, I. Bassi, P. Corradini, Makromol. Chem., 18/19, 455 (1955).

(33) A.A. Morton, Ind. Eng. Chem., 42, 1488 (1950).

(34) A.A. Morton, E.E. Magat and R.L. Letsinger, J. Am. Chem. Soc., 69, 950 (1947).

(35) G. Natta, J. Polym. Sci., 16, 143 (1955).

(36) J.L.R. Williams, J. Van DenBerghe, W.J. Dulmage, K.R. Dunham, J. Am. Chem. Soc., 78, 1260 (1956).

(37) R.C. Ferguson, Macromolecules, 4, 324 (1971).

(38) I. Pasquon, "Stereoregular Linear Polymers", in "Encyclopedia of Polymer Science and Technology", Vol. 13, H.F. Mark, N.G. Gaylord and N.M. Bikales, eds., Interscience Publishers, New York, 1970.

(39) U.W. Suter, S. Pucci and P. Pino, J. Am. Chem. Soc., 97, 1018 (1975), and references therein.

(40) P. Pino, U.W. Suter, Polymer, 17, 977 (1976).

(41) D.M. White, J. Am. Chem. Soc., 82, 5678 (1960).

(42) M. Kawasaki, T. Maekawa, K. Hayashi and S. Okamura, J. Macromol. Chem., 1, 489 (1966).

(43) D.J. Cram and K.R. Kopecky, J. Am. Chem. Soc., 81, 2748 (1959).

(44) W. Fowells, C. Schuerch, F.A. Bovey and F.P. Hood, J. Am. Chem. Soc., 89, 1396 (1967).

(45) T.J. Leiterg and D.J. Cram, J. Am. Chem. Soc., 90, 4019 (1968).

(46) G. Natta, L. Porri, G. Zanini and L. Fiore, Chim. Ind. (Milan), 41, 526 (1959).

(47) G. Natta, P. Pino, E. Mantica, F. Danusso, G. Mazzanti and M. Peraldo, Chim. Ind. (Milan), 38, 124 (1956).

(48) A. Zambelli, C. Wolfsgruber, G. Zannoni and F.A. Bovey, Macromolecules, 7, 750 (1974).

(49) F.A. Bovey, M.C. Sacchi and A. Zambelli, Macromolecules, 7, 752 (1974).

(50) E. Atteya and P. Pino, unpublished results.

(51) F. Patat and H. Sinn, Angew. Chemie, 70, 496 (1958).

(52) P. Pino, Adv. Polym. Sci., 4, 394 (1965).

(53) P. Cossee, Rec. Trav. Chim. Pays-Bas, 85, 1151 (1966)

(54) G. Henrici-Olivé and S. Olivé, Angew. Chem., 79, 764 (1967).

(55) P. Pino, G. Consiglio and J. Ringger, Justus Liebigs Ann. Chem., 509 (1975).

(56) K.J. Ivin, J.J. Rooney, C.D. Stewart, M.L.H. Green and R. Mahtab, J. Chem. Soc., Chem. Commun., 604 (1978).

(57) M. Löffelholz, M. Farina and U. Rossi, Makromol. Chem., 113, 230 (1968).

(58) J.F. Brown, Jr. and D.M. White, J. Am. Chem. Soc., 82, 5671 (1960).

(59) M. Farina, G. Natta, G. Allegra and M. Löffelholz, J. Polym. Sci., Part C, 16, 2517 (1967).

(60) Y. Chatani, S. Nakatami and H. Tadokoro, Macromolecules, 3, 481 (1970).

(61) J.D. D'Ianni, F.J. Naples and J.E. Field, Ind. Eng. Chem., 42, 95 (1950).

(62) E.W. Stavely and coworkers, Ind. Eng. Chem., 48, 778 (1956).

(63) Phillips Petroleum Co., Belg. Pat. 551,851 (October 17,1956).

(64) G. Natta, L. Porri and P. Corradini (to "Montecatini"), Ital. Pat. 566,940 (July 31, 1956).

(65) J.P. Durand, F. Dawans and Ph. Teyssié, J. Polym. Sci., Part B, 7, 111 (1969).

(66) G. Natta, L. Porri and G. Mazzanti (to "Montecatini"), Ital. Pat. 545,952 (March 12, 1955).

(67) J.P. Durand, F. Dawans and Ph. Teyssié, J. Polym. Sci., Part B, 6, 760 (1968).

(68) G. Allegra, F. Lo Giudice, G. Natta, U. Giannini, G. Fagherazzi and P. Pino, Chem. Commun., 1263 (1967).

(69) S. Otsuka and M. Kawakami, Kogyo Kagaku Zasshi, 68, 874 (1965).

(70) P. Cossee, "The Mechanism of Ziegler-Natta Polymerization. II.Quantum-Chemical and Crystal-Chemical Aspects", in "The Stereochemistry of Macromolecules", Vol. 1, A.D.Ketley, ed., M. Dekker, Inc., New York, 1967.

(71) L. Porri and M. Aglietto, Makromol. Chem., 177, 1465 (1976).

(72) I. Pasquon and L. Porri, in MTP Int. Rev. Sci., Phys. Chem., Ser. One, Vol. 8, Butterworths, London 1972, p. 159.

(73) J.C. Maréchal, F. Dawans and Ph. Teyssié, J. Polym. Sci., Part A-1, 8, 1993 (1970).

(74) S. Otsuka, K. Mori, T. Suminoe and F. Imaizumi, Eur. Polym. J., 3, 73 (1967).

(75) M. Ghalamkar-Moazzam and T.L. Jacobs, J. Polym. Sci., Polym. Chem. Ed., 16, 615 (1978).

(76) B.E. Davydov and B.A. Krentsel, Adv. Polym. Sci., 25, 1 (1977).

(77) G. Natta, G. Mazzanti and P. Corradini, Atti Accad. Nazl. Lincei, Rend., Classe Sci. Fis. Mat. Nat.,(8),25, 3 (1958).

(78) W.H. Watson, Jr., W.C. McMordie, Jr. and L.G. Lands, J. Polym. Sci., 55, 137 (1961).

(79) H. Shirakawa and S. Ikeda, Polym. J., 2, 231 (1971).

(80) H. Shirakawa, T. Ito and S. Ikeda, Polym. J., 4, 460 (1973).

(81) T. Ito, H. Shirakawa and S. Ikeda, J. Polym. Sci., Polym. Chem. Ed., 12, 11 (1974).

(82) M.M. Maricq, J.S. Waugh, A.G. MacDiarmid, H. Shirakawa and A.J. Heeger, J. Am. Chem. Soc., 100, 7729 (1978).

(83) C. Simionescu, S.V. Dumitrescu and V. Percec, J. Polymer Sci., Polym. Symp., 64, 209 (1978).

(84) R.H. Baughman and K.C. Yee, J. Polym. Sci., Macromol. Rev., 13, 219 (1978).

(85) G. Wegner, Chimia, 28, 478 (1974).

(86) P. Kubisa, I. Negulescu, K. Hatada, D. Lipp, J. Starr, B. Yamada and O. Vogl, Pure Appl. Chem., 48, 275 (1976).

(87) H. Yasuda and H. Tani, Macromolecules, 6 17 (1973).

(88) V.A. Kabanov, V.P. Zubov, V.P. Kovaleva and V.A. Kargin, J. Polym. Sci., Part C, 4, 1009 (1964).

(89) F. Millich, Adv. Polym. Sci., 19, 118 (1975).

(90) P. Pino, F. Ciardelli and G. Montagnoli, J. Polym. Sci., Part C, 16, 3265 (1968).

(91) P. Pino, F. Ciardelli and G.P. Lorenzi, J. Am. Chem. Soc., 85, 3888 (1963).

(92) T. Tsuruta, J. Polym. Sci., Part D, 179 (1972).

(93) W.H. Carothers and G. Arvin, J. Am. Chem. Soc., 51, 2560 (1929).

(94) V. Guidotti, M. Russo and L. Mortillaro, Makromol. Chem., 147, 111 (1971).

(95) L. Credali, P. Parrini, L. Mortillaro, M. Russo and T. Simonazzi, Angew. Makromol. Chem., 19, 15 (1971).

(96) I. Vancso-Szmercsanyi, K. Maros-Greger and E. Makay-Bodi, J. Polym. Sci., 53, 241 (1961).

(97) M. Goodman and J.S. Schulman, J. Polym. Sci., Part C, 12, 23 (1966).

(98) Zh.S. Sogomonyants and M.V. Vol'kenshtein, Izvest. Akad. Nauk SSSR, Otdel. Khim. Nauk, 611 (1957); C.A., 51, 15430c.

(99) P. Pino, G.P. Lorenzi, U.W. Suter, P.G. Casartelli, A. Steinmann, F.J. Bonner and J.A. Quiroga, Macromolecules, 11, 624 (1978).

(100) C.C. Price and M. Osgan, J. Am. Chem. Soc., 78, 4787
      (1956).

(101) Y. Kumata, N. Asada, G.M. Parker and J. Furukawa,Makromol.
      Chem., 136, 291 (1970).

(102) K. Tsuji, T. Hirano and T. Tsuruta, Makromol. Chem.,Suppl.1,
      55 (1975).

(103) N. Spassky, A. Pourdjavadi and P. Sigwalt, Eur. Polym. J.,
      13, 467 (1977).

(104) M. Sepulchre, A. Khalil, N. Spassky and M. Vert, Makromol.
      Chem., 180, 131 (1979).

(105) N. Spassky, P. Dumas, M. Moreau and J-P. Vairon, Macromole-
      cules, 8, 956 (1975).

(106) N. Spassky, P. Dumas, M. Sepulchre and P. Sigwalt, J.
      Polym. Sci., Part C, 52, 327 (1975).

(107) Y. Minoura, M. Takebayashi and C.C. Price, J. Am. Chem.
      Soc., 81, 4690 (1959).

(108) K. Tsuboyama, S. Tsuboyama and M. Yanagita, Bull. Chem.
      Soc. Jpn., 40, 2954 (1967).

(109) T. Yamashita, H. Mitsui, H. Watanabe and N. Nakamura,
      Polymer Preprints, 20, 828 (1979).

(110) W. Dittrich and R.C. Schulz, Angew. Makromol. Chem., 15,
      109 (1971).

(111) J. Kleine and H.H. Kleine, Makromol. Chem., 30, 23 (1959).

(112) R.C. Schulz and J. Schwaab, Makromol. Chem., 87, 90 (1965).

(113) C.G. Overberger and H. Kaye, J. Am. Chem. Soc., 89, 5649
      (1967).

(114) Y. Iwakura, K. Iwata, S. Matsuo and A. Tohara, Makromol.
      Chem., 146, 21 (1971).

(115) J.R. Shelton, D.E. Agostini and J.B. Lando, J. Polym. Sci.,
      Part A-1, 9, 2789 (1971).

(116) C.G. D'Hondt and R.W. Lenz, J. Polym. Sci., Polym. Chem.
      Ed., 16, 261 (1978).

(117) C.G. Overberger and J.K. Weise, J. Am. Chem. Soc., 90,
      3533 (1968).

(118) C.G. Overberger and H. Jabloner, J. Am. Chem. Soc., 85,
      3431 (1963).

(119) C.G. Overberger and G.M. Parker, J. Polym. Sci., Part C,
      22, 387 (1968).

(120) C.G. Overberger and T. Takeoshi, Macromolecules, 1, 1 (1968).

(121) M. Okada, H. Sumitomo and H. Komada, Polymer Preprints, 20,
      809 (1979).

(122) M. Szwarc, Adv. Polym. Sci., 4, 1 (1965).

(123) C.G. Overberger, T. Yoshimura, A. Ohnishi and A.S. Gomes,
      J. Polym. Sci., Part A-1, 8, 2275 (1970).

(124) R.C. Schulz and R.H. Jung, Makromol. Chem., 116, 190 (1968).

(125) A.P. Terentev, V.V. Dunina and E.G. Rukhadze, Vysokomol.
      Soedin., Ser. A, 9, 599 (1967).

(126) M. Goodman and Shih-chung Chen, Macromolecules, 3, 398 (1970).

(127) T. Tsuruta, S. Inoue and K. Matsuura, Makromol. Chem., 63, 219 (1963).

(128) M. Ishimori and T. Tsuruta, Makromol. Chem., 64, 190 (1963)

(129) T. Tsuruta, S. Inoue and I. Tsukuma, Makromol. Chem., 84, 298 (1965).

(130) H. Sakakibara, Y. Takabashi, H. Tadokoro, P. Sigwalt and N. Spassky, Macromolecules, 2, 515 (1969).

(131) K.J. Ivin, E.D. Lillie, P. Sigwalt and N. Spassky, Macromolecules, 4, 345 (1971).

(132) D.E. Agostini, J.B. Lando and J.R. Shelton, J. Polym. Sci., Part A-1, 9, 2775 (1971).

(133) K. Teranishi, T. Azaki and H. Tani, Macromolecules, 5, 660 (1972).

(134) T. Hirano, Pham Huu Khan and T. Tsuruta, Makromol. Chem., 153, 331 (1972).

(135) H. Tani, S. Yamashita and K. Teranishi, Polym. J., 3, 417 (1972).

(136) S. Yamashita, K. Wachi, N. Yamawaki and H. Tani, Macromolecules, 7, 410 (1974).

(137) K. Teranishi, M. Iida, T. Araki, S. Yamashita and H. Tani, Macromolecules, 7, 421 (1974).

(138) H.-G. Elias, H.G. Bührer and J. Semen, Appl. Polym. Symp., 26, 269 (1975).

(139) J. Semen and H.-G. Elias, Makromol. Chem., 179, 463 (1978).

(140) W.E. Hull and H.R. Kricheldorf, J. Polym. Sci., Polym. Letters Ed., 16, 215 (1978).

(141) C.C. Price, M.K.Akkapeddi, B.T. DeBona and B.C. Furie, J. Am. Chem. Soc., 94, 3964 (1972).

(142) S. Inoue, T. Tsuruta and J. Furukawa, Makromol. Chem., 53, 215 (1962).

(143) C. Coulon, N. Spassky and P. Sigwalt, Polymer, 17, 821 (1976).

(144) N. Spassky, A. Pourdjavadi and P. Sigwalt, Eur. Polym. J., 13, 467 (1977).

(145) Y. Tezuka, M. Ishimori and T. Tsuruta, Polymer Preprints, 20, 798 (1979).

(146) A.D. Aliev, B.A. Krentsel, G.M. Mamediarov, I.P. Solomatina and E.P. Tiurina, Eur. Polym. J., 7, 1721 (1971).

(147) A. Deffieux, M. Sepulchre, N. Spassky and P. Sigwalt, Makromol. Chem., 175, 339 (1974).

(148) Ph. Dumas, N. Spassky and P. Sigwalt, J. Polym. Sci., Polym. Chem. Ed., 12, 1001 (1974).

(149) N. Spassky, A. Leborgne, M. Reix, R.E. Prud'homme, E.Bigdeli and R.W. Lenz, Macromolecules, 11, 716 (1978).

(150) S. Yamashita, N. Yamawaki and H. Tani, Macromolecules, 7, 724 (1974).

(151) K. Matsura, S. Inoue and T. Tsuruta, Makromol. Chem., 80, 149 (1964).

(152) H.G. Bührer and H.-G. Elias, Makromol. Chem., 169, 145 (1973).

(153) M. Sepulchre, N. Spassky and P. Sigwalt, Macromolecules, 5, 92 (1972).

(154) T. Tsuruta, "Stereospecific Polymerization of Epoxides", in "Stereochemistry of Macromolecules", (A.D. Ketley, Ed.), M. Dekker, New York, 1967, Vol. 2, p. 177.

(155) P. Sigwalt, Pure Appl. Chem., 48, 257 (1976).

(156) K. Matsuura, T. Tsuruta, Y. Terada and S. Inoue, Makromol. Chem., 81, 258 (1965).

(157) Y. Kumata, J. Furukawa and T. Saegusa, Makromol. Chem., 105, 138 (1967).

(158) H. Yamaguchi, M. Nagasawa and Y. Minoura, J. Polym. Sci., Part A-1, 10, 1207 (1972).

(159) I.N. Topchieva, U.K. Zlobin, V.M. Potapov, R.Ya. Levina, B.A. Kabanov and V.A. Kargin, Vysokomol. Soedin., 6, 512 (1964).

(160) G. Natta, L. Porri and G. Sovarzi, European Polymer J., 1, 81 (1965).

(161) P. Walden, Zeit phys. Chem., 20, 383 (1896).

(162) P. Pino, F. Ciardelli, G.P. Lorenzi and G. Montagnoli, Makromol. Chem., 61, 207 (1963).

(163) W.J. Bailey and E.T. Yates, J. Org. Chem., 25, 1800 (1960).

(164) S. Nozakura, S. Takeuchi, H. Yuki and S. Murahashi, Bull. Chem. Soc. Jpn., 34, 1673 (1961).

(165) M. Goodman, K.J. Clark, M.A. Stake and A. Abe, Makromol. Chem., 72, 131 (1964).

(166) C. Carlini, F. Ciardelli, L. Lardicci and R. Menicagli, Makromol. Chem., 174, 24 (1973).

(167) G.P. Lorenzi, E. Benedetti and E. Chiellini, Chim. Ind. (Milan), 46, 1474 (1964).

(168) P. Pino, G.P. Lorenzi and E. Chiellini, J. Polym. Sci., Part C, 16, 3279 (1968).

(169) A.M. Liquori and B. Pispisa, Chim. Ind. (Milan), 48, 1045 (1966).

(170) O. Pieroni, F. Ciardelli, C. Botteghi, L. Lardicci, P. Salvadori and P. Pino, J. Polym. Sci., Part C, 22, 993 (1969).

(171) A. Allio and P. Pino, Helv. Chim. Acta, 57, 616 (1974).

(172) R.C. Schulz and H. Hilpert, Makromol. Chem., 55, 132 (1962).

(173) E.I. Klabunovskii, M.I. Shvartsman and Yu.I. Petrov, Izv. Akad. Nauk SSSR, Ser. Khim., 223 (1966); C.A., 64, 17734f.

(174) K.-J. Liu, J.S. Lignowski and R. Ullman, Makromol. Chem., 105, 8 (1967).

(175) K.-J. Liu, J.S. Lignowski and R. Ullman, Makromol. Chem., 105, 18 (1967).

(176) M. Imoto and Y. Kumata, Bull. Chem. Soc. Jpn., 38, 1615 (1965).

(177) E. Kaiser and R.C. Schulz, Makromol. Chem., 81, 273 (1975).

(178) A. Abe and M. Goodman, J. Polym. Sci., Part A, 1, 2193 (1963).

(179) R. Lazzaroni, P. Salvadori and P. Pino, J. Chem. Soc. D, 1164 (1970).

(180) G. Wulff, K. Zabrocki and J. Hohn, Angew. Chem., 90, 567 (1978).

(181) F. Ciardelli, E. Benedetti and O. Pieroni, Makromol. Chem., 103, 1, (1967).

(182) F. Millich and G.K. Baker, Macromolecules, 2, 122 (1969).

(183) A.J.M. Van Beijnen, R.J.M. Nolte, W. Drenth and A.M.F. Hezemans, Tetrahedron, 32, 2017 (1976).

(184) P. Pino, F. Ciardelli, G.P. Lorenzi and G. Natta, J. Am. Chem. Soc., 84, 1487 (1962).

(185) P. Pino, A. Oschwald, F. Ciardelli, C. Carlini and E. Chiellini, in "Coordination Polymerization", J.C.W. Chien, ed., Academic Press. Inc., New York, 1975, p. 25.

(186) E. Chiellini, G. Montagnoli and P. Pino, Polym. Letters, 7, 121 (1969).

(187) O. Pieroni, G. Stigliani and F. Ciardelli, Chim. Ind. (Milan), 52, 289 (1970).

(188) F. Ciardelli, P. Locatelli, M. Marchetti and A. Zambelli, Makromol. Chem., 175, 923 (1974).

(189) F. Ciardelli, E. Benedetti, G. Montagnoli, L. Lucarini and P. Pino, Chem. Commun., 285 (1965).

(190) F. Ciardelli, G. Montagnoli, D. Pini, O. Pieroni, C. Carlini and E. Benedetti, Makromol. Chem., 147, 53 (1971).

(191) C. Carlini, R. Nocci and F. Ciardelli, J. Polym. Sci., Polym. Chem. Ed., 15, 767 (1977).

(192) T. Higashimura and Y. Hirokawa, J. Polym. Sci., Polym. Chem. Ed., 15, 1137 (1977).

(193) E. Chiellini, Macromolecules, 3, 527 (1970).

(194) Y. Okamoto, K. Urakawa, K. Ohta and H. Yuki, Macromolecules, 11, 719 (1978).

(195) Y. Okamoto, K. Ohta and H. Yuki, Macromolecules, 11, 724 (1978).

(196) G. Bruno, G. Montaudo, N.Y. Hien, R.H. Marchessault, P. R. Sundararajan, J.E. Chandler and R.W. Lenz, J. Polym. Sci., Polym. Lett. Ed., 13, 559 (1975).

(197) G. Dall'Asta, G. Manetti, G. Natta and L. Porri, Makromol.
      Chem., 56, 224 (1962).
(198) G. Natta, G. Dall'Asta, G. Mazzanti and G. Motroni,
      Makromol. Chem., 69, 163 (1963).
(199) G. Natta, G. Dall'Asta and G. Mazzanti, Angew. Chem., 76,
      765 (1964).
(200) N. Calderon, Acc. Chem. Res., 5, 127 (1972).
(201) G. Dall'Asta and G. Motroni, Angew. Makromol. Chem., 16/17,
      51 (1971).
(202) G. Natta, G. Dall'Asta, I.W. Bassi and G. Carella, Makro-
      mol. Chem., 91, 87 (1966).
(203) G. Gianotti, G. Dall'Asta, A. Valvassori and V. Zamboni,
      Makromol. Chem., 149, 117 (1971).
(204) G. Natta, G. Dall'Asta, G. Mazzanti, I. Pasquon, A.
      Valvassori and A. Zambelli, J. Am. Chem. Soc., 83, 3343
      (1961).
(205) G. Dall'Asta and G. Mazzanti, Makromol. Chem., 61, 178
      (1963).
(206) G. Natta, M. Farina, M. Peraldo, P. Corradini, G. Bressan
      and P. Ganis, Accad. Nazl. Lincei, (8), 28, 442 (1960).
(207) G. Natta, M. Peraldo, M. Farina and G. Bressan, Makromol.
      Chem., 55, 139 (1962).
(208) T. Higashimura and Y. Hirokawa, J. Polym. Sci., Polym.
      Chem. Ed., 15, 1137 (1977).
(209) Chi-Yu Chen and I. Piirma, J. Polym. Sci., Polym. Symp.,
      65, 55 (1978).
(210) G. Natta, G. Dall'Asta, G. Mazzanti, I. Pasquon, A.
      Valvassori and A. Zambelli, Makromol. Chem., 54, 95 (1962).
(211) G. Natta, G. Dall'Asta and G. Mazzanti, Chim. Ind. (Milan),
      44, 1212 (1962).
(212) K. Kamio, K. Meyersen, R.C. Schulz and W. Kern, Makromol.
      Chem., 90, 187 (1966).
(213) G. Natta, M. Farina, M. Peraldo and G. Bressan, Makromol.
      Chem., 43, 68 (1961).
(214) G. Bressan, M. Farina and G. Natta, Makromol. Chem., 93,
      283 (1966).
(215) G.J. Schmitt and C. Schuerch, J. Polym. Sci., 49, 287 (1961).
(216) N. Beredjick and C. Schuerch, J. Am. Chem. Soc., 80, 1933
      (1958).
(217) G.J. Schmitt and C. Schuerch, J. Polym. Sci., 45, 313 (1960).
(218) H. Yamaguchi and Y. Minoura, J. Polym. Sci., Part A-1, 8,
      1467 (1970).
(219) N. Sakota, K. Kishiue, S. Shimada and Y. Minoura, J. Polym.
      Sci., Polym. Chem. Ed., 12, 1787 (1974).
(220) T. Doiuchi and Y. Minoura, Macromolecules, 10, 260 (1977).

(221) T. Doiuchi, K. Kubouchi and Y. Minoura, Macromolecules, 10, 1208 (1977).

(222) T. Doiuchi and Y. Minoura, Macromolecules, 11, 270 (1978).

(223) T. Doiuchi and Y. Minoura, Macromolecules, 11, 483 (1978).

(224) L. Porri, R. Rossi and G. Ingrosso, Tetrahedron Lett., 1083 (1971).

(225) G. Natta, L. Porri, P. Corradini, G. Zanini and F.Ciampelli, J. Polym. Sci., 51, 463 (1961).

(226) G. Natta, L. Porri, A. Carbonaro and G. Stoppa, Makromol. Chem., 77, 114 (1964).

(227) L. Porri and D. Pini, Chim. Ind. (Milan), 55, 196 (1973).

(228) G. Natta, M. Farina, M. Donati and M. Peraldo, Chim. Ind. (Milan), 42, 1363 (1960).

(229) M. Farina, M. Modena and W. Ghizzoni, Rend. Accad. Nazl. Lincei, (8), 32, 91 (1962).

(230) G. Natta, L. Porri and S. Valenti, Makromol. Chem., 67, 225 (1963).

(231) A.D. Aliev, B.A. Krentsel and T.N. Fedotova, Vysokomol. Soedin., 7, 1442 (1965).

(232) M. Farina, G. Audisio and G. Natta, J. Am. Chem. Soc., 89, 5071 (1967).

(233) G. Audisio and A. Silvani, J. Chem. Soc. Chem. Commun., 481 (1976).

(234) M. Miyata, K. Shinnen and K. Takemoto, Polymer Preprints, 20, 716 (1979).

(235) E.J. Vandenberg, J. Polym. Sci., Part A-1, 7, 525 (1969).

(236) E.J. Vandenberg, Pure Appl. Chem., 48, 295 (1976).

(237) E.J. Vandenberg, J. Polym. Sci., Part A-1, 10, 329 (1972).

(238) A. Momtaz, M. Reix, N. Spassky and P. Sigwalt,"New Developments in Ionic Polymerization", 1st Eur. Discussion Meeting on Polymer Science, Strasbourg, February 27 - March 2, 1978, Preprints, p. 84.

(239) E.J. Vandenberg, J. Polym. Sci., Part A-1, 10, 2903 (1972).

(240) C.C. Price and R. Spector, J. Am. Chem. Soc., 87, 2069 (1965).

(241) E.J. Vandenberg, Polym. Letters, 2, 1085 (1964).

(242) E. Field and M. Feller (to Standard Oil Co.), U.S. Pat. 2,791,576 (May 7, 1957).

(243) J.P. Hogan and R.L. Banks (to Phillips Petroleum Co.), U.S. Pat. 2,825,721 (March 4, 1958).

(244) G. Natta, P. Pino and G. Mazzanti (to Montecatini), Ital. Pat. 535,712 (November 17, 1955).

RECENT DEVELOPMENTS IN STEREOREGULATION BY ZIEGLER-NATTA POLYMER-
IZATION OF α-OLEFINS

F.Ciardelli[*] and A.Zambelli[**]

[*]   Istituto di Chimica Organica Industriale,University,
      and Centro di Studio del CNR per le Macromolecole
      Stereoordinate ed Otticamente Attive, Pisa, Italy
[**] Istituto di Chimica delle Macromolecole del CNR,
      Milano, Italy

1. INTRODUCTION

    Since its discovery a great deal of work has been done for
understanding the mechanism of Ziegler-Natta polymerization of
α-olefins. In spite of this almost 25 years old activity many
points do not appear entirely clarified and firmly established.
Indeed these catalysts join unique and very interesting properties
to a very large complexity. Thus the field is under constant
development and also the aspects generally accepted (1) are often
put in debate.For istance the α-olefin insertion in a transition
metal-carbon bond, considered as one of the fixed step of the mechanism, has been questioned very recently by proposing (2) a mechanism proceeding via a 1,2 hydrogen shift from the α-carbon of
the growing chain and formation of metallocycle and carbene
intermediates. Even if this last proposal does not explain end
group nature, particularly when using AlPh$_3$, large predominance
of primary insertion in isotactic polymerization and different
role of growing chain end and metal complex in the steric control,
it however indicates that several problems are still open as
discussed in other chapters.
    In the present chapter we shall limit ourself to discuss
the basic stereochemical features of α-olefins polymerization,
which appear rather well understood. Stereochemistry of iso- and
syndiotactic polymerization has been indeed interpreted not on
the ground of mechanistic speculations but on the basis of
polymer structure analysis, that is of firm experimental bases.
Reversely these conclusions can be used to check any chemical
mechanism which must explain also stereochemistry of resulting
macromolecules.

73

R. W. Lenz and F. Ciardelli (eds.), Preparation and Properties of Stereoregular Polymers, 73–83.
Copyright © 1979 by D. Reidel Publishing Company.

Steric interactions responsible for stereospecific polymerization
may occur between monomer and either catalytic complex or last
unit of the growing chain (3). The former has supposed to be
responsible for isotactic propagation because of its inherently
chiral structure as shown by the high crystallinity (4) and by
the separation in fractions with opposite optical activity (5)
of the polymers from racemic α-olefins. On the other side this
hypothesis and the determining role of growing chain end in
syndiotactic propagation have been demonstrated by copolymerizing
differently substituted α-olefins (6).

Propylene is the only α-olefin which can be polymerized
to either isotactic or syndiotactic polymers by means of
Ziegler-Natta catalyst. The stereochemistry of the two types of
stereospecific polymerization has been elucidated by structural
analysis of polymerization products (mainly propylene homopolymers
and ethylene-propylene copolymers) obtained in the presence of
different catalytic systems (7).

2. STEREOREGULATION IN POLYMERIZATION OF α-OLEFINS TO ISOTACTIC POLYMERS

High resolution ^{1}H- and ^{13}C-NMR of polypropylene and of
ethylene-propylene copolymers as well as determination of
enantiomers distribution in polymers from racemic α-olefins
demonstrated that the catalytic complex is responsible for the
isotactic enchainment. This implies the chiral (racemic) structure
of the catalyst which at present seems to be adequately proved as
discussed in the two following sections.

2.1 NMR Study of propylene and ethylene-propylene copolymers.

In the case of isotactic polymerization it has been observed
that the overal stereochemical mechanism of addition to the double
bond is cis (8,9). Indeed NMR analysis of polymers obtained in
the presence of isotactic catalysts from cis- and trans-1d$_1$-propene
indicated that erythro-diisotactic polymers are obtained from the
former and threo-diisotactic polymers from the latter.

The polymerization is highly regiospecific: only head-to-tail
einchained monomer units are detected in highly isotactic polymers
and only odd methylene sequences are detected in ethylene/propylene
copolymers (10-12). The insertion of all propylene units into
the reactive metal-carbon bond is primary (or antimarkownikow)

$$Mt - P + C_3H_6 \longrightarrow Mt-CH_2-\underset{\underset{CH_3}{|}}{CH}-P$$

as determined by detailed analysis of the end groups (13,14).

The steric control is due to the chirality of the catalytic

sites rather than to the chirality of the growing chain end.
This conclusion has been achieved on the basis of the following
considerations:

i) The statistic of the stereochemical sequences of the
    substituted carbon atoms in propylene homopolymers is
    substantially described by the "enatiomorphic-sites model"
    elaborated by Shelden and Furukawa (10,15). This model
    assumes that the configuration of adding monomer is determined
    primarily by the configuration of the catalyst, whereas the
    configuration of the growing chain has only a negligible
    effect.

ii) The steric control is transmitted across isolated ethylene
    units as shown by ^{13}C-NMR analysis of the stereochemical
    environment of ethylene units in ethylene-propylene copolymers
    of low ethylene content. Indeed copolymerization of enriched
    ethylene-1 ^{13}C with propylene in the presence of catalysts
    capable of highly isotactic control gives polymers showing
    only two ethylene resonances for the $CH_2-CH_2-CH_2$-groups,
    thus suggesting the presence of only <u>meso</u> (A) or racemic
    (B) situation (11,12,16).

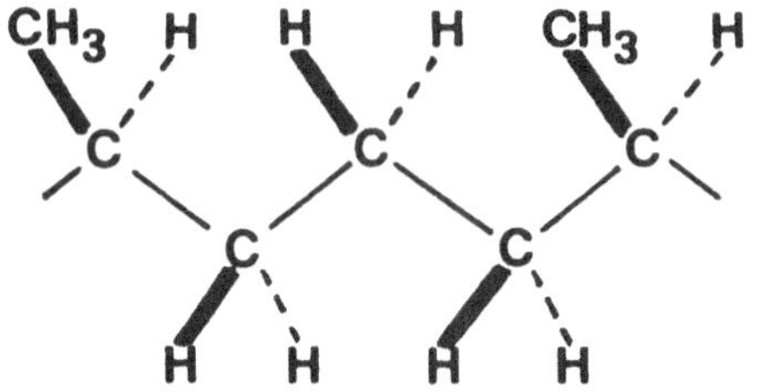

(A)

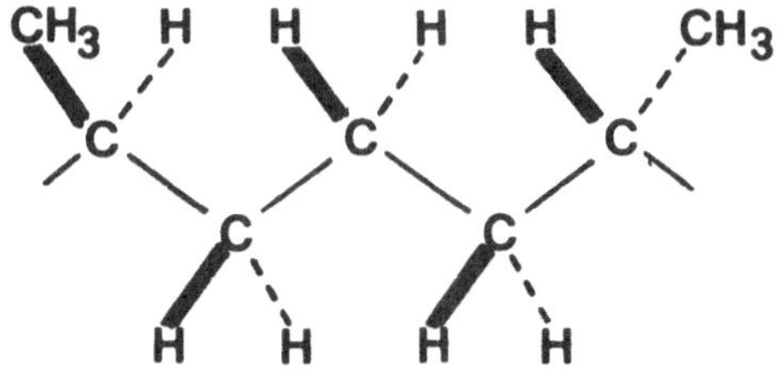

(B)

The two peaks detected in sample obtained with isotactic-specific
catalyst were assigned to <u>meso</u> ethylene units. This result clearly
indicates that insertion of ethylene in the growing chain does not

change configuration of propylene units according to control
of the stereochemistry of the addition by the catalytic metal
complex; only in this situation in fact the stereoregulation
can be transmitted across the achiral ethylene unit.

iii) Determination of the stereochemistry of end groups is a
third manner of grasping the problem of origin of steric
control in isotactic polymerization. As observed by Carman
and coworkers (17) the carbon atoms of the two methyls of
the isopropyl group of branched hydrocarbons are diastereo-
topic and show different ^{13}C shifts . Work in progress (14)
suggests that when 3-^{13}C enriched propylene ($^{*}C_3H_6$) is
polymerized in the presence of $TiCl_3/Al(CH_3)_3/Zn(CH_3)_2$
catalytic system, the ^{13}C enriched carbon atoms of the
two isopropyl end groups of the resulting isotactic
macromolecules are both in an __erythro__ relationship.

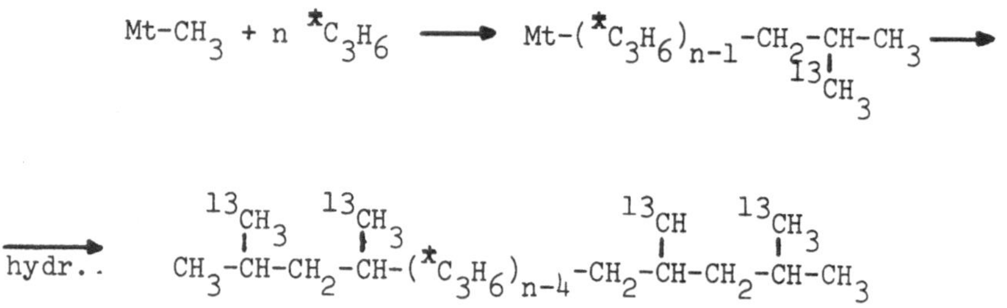

$$Mt-CH_3 + n\ ^{*}C_3H_6 \longrightarrow Mt-(^{*}C_3H_6)_{n-1}-CH_2-\underset{\underset{CH_3}{\overset{13}{|}}}{CH}-CH_3 \longrightarrow$$

$$\xrightarrow{hydr..}\ CH_3-\underset{\underset{CH_3}{\overset{13}{|}}}{CH}-CH_2-\underset{\underset{CH_3}{\overset{13}{|}}}{CH}-(^{*}C_3H_6)_{n-4}-CH_2-\underset{\underset{CH}{\overset{13}{|}}}{CH}-CH_2-\underset{\underset{CH_3}{\overset{13}{|}}}{CH}-CH_3$$

(as defined in references 18 and 19) with respect to the
methyl substituents of the respective first neighbouring
monomer unit. In the above scheme the isopropyl group arising
from Mt-chain hydrolysis in on the left and the one formed
by insertion of propylene in the Mt-CH_3 bond, at the beginning
of the polymerization, is on the right. Thus the first and
second units inserted are subjected to the same isotactic
steric control, despite nor the starting methyl group nor
the isobutyl group formed by insertion of the first propylene

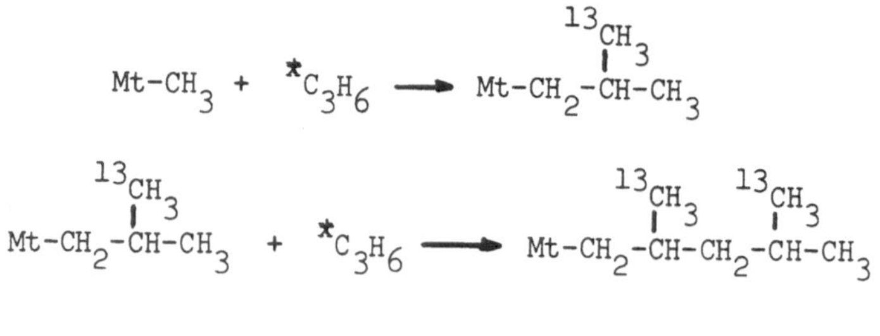

$$Mt-CH_3 + \ ^{*}C_3H_6 \longrightarrow Mt-CH_2-\underset{\underset{CH_3}{\overset{13}{|}}}{CH}-CH_3$$

$$Mt-CH_2-\underset{\underset{CH_3}{\overset{13}{|}}}{CH}-CH_3 + \ ^{*}C_3H_6 \longrightarrow Mt-CH_2-\underset{\underset{CH_3}{\overset{13}{|}}}{CH}-CH_2-\underset{\underset{CH_3}{\overset{13}{|}}}{CH}-CH_3$$

molecule are chiral.
The driving force of 5 Kcal/mole evaluated for the isotactic

propagation (20) is to assign to the asymmetric structure
of the active sites in the catalytic complex.

2.2. Polymerization of chiral α-olefins

Results obtained by polymerizing racemic α-olefins having
the asymmetric C-atom in α to the bouble bond in the presence
of the same type of catalysts are consistent with the above
conclusion.

Isotactic polymers of racemic 3,7-dimethyl-1-octene can be
separated in fractions having optical activity of opposite
sign by elution on an optically active crystalline support
consisting of poly-[(S)-3-methyl-1-pentene] (21). The separation
degree F even for highly isotactic fractions is not higher

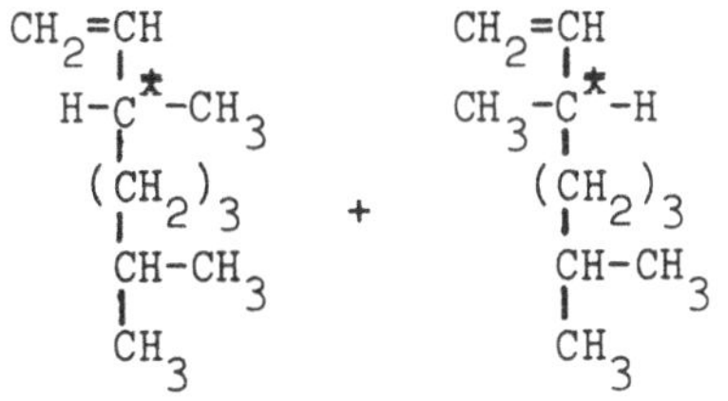

$$CH_2=CH \quad H-C^*-CH_3 \quad (CH_2)_3 \quad CH-CH_3 \quad CH_3$$

$$CH_2=CH \quad CH_3-C^*-H \quad (CH_2)_3 \quad CH-CH_3 \quad CH_3$$

+

racemic 3,7-dimethyl-1-octene

$$\sim CH_2-\underset{\underset{C_2H_5}{|}}{\overset{\overset{H}{|}}{C}}-CH_2-\underset{\underset{C_2H_5}{|}}{\overset{\overset{H}{|}}{C}} \sim$$
$$H_3C-C^*-H \quad H_3C-C^*-H$$

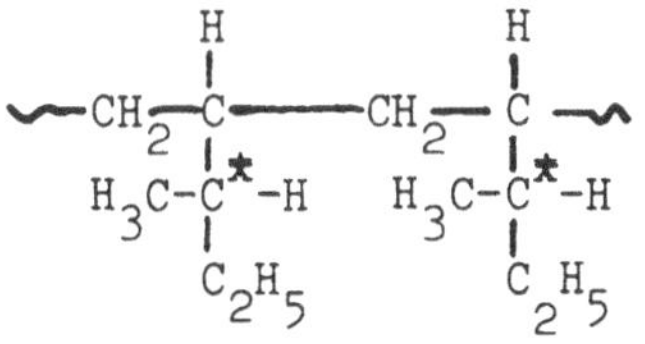

isotactic poly-[(S)-3-methyl-1--pentene]

than 40% (22),but a very similar value has been obtained by
submitting to elution under the same conditions a 1/1 mixture
of isotactic poly-[(S)-and poly-[(R)-3,7-dimethyl-1-octene](Table
1)(23). As in this last case the intrinsic separability D is
100%, the separation efficiency of the method E=F/D is lower
than 40%. Accordingly the intrinsic separability of the
highly isotactic polymer from racemic 3,7-dimethyl-1-octene
must be higher than 90%, indicating for each macromolecule
an average enantiomeric purity of the same order of magnitude.
These results exclude the formation of statistical and block
copolymers of the two antipodes and are rather consistent with
the formation of macromolecules consisting of long sections
of units derived from the same antipode connected by isolated
mistakes as shown below. Such a structure can be formed only
if the catalytic complex is able to distinguish between the to

n (R + S) ⟶ R R R S R R R R R R + S S S R S S S S S S

[R and S represent (R)- and (S)-3,7-dimethyl-1-octene,respectively]

Table 1

Separation in Fractions with Opposite Optical Rotation of
Polymers from Racemic 3,7-Dimethyl-1-octene and of an Equimolar
Mixture of Homopolymers of the Two Antipodes (23).

Sample submitted to elution on poly-(S)--3-methyl-1-pentene	Separation degree,% F	Intrinsic Separability,% D
Poly-[(R)(S)-3,7-dimethyl--1-octene]	33.6	91
Equimolar mixture of Poly-[(R)-and Poly-[(S)--3,7-dimethyl-1-octene]	37.0	100

antipodes, its capability being maintained even if a "wrong"
monomer insertion occurs. Moreover the random copolymers of
3,7-dimethyl-1-octene with ethylene have been separated in
fractions having optical rotation with opposite sign (24).
The separation degree F attained (25) was only slightly lower
than for the polymers from racemic 3,7-dimethyl-1-octene
(Table 2).

Table 2

Separation Degree F in Fractions Having Opposite Optical Rotation
for Polymers of Racemic 3,7-Dimethyl-1-octene and for its
Copolymer with Ethylene.

Polymer	F%
Poly-[(R)(S)-3,7-dimethyl-1-octene]	27 - 34
(R)(S)-3,7-dimethyl-1-octene/ Ethylene 56/44 copolymer	23 - 29

It is reasonable to expect that the efficiency E of the separation is lower for the ethylene copolymer; thus the intrinsic separability D of this last is probably higher than that observed indicating that insertion of ethylene in the growing chain does not affect the stereoselectivity of a single site towards the two monomer antipodes. The process can be represented as below, where R and S indicate (R)- and (S)-3,7-dimethyl-1-octene, respectively:

$$R + S + CH_2=CH_2 \Big[\begin{array}{l} \longrightarrow RRR(CH_2CH_2)_x RRSRR(CH_2CH_2)_y RR \\ \\ \longrightarrow SSS(CH_2CH_2)_x SSRSS(CH_2CH_2)_y SS \end{array}$$

R   +   S   racemic monomer

In this case also,the only explanation possible is that the catalytic site is able to distinguish between the R- and S-monomer,the assistance from the growing chain being vanishing small if any.

According to the previous presentation heterogenous Ziegler-Natta catalysts capable of isotactic propagation should have a racemic structure with chiral site of both levo and dextro configuration. It appeared very attractive to attempt the preparation of catalytic systems containing sites of a single chirality only by using optically active components. We have to say that up to now this goal has been only partially achieved starting with optically active metal alkyls (26) or using optically active electron donors (27) for catalyst preparation. These systems are indeed able to induce preferential polymerization of a monomer antipode but the"stereoelectivity" is not very high, the enantiomeric prevalence in the polymer being lower than 20%, while from the separation data we could expect that a real chiral site is able to distinguish between the two antipodes up to more than 90%.

Rather an interesting application of the chiral structure of active sites is obtained by copolymerizing a racemic α-olefin with an optically active one (28). In this case we have actually a terpolymerization with two monomers having the same absolute configuration (for instance R) and a third monomer having opposite configuration (S). As the two (R)-monomer will polymerize on the centers of the same chirality they yield a copolymer, whereas the (S)-monomer forms a homopolymer:

R + S + R' $\longrightarrow$ SSSSSS + RR'RRRR'R'R

racemic   optically   (S)-homopolymer   (R)(R')-copolymer
monomer   active
               monomer

| R and S represent (R)- and (S)-3,7-dimethyl-1-octene respec-
tively and R' is (R)-3-methyl-1-pentene |

The final product, because of different solubility of homopolymer
and copolymer can be separated in fractions of opposite optical
rotation by simple solvent extraction.

3. Polymerization of propylene to syndiotactic polymers

The best way to obtain syndiotactic polypropylene is the
low temperature polymerization in the presence of homogeneous
vanadium based catalysts (29). Also, in this case the
overall stereochemical mechanism of addition to the double bond
is cis (30) as for isotactic propagation, but this seems to be the
only common feature of the two types of stereospecific
polymerization of propylene.

The presence of "head-to-head" and "tail-to-tail" units in
addition to the head-to-tail ones in propylene homopolymers
(11,32), and of a number of even methylene sequences in
ethylene/propylene copolymers (11,31,33-35) denotes lack of
regiospecificity. As a consequence the syndiotactic polymerization
of propylene is conveniently described as a binary copolymerization
(11, 31, 36) between "head-to-tail" and "tail-to-head" oriented
monomer molecules:

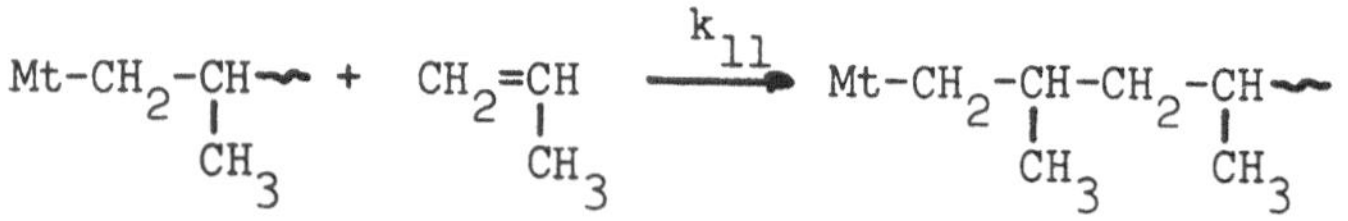

Mt-CH$_2$-CH$\sim$ + CH$_2$=CH $\xrightarrow{k_{11}}$ Mt-CH$_2$-CH-CH$_2$-CH$\sim$ [11]
       | | | |
       CH$_3$ CH$_3$ CH$_3$ CH$_3$

Mt-CH$_2$-CH$\sim$ + CH=CH$_2$ $\xrightarrow{k_{12}}$ Mt-CH-CH$_2$-CH$_2$-CH$\sim$ [12]
       | | | |
       CH$_3$ CH$_3$ CH$_3$ CH$_3$

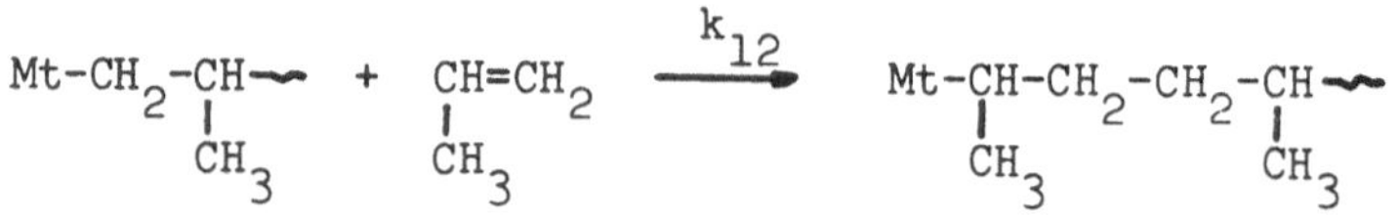

$$\text{Mt-CH-CH}_2\text{~} + \text{CH}_2\text{=CH} \xrightarrow{k_{21}} \text{Mt-CH}_2\text{-CH} - \text{CH-CH}_2\text{~} \qquad [21]$$
$$\underset{\text{CH}_3}{|} \qquad \underset{\text{CH}_3}{|} \qquad\qquad \underset{\text{CH}_3}{|}\ \underset{\text{CH}_3}{|}$$

$$\text{Mt-CH-CH}_3\text{~} + \text{CH=CH}_2 \xrightarrow{k_{22}} \text{Mt-CH-CH}_2\text{-CH-CH}_2\text{~} \qquad [22]$$
$$\underset{\text{CH}_3}{|} \qquad \underset{\text{CH}_3}{|} \qquad\qquad \underset{\text{CH}_3}{|}\ \underset{\text{CH}_3}{|}$$

The step [22] has been shown to be the only one syndiotactic specific (11,31,32), the prevailingly syndiotactic structure of the chains indicating that:

$$k_{11}/k_{12} \ < \ k_{22}/k_{21}$$

and

$$k_{11}k_{22}/k_{12}k_{21} \ > \ 1$$

The stereochemical sequences of the prevailingly syndiotactic polymers follow a perturbed first order Markov model (a strictly first order Markov model should be expected from the propagation step [22]) (20). The secondary (or Markownikow) regiospecificity of the step [22] turns primary across ethylene units with contemporary decrease of the syndiotactic content. Syndiotactic control is therefore lost whenever the last unit is achiral.

These facts together with the alternating tendence of syndiotactic specific catalysts in the copolymerization of ethylene with propylene and 1-butene as well as the higher syndiotactic content of propylene/1-butene copolymers than the corresponding propylene/ethylene copolymers (6) indicate that the syndiospecific propagation is controlled by the steric interactions between the last chiral unit of the growing chain end and the incoming monomer. These conclusions are confirmed by the results obtained by detailed analysis of the macromolecules end groups (14). It is of interest to remark that a driving force of less than 2 Kcal/mole has been estimated for syndiotactic polymerization control (20), appreciably lower than for the isotactic polymerization.

Recently a polymerization mechanism assuming pentacoordinate catalytic complexes has been proposed which accounts for a number of stereochemical details concerning this type of polymerization (37).

REFERENCES

1.   See for instance "Coordination Polymerization: a
     Memorial to Karl Ziegler", J.W.C.Chien Ed., Academic Press,
     1975.
2.   R.J.Ivin, J.J.Rooney, C.D.Stewart, M.L.H.Green, and
     R.Mahtab, Chem.Commun., 604 (1978).
3.   G.Natta, J.Inorg.Nuclear Chem., 8, 589 (1958).
4.   G.Natta, P.Pino, G.Mazzanti, P.Corradini, and U.Giannini,
     Rend.Accad.Naz.Lincei, 19 (8), 397 (1955).
5.   P.Pino, F.Ciardelli, G.P.Lorenzi, and G.Natta, J.Am.Chem.
     Soc., 84, 1487 (1962).
6.   A.Zambelli, A.Léty, C.Tosi, and I.Pasquon, Makromol.Chem.,
     115, 73 (1968).
7.   See e.g.: A.Zambelli and C.Tosi, Adv.Polymer Sci.
     (Fortschritte der Hochpolymeren Forschung) 15, 31 (1974).
8.   G.Natta, M.Farina, and M.Peraldo, Chim.Ind.(Milan), 42,
     255 (1960).
9.   T.Miyazawa and T.Ideguchi, J.Polymer Sci., B 1, 389 (1963).
10. A.Zambelli in "NMR Basic Principles and Progress", P.Diehl,
     E.Fluck and K.Kosfeld Eds., Springer Verlag, Berlin, 1971,
     pp.101-108.
11. A.Zambelli, G.Bajo, and E.Rigamonti, Makromol.Chem., 179,
     1249 (1978).
12. J.M.Sanders and R.A.Komoroski, Macromolecules, 10, 1214
     (1977).
13. G.Natta, P.Pino, E.Mantica, F.Danusso, G.Mazzanti, and
     M.Peraldo, Rend.Ist.Lomb.Sci.Lett., 91, 755 (1957).
14. A.Zambelli, P.Locatelli, and E.Rigamonti, Macromolecules,
     in press.
15. R.A.Shelden, T.Fueno, T.Tsunetsugu, and J.Furukawa,
     Polymer Letters, 3, 23 (1965).
16. G.J.Ray, P.E.Johnson, and J.R.Knox, Macromolecules, 10,
     773 (1977).
17. C.J.Carman, A.R.Tarpley Jr., and J.H.Goldstein, Macromolecules
     6, 719 (1973).
18. A.Zambelli and G.Gatti, Macromolecules, 11, 485 (1978).
19. A.Zambelli, P.Locatelli, and G.Bajo, Macromolecules in
     press.
20. A.Zambelli, P.Locatelli, G.Zannoni, and F.A.Bovey,
     Macromolecules, 11, 923 (1978).
21. P.Pino, F.Ciardelli, and G.Montagnoli, J.Polymer Sci., C,
     16, 3265 (1968).
22. P.Pino, G.Montagnoli, F.Ciardelli, and E.Benedetti,
     Makromol.Chem., 93, 158 (1966).
23. G.Montagnoli, D.Pino, A.Lucherini, F.Ciardelli, and P.Pino,
     Macromolecules, 2, 684 (1969).
24. O.Pieroni, G.Stigliani, and F.Ciardelli, Chim.Ind.(Milan),

52, 289 (1970).

25. F.Ciardelli, P.Locatelli, M.Marchetti, and A.Zambelli, Makromol.Chem., 175, 923 (1974).

26. F.Ciardelli, C.Carlini, G.Montagnoli, L.Lardicci, and P.Pino, Chim.Ind.(Milan), 50, 860 (1968).

27. C.Carlini, R.Nocci, and F.Ciardelli, J.Polymer Sci., Polymer Chem.Ed., 15, 767 (1977).

28. F.Ciardelli, C.Carlini and G.Montagnoli, Macromolecules, 2, 296 (1969).

29. G.Natta, I.Pasquon, and A.Zambelli, J.Am.Chem.Soc., 84, 1488 (1962).

30. A.Zambelli, M.G.Giongo, and G.Natta, Makromol.Chem., 112 , 183 (1968).

31. A.Zambelli, C.Tosi, and C.Sacchi, Macromolecules, 5, 649 (1972).

32. T.Asakura, I.Ando, A.Nishioka, Y.Doi, and T.Keii, Makromol. Chem., 178, 791 (1977).

33. W.O.Crain Jr., A.Zambelli, and J.D.Roberts, Macromolecules, 4, 330 (1971).

34. C.J.Carman, R.A.Harrington and C.E.Wilkes, Macromolecules, 10, 536 (1977).

35. J.C.Randall, Macromolecules 11, 33 (1978).

36. F.A.Bovey, M.C.Sacchi, and A.Zambelli, Macromolecules, 7, 752 (1974).

37. A.Zambelli and G.Allegra, submitted to Macromolecules.

KINETIC CONSIDERATIONS AND ACTIVE CENTRE DETERMINATION IN ZIEGLER-NATTA POLYMERIZATION.

Peter J.T. Tait

Department of Chemistry, U.M.I.S.T.
Manchester, United Kingdom.

Recent developments of commercial processes (1) for the polymerization of ethylene and propylene and which involve the use of second generation Ziegler-Natta catalysts have highlighted the need of reliable methods for the evaluation of catalyst activity. Comparative studies of catalyst activities are only useful, however, provided the factors which control these activities are understood. Such an understanding requires a fuller appreciation of the kinetic behaviour of these fascinating polymerization systems and the development of more accurate methods for the determination of active centre concentration.

1. KINETIC ASPECTS OF ZIEGLER-NATTA POLYMERIZATION

For the purpose of this paper only recent studies will be considered and their consideration further limited to an analysis of their relevance to catalyst activity and active centre concentration.

1.1 Phenomenological kinetic behaviour

The typical rate-time behaviour exhibited by conventional heterogeneous and homogeneous catalyst systems has been detailed by Keii (2) and Tait et al (3). Typical rate-time curves are shown in Figure 1.

R. W. Lenz and F. Ciardelli (eds.), Preparation and Properties of Stereoregular Polymers, 85-112.

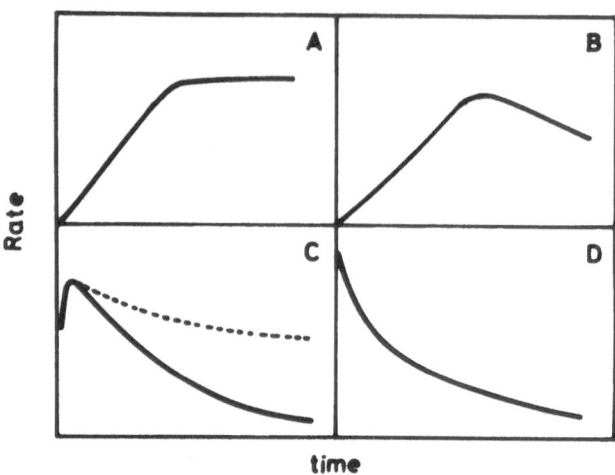

Figure 1. Typical rate–time plots.

Most conventional first generation systems, e.g.,
α-TiCl$_3$,VCl$_3$, δ-TiCl$_3$.0.33 AlCl$_3$, etc., show A or B type
behaviour depending on the particular concentration and type
of cocatalyst used and also on the temperature of the
polymerization. In both cases there is a definite settling
period during which the rate increases to a maximum value
after which it either remains constant or gradually decreases.
Many high activity catalyst systems exhibit C type behaviour
where the rate very quickly, in some cases almost
instantaneously, reaches a maximum value and then decreases
either slowly, as in the case of ether treated Solvay & Cie
catalysts, or in some cases very quickly, e.g. magnesium
alkyl reduced TiCl$_4$/AlEt$_3$ systems at higher aluminium alkyl
concentrations (4). Many homogeneous systems, e.g.,
Cp$_2$TiEtCl/AlEtCl$_2$/ethylene (5), and certain modified catalyst
systems, e.g., VOCl$_2$.2THF/AlR$_3$/vinyl chloride (6) are of type
D where a bimolecular decomposition of the active centres leads
to a very rapid decrease in the polymerization rate. Some
transition metal alkyl systems supported on a particular support
may also show D type behaviour although when supported on a
different support they will show C type behaviour. The
behaviour of most systems is very sensitive to the actual
conditions of polymerization and the concentrations of the
catalyst and cocatalyst components. Hence in comparing catalyst
activities it is essential to specify the exact polymerization
systems, i.e. catalyst, cocatalyst, monomer, additive, diluent,
and experimental conditions, and in addition the appropriate

time intervals from the beginning of the polymerization,
otherwise much confusion may result. These same
considerations also have important implications when
concentrations of active centres are reported.

1.2 Formulation of polymerization sequence

     The various steps in the overall polymerization process,
i.e. chain initiation, chain propagation and chain transfer
were first detailed by Natta (7), and have recently been
discussed in detail by Cooper (8). Whilst there is still no
general agreement concerning the finer details of these steps
the more generally accepted aspects of the polymerization
sequence may be summarized as follows.
     (a) The active centres are located on the lateral faces
of the crystalline metal halide crystals. This belief is
supported by the electron microscopic studies of Rodriguez
& van Looy (9), Kollar et al (10) and Guttman & Guillet
(11,12) on gas phase polymerization systems, although Kissin
& Chirkov (13,14) have concluded from their active centre
concentration determinations (methanol inhibition) that the
active centres cover all the surface of the catalyst. Similar
conclusions were reached by Burfield & Tait (15) using a
tritiated quenching technique. Unfortunately all these
investigations are complicated by breakdown of the catalyst
aggregates. Also it must be remembered that many of the
newer types of Ziegler-Natta catalysts have highly disordered
structures which further complicate this issue.
     (b) The growth of the polymer chain results from a primary
insertion reaction of the olefin into the transition metal
carbon bond and this insertion reaction occurs with _cis_
stereochemistry, Evidence for believing that the propagating
chain is attached to the transition metal atom comes from
quenching studies using tritiated alcohols (16), copolymerization
studies (17), and the development of alkyl free catalysts
(18), and the independence of the propagating rate constant on
the particular type of aluminium alkyl used (15). The spectral
studies of Longi et al (19), Takegami and Suzuki (20,21)
and Zambelli (22) support the belief that chain propagation
involves insertion of the monomer into a primary carbon-metal
bond with the formation of another primary carbon-metal bond.
Studies on the polymerization of $1-d_1$-deuteropropane
have indicated that the double bond opens in the _cis_
direction (23). A different sequence has recently been
proposed by Green et al (24,25) which involves a 1,2
hydrogen shift.

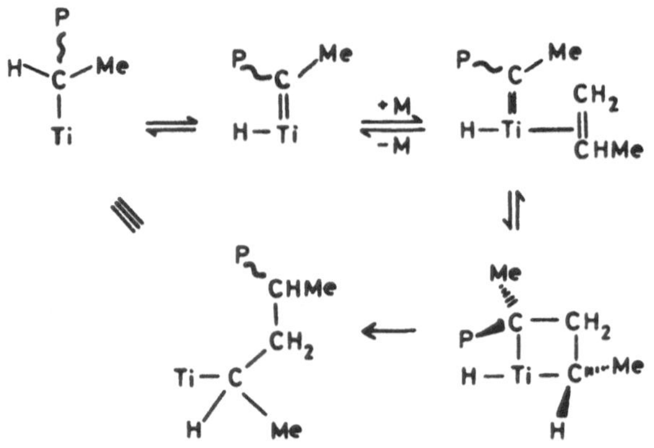

Figure 2. Polymerization sequence involving
a 1,2 hydrogen shift.

However positive evidence for this type of sequence in
Ziegler-Natta systems has still to be produced.

(c) Olefin coordination by means of π-complex formation
is a preliminary step to the insertion reaction. Although
there is as yet little direct evidence for prior coordination
of the olefin at the transition metal atom such a hypothesis
is supported by considerable kinetic evidence and has been
considered by Tait et al (15,26-30) and Zakharov et al (31).

(d) The isospecific arrangement of the monomer units
arises from steric hindrance between the monomer unit and the
halogen atoms joined to the transition metal atom. This
proposal which is an essential postulate of the Cossee-Arlman
theory (32,33) is in agreement with results obtained by
Pino (34) for the stereoselective polymerization of racemic
α-olefins and with n.m.r. data obtained by Zambelli (35).
Doi and Asakura have reached similar conclusions (36,37).

(e) Chain transfer reactions take place with both
aluminium alkyl and monomer. Conclusive evidence for these
reactions comes from the many investigations of the molecular
weight dependence on both the aluminium alkyl and monomer
concentrations (e.g. 2,7,8,38).

1.3 Kinetic Models
    Very many kinetic models have been proposed for both
heterogeneous and homogeneous Ziegler-Natta polymerization
(c.f. 2,8,18) but for the purposes of this paper only the
recent models proposed by Tait et al (26,29,30), Yermakov
et al (31) and Bohm (39) will be considered. These three
kinetic treatments have the common features that chain
propagation is treated as a two-stage reaction, monomer

coordination at an active centre being followed by subsequent insertion of the adsorbed (complexed) monomer molecule into an active transition metal-carbon bond and that complexation of aluminium alkyl with active sites is envisaged.

1.3.1 Kinetic model of Tait et al. The main features of the kinetic scheme proposed by Tait et al (26,29,30) are as follows.

(a) The active centres are formed by the interaction of the metal alkyl compound with the transition metal halide. These centres are considered to be basically monometallic in nature in the sense that insertion of monomer takes place into active transition metal-carbon bonds, but this does not exclude more highly complexed species. During the steady state polymerization conditions and for any given system a maximum limiting number of these centres are considered to be present for a given concentration of heterogeneous catalyst.

(b) Chain propagation takes place between an active centre and adsorbed monomer, i.e. the propagation reaction is a two-stage reaction.

(c) The overall rate of polymerization, R_p is given by

$$R_p = k_p \Theta_M C^* \qquad (1)$$

where k_p is the propagation rate constant with respect to adsorbed monomer, Θ_M is the fraction of surface sites covered by adsorbed monomer, and C^* is the concentration of active centres.

(d) Monomer and aluminium alkyl compounds are involved in competitive reversible adsorption reactions with the active centres, viz.,

$$Cat - P + M \xrightleftharpoons{K_M} M. Cat - P \qquad (2)$$

$$Cat - P + A \xrightleftharpoons{K_A} A. Cat - P \qquad (3)$$

The adsorption of monomer and aluminium alkyl compounds are thus regarded as the equivalent of complexation reactions and it is an important feature of this model that active centres can become inactivated by complexation with metal alkyl compounds or derivatives. Hence it is essential to distinguish between potential and propagating centres (40, 41).

(e) Adsorption of both monomer and metal alkyl are described by Langmuir - Hinshelwood isotherms such that

$$\Theta_M = \frac{K_M [M]}{1 + K_M[M] + K_A[A]} \qquad (4)$$

$$\Theta_A = \frac{K_A [A]}{1 + K_M[M] + K_A[A]} \tag{5}$$

where $[M]$ and $[A]$ are the concentrations of M and metal alkyl and K_M and K_A are the respective equilibrium constants.

(f) The overall rate of polymerization is given by the expression

$$R_p = \frac{k_p K_M [M] C^*}{1 + K_M[M] + K_A[A]} \tag{6}$$

(g) Chain transfer reactions take place between adsorbed metal alkyl and the growing polymer chain and also between adsorbed monomer and the growing polymer chain.

This model has been successfully applied to an analysis of the kinetic behaviour and molecular weight and active centre concentration dependencies in the system $VCl_3/AlR_3/4$-methyl pent-1-ene (4-MP-1) (27,29,30).

1.3.2 <u>Kinetic model of Yermakov et al</u>. More recently Yermakov and Zakharov et al (31) have used a somewhat similar model to interpret their extensive kinetic and active centre studies on the polymerization of ethylene and propylene by the catalyst systems δ-$TiCl_3.0.3.AlCl_3/AlEt_3$ or $AlEt_2Cl$. This model is based on the following assumptions.

(a) The propagating centres are monometallic and involve an active titanium-carbon bond. Aluminium alkyl cocatalysts are not considered to be incorporated in the propagating centres.

(b) The propagation reaction is a two-stage reaction where prior coordination of monomer on the titanium ion in the active centre is followed by subsequent insertion into the titanium bond.

(c) The aluminium alkyl affects the rate of polymerization and the stereoregularity of the polymer formed due to a change in the number of active centres and in the ratio of centres of different stereospecificities.

(d) Reversible adsorption of aluminium alkyl occurs at the active centres with the formation of temporarily inactive centres as proposed earlier by Tait et al (26, 27).

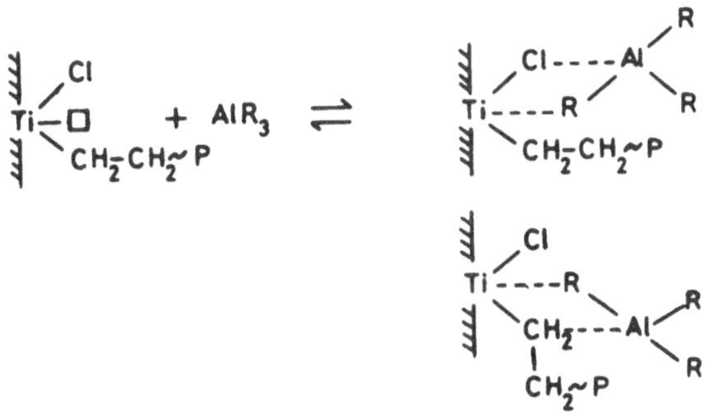

Figure 3. Reversible adsorption of aluminium alkyl.

(e) Aluminium alkyl compounds participate in a chain transfer reaction.

(f) The overall rate of polymerization is given by the expression

$$R_p = k_p C_p [M] \qquad (7)$$

where C_p is the number of propagating centres.

The total number of active centres, C_o, is given by the equation

$$C_o = C_p + C_A \qquad (8)$$

where C_A is the number of centres complexed by aluminium alkyl.

1.3.3 **Kinetic model of Bohm.** A very comprehensive reaction model has been proposed by Bohm (39) which is claimed to explain, at least in a qualitative manner, the known kinetic phenomena for both heterogeneous and homogeneous Ziegler–Natta polymerization reactions. This model has the following characteristics.

(a) It is based on the Rideal model for adsorption, i.e. catalytic reaction takes place between a radical or atom fixed at an active centre by a covalent bond and an adsorbed molecule. The adsorbed molecule is held in a deep van der Waals trough which is regarded as the free coordination site.

(b) Complex formation between olefin and transition metal atom is assumed in keeping with the concepts of Arlman and Cossee (33) and Henrici–Olive and Olive (42).

(c) The model assumes reversible complexation between alkyl aluminium and an active centre as in the Tait and

Yermakov models.

(d) It allows for chain transfer with complexed aluminium alkyl.

(e) It is comprehensive in that it makes use of most of the reactions which have been previously formulated for this type of polymerization, e.g. spontaneous termination, etc., (8).

Bohm has usefully applied this model to a kinetic and molecular weight analysis of the polymerization of ethylene by a highly active catalyst derived from $Mg(OEt)_2$ and $TiCl_4$ (43, 44).

It is the opinion of the author that it is unwise to expect any one model to fit all cases because of the very wide range of Ziegler-Natta and related catalysts which have been developed. Nevertheless it is essential in order to interpret the accumulated active centre concentration data to have fairly well defined kinetic models. This requirement will become evident in the next section of this paper and it is for this reason that these kinetic models have been presented in some detail.

2. ACTIVE CENTRE DETERMINATION

These past few years have witnessed an increasing interest in efforts to determine the concentrations of active centres, C^*, in Ziegler-Natta polymerization systems. This interest arises both from academic and industrial considerations and is of considerable importance. Indeed the development of more accurate and less ambiguous methods for the determination of C^* must now be given high priority and for the following reasons. Firstly, the determination of active centre concentrations provides the sole route for the evaluation of the absolute values of rate constants for chain propagation and also for the transfer reactions which occur in these polymerization systems. Secondly, a knowledge of the concentrations of active centres is necessary before any meaningful and direct comparison between the activities of various polymerization systems can be made. Such comparisons have been useful in the development of second generation catalyst systems. However as has been remarked earlier the overall kinetic behaviour of these systems must be considered in parallel with such data. Thirdly, there is increasing evidence that the nature of the procedures adopted for the determination of C^* together with the character and numerical values obtained can lead to useful information concerning both the nature and types of the active centres which are present in various catalyst systems.

A considerable number of indirect methods have been developed which allow C* to be determined and a consideration of the validity of some of these has already been published by Schnecko and Kern (45-47). For the purposes of this paper it is convenient to classify these methods into five main categories.

2.1 Kinetic, molecular weight and adsorption methods.

A number of different techniques for the determination of C* have been devised which rely on kinetic studies or either a combination of kinetic observations and molecular weight determinations, or simultaneous kinetic and adsorption studies.

2.1.1 <u>Variation of molecular weight with time</u>. This method was first devised by Natta et al (7,48) whereby C* was obtained from the variation of the number average degree of polymerization, $\bar{P}_n$, with time. and has been quite widely used (7,48-51). Natta showed that the following equations could be used to evaluate C*.

$$\bar{P}_n = \frac{R_p \tau}{C^*} \tag{9}$$

$$\tau = \frac{d(1/\bar{P}_n)}{d(1/t)} \cdot \bar{P}_n \tag{10}$$

Consequently τ may be determined from a plot of $1/\bar{P}_n$ versus $1/t$ together with a knowledge of $\bar{P}_n$ (asymptotic value) and used in equation (9) to evaluate C*.

Schnecko and Kern (45) have pointed out that τ and C* as determined above represent average values over the time of polymerization. The derivation requires that there is no change in molecular weight distribution with time. Additionally there is the problem that when $\bar{M}_n$ reaches a constant value $d(1/\bar{P}_n)/d(1/t)$ will tend to zero as will τ . To overcome this problem it is necessary to use $\bar{M}_n$ values from the early stages of the polymerization, but in this case C* may be changing due to complications arising from any settling period.

Haward et al (4) have made use of molecular weight data to determine C* by using the equation

$$\frac{Y}{\bar{M}_n} = (1 + k_{tr} [X] t) C^* \tag{11}$$

$$= C^* \text{ when } t = 0$$

where y is the yield of polymer at time t and k_{tr} is the rate constant for chain transfer with agent X (e.g., hydrogen). A similar procedure has been used by Bohm (39,43). As before only average values of C* can be obtained and there must be no change in molecular weight distribution.

2.1.2 <u>Non-stationary kinetics</u>. A complex kinetic scheme has been developed by Chirkov et al (52) and used to determine C* for the $VCl_3/Al(i-Bu)_3$/propylene system. The settling period is attributed to a slow initiation step and hence C* is evaluated from the initial non-stationary period of the polymerization. The method is limited in that it depends on this particular kinetic interpretation and in any case many polymerization systems show no settling period.

2.1.3 <u>Inhibition by alcohols</u>. The inhibition effects have been used by Chirkov et al (13) to evaluate C* for the polymerization system $TiCl_3/AlEt_3$/propylene. The method depends on the assumptions that all the alcohol added is adsorbed on the catalyst surface and than an active site is blocked by one alcohol molecule. Unfortunately with this type of technique there is always the uncertainty of exactly what the added alcohol reacts with, and to what extent. In the case of heterogeneous systems the problem is more acute since the alcohol may not only react with active centres but may even be adsorbed over the entire catalyst surface. Consequently values of C* determined by this method can be expected to give only the upper limit.

2.1.4 <u>Simultaneous kinetic and adsorption studies</u>. A promising new method which involves simultaneous rate and adsorption measurements has been devised by Caunt (53) and further developed by Burns, Caunt and Tait (54,55). When a known amount of allene is injected into a Ziegler-Natta polymerization reaction there is an immediate and dramatic decrease in the overall rate of polymerization which then slowly increases to reach its original value. This effect is shown in Figure 4. On injection the allene partitions itself between the diluent and the gaseous phase whilst in addition a small amount is adsorbed on the active centres. The essence of this method is the determination of the amount of allene adsorbed on the catalyst surface at a time t_o, when the maximum number of polymerization centres have been complexed, but before the allene is itself polymerized. $\lfloor$ N.B., $R_p(\text{allene}) = 10^{-3} R_p (\text{propylene}) \rfloor$ and to relate this amount to the number of active centres. In practice it is convenient to determine the total number of active centres by plotting the % drop in rate versus mole % allene adsorbed per mole $TiCl_3$. A typical plot is shown in Figure 5. More recently a similar procedure has been carried out by Ajayi, Caunt and Tait (56) using CO when the use of a different adsorbent was found to yield the same value for C* for the system $\delta-TiCl_3.0.33AlCl_3$ / $AlEt_2Cl$ / propylene. These results

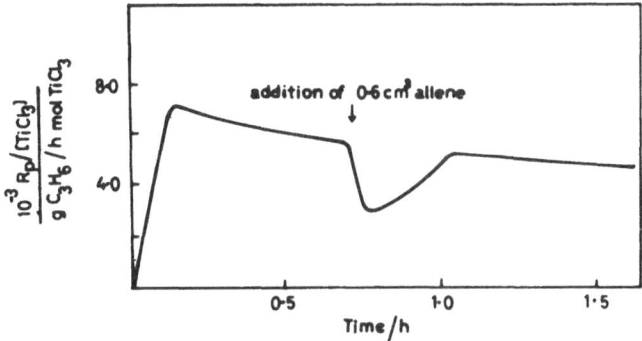

Figure 4. Effect of injection of allene on rate of
polymerization of propylene. Solvay & Cie ether treated
catalyst; $AlEt_2Cl:TiCl_3$ = 2 : 1 ; P = 1 atm,
temperature = 60 °C.

have also been incorporated into Figure 5. The agreement
between these two adsorbates is considered good proof for the
validity of this method.

Obviously certain operational criteria must be fulfilled
before this type of method can be successfully applied for the
evaluation of C*.

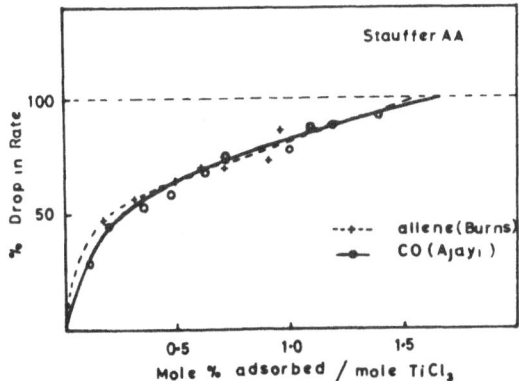

Figure 5. Plot of % drop in rate versus mole % allene or
CO adsorbed. Stauffer – AA $TiCl_3$ catalyst;
$AlEt_2Cl:TiCl_3$ = 2 : 1 ; P = 1 atm; temperature = 60 °C

(a) The adsorbate must remain on the catalyst surface long enough for its concentration to be measured or determined.

(b) All centres must be complexed and the system must have reached equilibrium.

(c) The adsorbate must be of a similar chemical nature and size to the monomer so that adsorption only takes place on the polymerization sites.

(d) Only one molecule of adsorbate should be adsorbed per active centre, or else the stoichiometry must be known.

From an analysis of the data obtained from a number of Ziegler-Natta catalyst systems it is believed that these criteria have been met in the use of allene and CO.

Figure 5, which is typical of many catalyst systems, is of considerable interest because it clearly and directly demonstrates, perhaps for the first time, that some centres are more reactive than others. For the simplest case where only two different types of centres are assumed to be present then their activities differ by about a factor of ten. However the true situation may be more complex than this.

Plots of the type of Figure 5 may be used to "finger-print" the type of catalyst system in use. Some systems would seem to be more uniform as far as site activity is concerned as is shown in Figure 6.

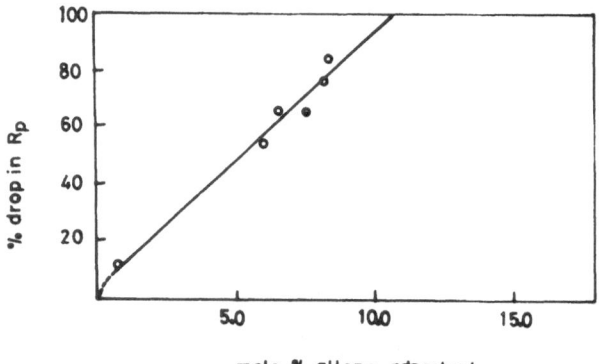

mole % allene adsorbed

Figure 6. Plot of % drop in rate versus mole % allene adsorbed. Solvay & Cie ether treated catalyst; $AlEt_3$: $TiCl_3$ = 2 : 1, P = 1 atm; temperature = 60 °C.

The concept that there are centres of different activities has been postulated before; by Baker & Tait (57) from copolymerization studies, by Fuji (58) and by Keii (59) amongst others. Such a situation may arise because of the actual location of centres, e.g., centres on exposed crystal edges, centres located in catalyst pores, etc., and/or because of the existence of centres of different chemical composition, e.g., Cossee type centres (32,33) or van Looy type centres (9). Both Fuji (58) and Keii (59) have detailed some possible structures. There is in addition the possible effects of centre clustering which has to be considered, i.e., whether or not isolated centres have the same activity as those in clusters. Electron microscopic studies would seem to indicate that under certain situations active centre clustering occurs (c.f.,11).

Finally it is evident that these results must lead to the concept of <u>average</u> k_p <u>values</u> and this must be recognized when comparing k_p values derived from C* values where these have been determined using different techniques.

Table I lists some typical C* values obtained using kinetic, molecular weight and adsorption techniques. It is quite evident that a wide range of results have been obtained by different workers, although in some cases the catalyst systems employed are different.

2.2 Quenching methods

Quenching procedures have been until recently very widely used for the determination of C* and are based on the reaction.

$$Cat^+ \ \overline{C}H_2-CH_2 \sim P + QL \rightarrow Cat-Q + L-CH_2-CH_2 \sim P$$

Typical quench reagents have included radioactive iodine, inactive iodine, tritiated alcohols, tritiated water, and deuterated methanol and water.

Necessary criteria for the successful use of any quenching technique include the following.

(a) The quenching agent should completely stop the polymerization, i.e., the reaction with the active centres must quantitative.

(b) The quenching agent should interact only with active metal-polymer bonds or if not the other reactions must be allowed for.

(c) Isotopic exchange or side reactions with polymer should be absent.

(d) Kinetic isotopic effects should be absent or directly measurable.

Chain transfer processes with aluminium alkyl compounds in Ziegler-Natta polymerization means that there can exist at least two types of polymer chain both of which will react with the quenching agent, i.e.

TABLE I

Active Centre Concentrations – Molecular weight, kinetic and adsorption methods.

Author	Catalyst	Monomer	$T/^{\circ}C$	$C^* \times 10^3/$ mol/mol	Notes
Natta	α-TiCl$_3$/AEt$_3$	P	70	2-6.3	Molecular weight/ kinetic
Tanaka and Morikava	δ-TiCl$_3$0.33AlCl$_3$/ AlEt$_3$	P	70	11	Molecular weight/ kinetic
Grievson	TiCl$_3$/AlEt$_2$Cl	E	40	15	Molecular weight/ kinetic
Chirkov	VCl$_3$/Al(i-Bu)$_3$	P	70	5.4	Kinetic, non-stationary states
	VCl$_3$/Al(i-Bu)$_3$	E	70	5.4	
Schnecko and Kern	TiCl$_3^*$ /AlEt$_2$Cl	P	60	7.5	Molecular weight/ kinetic
	TiCl$_3^{**}$/AlEt$_2$Cl	P	60	19	
Kissin and Chirkov	α-TiCl$_3$/AlEt$_3$	P	70	5.2	Inhibition
Burns et al	γ-TiCl$_3$/AlEt$_2$Cl	P	60	15.3	Allene adsorption/ kinetic
Ajayi et al	γ-TiCl$_3$/AlEt$_2$Cl	P	60	16.2	CO adsorption/ kinetic
Haward et al	TiCl$_4$/MgR$_2$/AlEt$_2$Cl	E	50	600	Molecular weight
	TiCl$_4$/MgR$_2$/AlEt$_2$Cl	E	30	150	(Magnesium reduced catalyst)
Bohm	Mg(OEt)$_2$/TiCl$_4$/ AlR$_3$	E	85	700	Molecular weight (High activity catalyst)

*TiCl$_4$ reduced by Ti; **TiCl$_4$ reduced by Al.

$$Cat^+ \ \bar{C}H_2\text{-}CH_2 \diagdown P \quad \text{and } R_2 \ AlCH_2\text{-}CH_2 \diagdown P$$

Thus the total number of metal-polymer bonds, MPB, in the polymerization system will increase with time and may be described by the equation

$$[MPB]_t = C^* + N_t \qquad (12)$$

where $[MPB]_t$ is the total concentration of metal-polymer bonds at time t and N is the concentration of transferred metal-polymer bonds (26).

In order to determine C^* it is essential to distinguish between labelled polymer molecules resulting from active centres and from transferred polymer chains now bound to aluminium and non-propagating. The method usually adopted is to extrapolate the linear region of the plot of $[MPB]$ versus time to zero time, or to zero conversion as is illustrated in Figure 7 (15). The rate of transfer with aluminium alkyl may be obtained from such a plot and becomes an added bonus when using this method.

2.2.1 <u>Use of iodine</u>. Chien (60) was the first to report on the use of a quench technique using $^{131}I_2$ for the determination of C^* in the polymerization of ethylene by the soluble catalyst system Cp_2TiCl_2 / $AlEt_2Cl$, and additionally he demonstrated the presence of a non-propagation species. Caunt (61) and Schnecko et al (62) have used inactive I_2 as a quench reagent. Both Caunt and Schnecko observed an increase in $[MPB]$ with time and therefore C^* was determined by extrapolation to zero time. Unfortunately the reaction between the active centres and the iodine molecules may not be quantitative and additionally the technique suffers from other serious complications. These drawbacks have been discussed in detail by Schnecko et al (45).

2.2.2 <u>Use of tritiated alcohol</u>. One of the foundation studies in this area was that by Feldman and Perry (16) who determined values for C^* for the polymerization system $TiCl_4$ / $Al(i\text{-}Bu)_3$ / ethylene using tritiated methanol as a quench reagent. Since then the method has been successfully used by many other workers including Kohn et al (63), Coover et al (64), Bier et al (65), Schnecko and Kern (45) for the system $TiCl_3$ / $AlEt_2Cl$ / propylene, Cooper et al (66) for the VCl_3 / $AlEt_3$ / isoprene system, and by Burfield and Tait (15) for their studies on the VCl_3 / AlR_3 / 4-MP-1 system. More recently the method has been used by Yermakov and Zakharov (67,68) for the system $TiCl_2$ / $AlEt_2Cl$ / ethylene and $TiCl_3$ 0.3 $AlCl_3$ / $AlEt_2Cl$ / propylene and by Baulin et al (69) for the supported system $TiCl_4$ / MgO / $AlEt_3$ and $TiCl_4$ / $Al_2O_3.SiO_2$ / $AlEt_3$.

Provided care is taken to extrapolate the measured $[MPB]$

TABLE II

Active Centre Concentration – Quenching Methods

Author	Catalyst	Monomer	T/°C	$C^* \times 10^3$ mol/mol	Notes
Schnecko et al	$TiCl_3/AlEt_2Cl$	P	60	3	I_2 quenching
Caunt	$\alpha\text{-}TiCl_3/AlEt_2Cl$	P	60	4	I_2 quenching
Feldman and Perry	$TiCl_4Al(i\text{-}Bu)_3$	E	55	11	Tritium quenching
Coover et al	$TiCl_3(StaufferHA)/$ $AlEt_2Cl$	P	70	1	Tritium quenching
Schnecko et al	$\delta\text{-}TiCl_3.0.33AlCl_3/$ $AlEt_2Cl$	E P B	60 60 60	6.9 5.3 3.5	Tritium quenching
Burfield and Tait	$VCl_3/Al(i\text{-}Bu)_3$ $VCl_3/Al(n\text{-}Bu)_3$ $VCl_3/Al(n\text{-}Hex)_3$ $VCl_3/AlEt_3$	4-MP-1 4-MP-1 4-MP-1 4-MP-1	30 30 30 30	0.38 0.33 0.23 0.61	Tritium quenching
Baulin et al	$TiCl_4/MgO/AlEt_3$ $TiCl_4/Al_2O_3,SiO_2/$ $AlEt_3$	E E	70 70	210 235	Tritium quenching
Yermakov et al	$TiCl_2$ $\delta\text{ -}TiCl_3.0.3AlCl_3/$ $AlEt_2Cl$	E P	75 70	0.036α 60α	Tritium quenching $\alpha \equiv$ isotope factor

values back to zero time or zero conversion and to determine
the kinetic isotope effect appropriate for the polymerization
system under examination, tritium quenching can be expected
to provide reliable methods for C*. However, there is always
the possible complications of isotopic exchange reactions (70)
and contamination by active catalyst residues. Further
development work is needed in this area, especially when
using supported type catalysts. The significance of C* values
obtained using this technique will be discussed later.

Some typical C* values are listed in Table II.

2.3 Radiotagging methods

Due to the very low values of C* present in many of the
earlier Ziegler-Natta catalyst systems the use of radiolabelling
techniques has received considerable attention by a number
of workers.

2.3.1 <u>Use of labelled cocatalysts</u>. This particular
technique relies on the now well substantiated assumption that
the active centres are derived from alkylation reactions
such as

$$TiCl_3 \; + \; Al\overset{*}{R}_3 \longrightarrow Cat — R^*$$

so that an alkylated transition metal species is produced from
the labelled aluminium alkyl. The value of C* may thus be
obtained from the number of labelled groups present in the
polymer. Values of C* have been determined by Natta et al
(7,48) from both adsorption studies and from the ^{14}C labelled
alkyl groups incorporated into the polymer. Chien (60) also
used ^{14}C labelled $AlEt_2Cl$ when studying the homogeneous
polymerization system $Cp_2TiCl_2/AlMe_2Cl$ ethylene. More recently
Ayrey and Mazza (71) have carried out some detailed studies on
the validity of using labelled cocatalysts and have concluded
that olefin displacement reactions can give rise to contamination.

Also serious complications may arise because of the
adsorption (or complexation) of metal alkyl molecules and from
chain transfer reactions leading to the incorporation of
additional labelled groups into the polymer chain. Hence the
values of C* obtained must be regarded as representing an
upper limit.

2.3.2 <u>Carbon monoxide and carbon dioxide radiolabelling</u>.
Most of the recent work using radiotagging techniques has
involved the use of catalyst poisons such as ^{14}CO and $^{14}CO_2$
rather than ^{14}C labelled cocatalysts. Although this
procedure is sometimes regarded as a quenching reaction this
is not strictly correct since the polymerization is first
treated with either ^{14}CO or $^{14}CO_2$ and then quenched after
an appropriate time interval by the addition of alcohol.

TABLE III

CARBON MONOXIDE RADIOLABELLING

Catalyst	$10^3 C*/mol/mol$ $TiCl_3$		$k_p /dm^3 /mol$ sec	
	Soluble	Insoluble	Soluble	Insoluble
Stauffer AA/ AlEt$_2$Cl	2.2	7.5	1.1	10.5
Solvay & Cie/AlEt$_2$Cl (Ether treated)	1.9	7.5	1.5	12.9
Solvay & Cie/AlEt$_3$ (Ether treated. Rapid decay in rate with time. Maximum rate values)	5.7	13.2	38.2	58.4

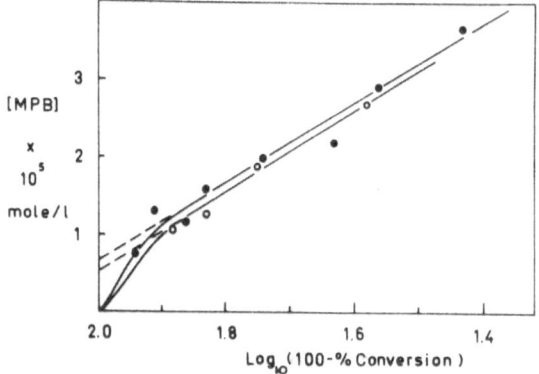

Figure 7. Variation of MPB concentration with $\log_{10}$ (100-% conversion) for Al(i-Bu)$_3$. [4-MP-1$_o$]= 2.00 mol/dm^3 ; VCl$_3$= 18.5 mmol/dm^3 ●, 30 °C ; 0, 50 °C.

The overall reaction sequence may be represented as

$$
\begin{array}{ccc}
& & ^{14}CO \\
& & \downarrow \\
L_xM\text{-}CH_2\!\!\sim\!\!P \;+\; ^{14}CO \;\rightleftharpoons\; & & L_xM\text{-}CH_2\!\!\sim\!\!P \\
& & \downarrow \\
\underset{O}{\overset{\parallel}{H\text{-}C\text{-}CH_2\!\!\sim\!\!P}} \quad\xleftarrow{\;ROH\;}\quad & & \underset{O}{\overset{\parallel}{L_xM\text{-}C\text{-}CH_2\!\!\sim\!\!P}}
\end{array}
$$

The above reaction corresponds to the known insertion reaction of CO into 6-bonded metal alkyls for organometallic compounds of transition metals (72).
    The development of ^{14}CO radiotagging has been pioneered by Yermakov et al who have produced numerous publications in this field (e.g., 67,68, 73–77). Additionally the technique has been used by Fuji(58), Chien & Hsieh (78) and by Burns, Caunt & Tait (54,55).
    One of the complications of this method is that an increase in polymer radioactivity is observed as the time of contact between the CO and the polymerization system is increased, i.e., with increase in the duration of the interval after injection of CO and quenching with alocohol. This effect is shown in Figure 8 (54, 55).

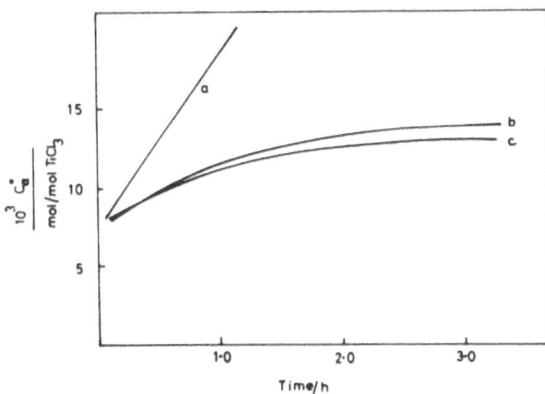

Figure 8. Variation of Apparent active centre concentration with interval between CO injection and quenching by alcohol.

The rapid increase in ^{14}C in the polymer when propylene is present in the system (plot a) is believed to result from a copolymerization reaction. It is evident however that even in the absence of propylene (plots b and c, when the propylene has been replaced by N_2) that an increase in polymer radioactivity still takes place. The phenomena shown in Figure 8 is quite general and for many catalyst systems could lead to an increase in the apparent value of C* of up to 100% if comparison is made between the value corresponding to a 4 to 6 min interval (believed to be the time taken for the CO to be adsorbed on the active centres) between injection of CO and quenching and the limiting value of C* after some 180 min or so.

Mejzlik and Lesna (79) have concluded from their examination of this technique that the reaction of CO and CO_2 with active centres is very slow such that their retarding influence on olefin polymerization should be considered as a sorption equilibrium. This view is not however consistent with the results shown in Figure 8 which shows that a very large percentage of the CO molecules insert very rapidly.

The following possible explanations for the observed phenomena should be considered.

(a) The existence of two or more types of active centre, one of which is very inactive leading to slow CO insertion while the other is much more active and consequently allows the CO to insert more quickly.

(b) Initiation of new active bonds by transfer reaction such as

$$\text{Cat-}\overset{14}{\underset{\underset{O}{\|}}{C}}\text{P} \; + \; \text{AlR}_3 \longrightarrow \text{Cat-R} \; + \; \text{R}_2\text{Al-}\overset{14}{\underset{\underset{O}{\|}}{C}}\text{P}$$

(c) Slow displacement of complexed aluminium alkyl compounds from potential active centres, viz.,

However it is apparent that further fundamental work is required in this field.

Nevertheless in spite of these uncertainties the method has a number of advantages.

(a) The experimental procedures are relatively simple and less time consuming than many others.

(b) Complications due to chain transfer with aluminium alkyl compounds as happens with tritium quenching are absent.

(c) Isotope effects may be neglected.

(d) The method allows differentiation between centres producing soluble and insoluble polymer (i.e. polymer soluble and insoluble in hydrocarbon solvent at 60 °C in the present case). Data from some recent results (54, 55) is shown in Table III. As is evident from Table III useful information concerning average k_p values for centres producing "atactic" and "isotactic" polymer can thus be obtained.

Table IV contains some typical values of C* obtained from radiotagging procedures.

2.4 Electron Microscopy

The gas phase polymerization of propylene on the surface of α-TiCl$_3$ crystals activated by AlMe$_3$ has been investigated by Guttman and Guillet by means of electron microscopy, (11,12). By assuming that each fibril was attached to only one active centre values for C* were determined. However the values obtained were several orders of magnitude less than those obtained by other methods, a discrepancy which may be due to active centre clustering, or to the somewhat artificial conditions of these imaginative experiments (i.e. gas phase).

2.5 Spectroscopic methods

A paramagnetic probe technique has been developed by Chien (80) and used to characterize surface hydroxyl groups of materials such as Cab-O-sil and Mg(OH)Cl when reacted with VCl$_4$ (78). Although this method has not so far been developed to determine values of C* it could provide useful information on the maximum number and location of supported transition metal atoms which could be potential active centres.

3. COMPARATIVE STUDIES ON ACTIVE CENTRE CONCENTRATIONS

Meaningful comparisons between values of C* obtained by different workers are both difficult and hazardous due to the variation in experimental conditions which have been employed by various workers. There is also the problem of exactly what is determined by the different measuring techniques.

Chirkov (52) has compared values of C* obtained using non-stationary state kinetics with those obtained by Tanaka & Morikava (49). Schnecko & Kern (45) have discussed values of C* obtained by kinetic and molecular weight techniques and by

106   P. J. T. TAIT

TABLE IV
Active Centre Concentrations – Radiolabelling Methods.

Author	Catalyst	Monomer	$T/^{\circ}C$	$C^* \times 10^3/$ mol/mol	Notes
Natta	$\measuredangle$-TiCl$_3$	P	70	3–6.3	^{14}C
Chien	α-TiCl$_3$	P	70	36	^{14}C
		P	50	106	^{14}C
Zakharov et al	δ-TiCl$_3$.0.3AlCl$_3$/ AlEt$_3$	E	80	0.43	$^{14}CO/PH_3$
	TiCl$_4$ on poly-ethylene/AlEt$_3$	E	80	2.6	
	TiCl$_4$/MgCl$_2$/ AlEt$_3$	E	80	5.2–18	
	TiCl$_2$	P	70	0.0013	$^{14}CO_2$
	δ-TiCl$_3$.0.33AlCl$_3$/ AlEt$_2$Cl	P	70	1.4	$^{14}CO/PH_3$
Burns et al	δ-TiCl$_3$.0.3AlCl$_3$/ AlEt$_2$Cl	P	60	10	^{14}CO
Yermakov and	Cr(allyl)$_3$/SiO$_2$	E	80	1.5	$^{14}CO/PH_3$
Zakharov	Zr(allyl)$_3$/SiO$_2$	E	80	3.0	$^{14}CO/PH_3$
	TiBz$_4$/Al$_2$O$_3$	E	80	3.6	$^{14}CO/PH_3$
Chien and Hsieh	Bz$_4$Ti/Mg(OH)Cl/ AlEt$_2$Cl	E	50.	1.1	$^{14}CO/PH_3$
	Bz$_3$TiCl/Mg(OH)Cl/ AlEt$_2$Cl	P	50	0.9	$^{14}CO/PH_3$

tritium quenching. One of the problems associated with such comparisons (52) is the sensitivity of these catalyst systems to pretreatment and experimental handling, especially when this is carried out in different laboratories.

Recently Ajayi, Burns, Caunt and Tait (54,55,56) have carried out a series of comparative studies using selected catalyst systems and selected methods for the determination of $C*$. Some representative results are shown in Table V. From a study of Table V the following points are evident.

(a) For the catalyst system δ-TiCl$_3$0.33 AlCl$_3$ / AlEt$_2$Cl allene adsorption, CO adsorption and ^{14}CO radiotagging give about the same values for $C*$ although the ^{14}CO technique for $t_q = 4$ min gives a slightly lower value. Any final conclusion on this result is dependent on the interpretation placed on Figure 8.

(b) Although the actual values of $C*$ for different types of TiCl$_3$ - based catalyst systems vary the average values of k_p remain constant when $C*$ is determined from allene adsorption experiments. Values of k_p based on ^{14}CO radiotagging show more scatter and their interpretation is less certain especially when using Solvay & Cie ether treated catalysts and AlEt$_3$.

(c) Values of $C*$ obtained from tritium quenching are higher than those obtained from ^{14}CO radiotagging. This is evident from a comparison of the results obtained from the VCl$_3$/ AlEt$_3$/ 4-MP-1 system. Similar conclusions have been reached by Yermakov et at (31, 67).

It is however to be expected that these two methods might give rise to different values for $C*$. Complexed aluminium alkyl compounds (with at least one polymeric chain) should not be displaced from the active centres by CO. Such compounds would be quenched in the usual manner by a tritiated alcohol yielding polymer chains containing tritium. This situation is easily understood if the polymerization sequence (c.f. Tait kinetic model) detailed in Figure 9 is studied.

Thus

$$C_T = C_{\square} + C_M + C_A \qquad (13)$$

If C_p is the number of propagating centres at any given time then

$$C_p = C_{\square} + C_M \qquad (14)$$

Hence $\quad C_T = C_p + C_A \qquad (15)$

Tritium quenching will give C_T whilst allene and CO adsorption and ^{14}CO radiolabelling should give C_p. Indeed it can be shown that

$$\frac{(C*)_{CO}}{(C*)_{ROT}} = 1 - \Theta_A \qquad (16)$$

TABLE V

COMPARISON OF C* AND k_p VALUES FOR THE CATALYST SYSTEMS UNDER STUDY

	VCl$_3$/AlEt$_3$/4-MP-1*[3]	Solvay & Cie †/AlEt$_2$Cl/propylene**	Stauffer-AA/AlEt$_2$Cl/propylene**	Solvay & Cie ≠/AlEt$_2$Cl/propylene**	Solvay & Cie ≠/AlEt$_3$/propylene**
$10^3 \cdot C^*/(\text{mol/mol TiCl}_3)$					
Tritium	0.61	40α			
^{14}CO(t_q = 4 min)	0.17	9.6	10.0	7.9	20.8
^{14}CO(t_q = 180 min)		15.2	20.0	13.2	45.3
Allene (% rate drop)		27.2	15.3	22.4	106
Allene (Langmuir)		23.3	13.7	22.7	121
CO (% rate drop)			16.6		
$k_p/(\text{dm}^3/\text{mol sec})$					
^{14}CO (t_q = 4 min)		10.4	8.2	17.2	47.5
Allene (Langmuir)		7.6	5.9	7.6	10.2

t_q = time of quenching, α = isotope factor, *temp = 30 °C, **temp = 60 °C, † and ≠ different catalysts.

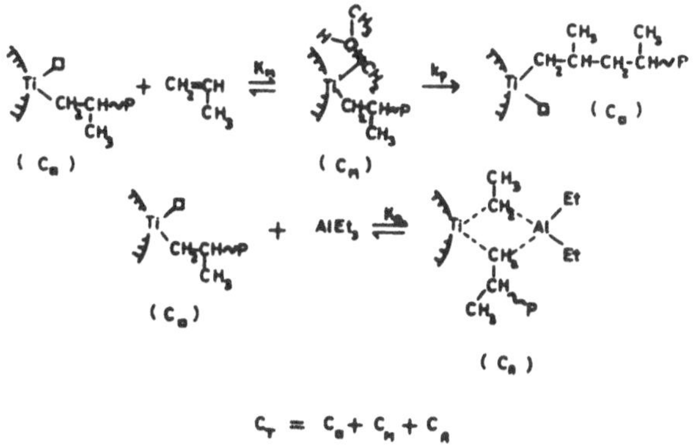

Figure 9. Polymerization sequence showing
centres complexed by monomer and aluminium
alkyl.

Although this treatment neglects more complicated reactions
and is based on a simplified reaction sequence values of
C^* obtained for the polymerization systems VCl_3/$AlEt_3$/4-MP-1
and δ-$TiCl_3$.033 $AlCl_3$/$AlEt_2Cl$/propylene are in good
agreement with equation 16.

In conclusion a final comparison of the types of
information which may be obtained from these methods for
evaluating C^* is made in Table VI.

TABLE VI

SUMMARY

Tritium Quenching

(i)   Gives C^*_{TOTAL}

(ii)   Also allows evaluation of R_{ta}

^{14}CO RADIOLABELLING/CO ADSORPTION (ALLENE ADSORPTION)

(i)   Gives C^*_p

(ii)   Gives distribution of active centres (adsorption)

(iii) Allows evaluation of $(C^*_p)_{sol}$ and $(C^*_p)_{insol}$, hence

       of $(k_p)_{sol}$ and $(k_p)_{insol}$

REFERENCES

1. B. Diedrich, <u>Applied Polymer Symposium</u>, <u>26</u>, 1, 1975.
2. T. Keii,"<u>Kinetics of Ziegler-Natta Polymerization</u>,"
    Chapman and Hall, London, 1972.
3. D.R. Burfield, I.D. McKenzie and P.J.T.Tait, <u>Polymer</u>,
    <u>17</u>, 130, 1976.
4. D.G. Boucher, I.W. Parsons and R.N. Haward,
    <u>Die Makromol.Chem.</u>, <u>175</u>, 3461, 1974.
5. L.F. Borisova, E.A. Fushman, E.I. Vizen and N.M.Chirkov,
    <u>Europ.Polymer J.</u>, <u>9</u>, 953, 1973.
6. A.G. Chesworth, R.N. Haszeldine and P.J.T.Tait,
    <u>Polymer</u>, <u>14</u>, 224, 1973.
7. G. Natta, <u>J. Polymer Sci.</u>, <u>34</u>, 21, 1959.
8. W. Cooper in "<u>Comprehensive Chemical Kinetics</u>",
    Ed. C.H. Bamford and C.F.H. Tipper, Amsterdam,1976 p.133.
9. L.A.M. Rodriguez and H.M. Van Looy, <u>J. Polymer Sci.</u>,
    <u>A-1</u>, <u>4</u>, 1971, 1966.
10. L. Kollar, A. Simon and D. Kallo, <u>Magy.Kem.Foly.</u>,
    <u>74</u>, 289, 1968.
11. J.Y. Guttman and J.E. Guillet, <u>Macromolecules</u>, <u>1</u>,
    461, 1970.
12. J.Y. Guttman and J.E. Guillet, <u>Macromolecules</u>, <u>3</u>,
    470, 1970.
13. Yu. V. Kissin, S.M. Mezhikorsky and N.M. Chirkov,
    <u>Europ. Polymer J</u>, <u>6</u>, 267, 1970.
14. Yu.V. Kissin and N.M. Chirkov, <u>Europ. Polymer J</u>,
    <u>6</u>, 525, 1970.
15. D.R. Burfield and P.J.T. Tait, <u>Polymer</u>, <u>13</u>, 315, 1972.
16. C.F. Feldman and E. Perry, <u>J. Polymer Sci.</u>,
    <u>46</u>, 217, 1960.
17. F.J. Karol and W.L. Carrick, <u>J. Am. Chem. Soc.</u> <u>83</u>,
    2654, 1961.
18. J. Boor, <u>Macromol. Rev.</u>, <u>2</u>, 115, 1967.
19. P. Longi, G. Mazzanti, A. Roggero and M.P.Lachi,
    <u>Makromol Chem</u>, <u>61</u>, 63, 1963.
20. Y. Takegami, T. Suzuki and T. Okazaki, <u>Bull Chem.Soc.
    Japan</u>, <u>42</u>, 1060, 1969.
21. T. Suzuki and Y. Takegami, <u>Bull Chem.Soc.Japan</u>, <u>43</u>,
    1484, 1970.
22. A. Zambelli, C. Tosi and C. Sacchi, <u>Macromolecules</u>, <u>5</u>,
    649, 1972.
23. A. Zambelli, M.G. Giongo and G. Natta, <u>Makromol Chem.</u>,
    <u>112</u>, 183, 1968.
24. M.L.H. Green, <u>Pure Appl. Chem</u>, <u>50</u>, 27, 1978.
25. K.J. Ivin, J.J. Rooney, C.D. Steward, M.L.H. Green and
    R. Mahtab, <u>J.C.S. Chem. Comm.</u>, 1978, 604.
26. D.R. Burfield, I.D. McKenzie and P.J.T. Tait,
    <u>Polymer</u>, <u>13</u>, 302, 1972.
27. I.D. McKenzie, P.J.T. Tait and D.R.Burfield,<u>Polymer</u>, <u>13</u>

307, 1972.

28. D.R. Burfield, P.J.T. Tait and I.D. McKenzie, Polymer, 13, 321, 1972.

29. P.J.T. Tait, Chem.Tech., 5, 688, 1975.

30. P.J.T. Tait, in "Coordination Polymerization", Ed., J.C.W. Chien, Academic Press, 1975, p155.

31. V.A. Zakharov, G.D. Bukatov, N.B. Chumaeskii and Yu. I. Yermakov, Makromol Chem., 178, 967, 1977.

32. P. Cossee, J. Catal., 3, 80, 1964.

33. E.J. Arlman and P. Cossee, J. Catal., 3, 99, 1964.

34. P. Pino, Adv.Polymer Sci., 4, 393, 1965.

35. A. Zambelli, Seventh Colloquium on NMR Spectroscopy, Aachen, 1970.

36. Y. Doi and T. Asakura, Makromol. Chem., 176, 507, 1975.

37. Y. Doi, T. Kohara, H. Koiwa and T. Keii, Makromol Chem., 176, 2159, 1975.

38. I.D. McKenzie and P.J.T. Tait, Polymer, 13, 510, 1972.

39. L.L. Bohm, Polymer, 19, 545, 1978.

40. D.R. Burfield and P.J.T. Tait, Polymer, 15, 87, 1974.

41. D.R. Burfield, Polymer, 16, 384, 1975.

42. G.Henrici-Olive and S. Olive, Angew. Chem., 79, 764, 1967.

43. L.L. Bohm, Polymer, 19, 553, 1978.

44. L.L. Bohm, Polymer, 19, 562, 1978.

45. H. Schnecko and W. Kern, IUPAC Internat.Macromolecular Symposium, Budapest, 1969, p.365.

46. H. Schnecko and W. Kern, Chemiker Z., 94, 229, 1970.

47. H. Schnecko, K.A. Jung and W. Kern, in "Coordination Polymerization", Ed. J.C.W.Chien, Academic Press, New York, 1975, p.91.

48. G. Natta and I. Pasquon, Adv.Catal., 11, 1, 1959.

49. S. Tanaka and H. Morikava, J. Polymer Sci., A-3, 3147, 1965.

50. I.D. McKenzie, Ph.D. thesis, Manchester, 1967.

51. H. Schnecko, W. Dost and W. Kern, Makromol.Chem., 121, 159, 1969.

52. N.M. Chirkov, IUPAC Internat.Macromolecular Symposium, Budapest, 1969, p.297.

53. A.D. Caunt (in publication).

54. A.L. Burns, Ph.D. Thesis, Manchester, 1976.

55. A.L. Burns, A.D. Caunt and P.J.T.Tait, (in publication).

56. T.T. Ajayi, A.D. Caunt and P.J.T.Tait (in publication)

57. B. Baker and P.J.T. Tait, Polymer, 8, 225, 1967.

58. S. Fuji, in "Coordination Polymerization", Ed., J.C.W. Chien, Academic Press, 1975, p.135.

59. T. Keii, in "Coordination Polymerization", Ed., J.C.W. Chien, Academic Press, 1975, p.263.

60. J.C.W. Chien, J.Am.Chem.Soc., 81, 86, 1959.

61. A.D. Caunt, J. Polymer Science, C-4, 49, 1963.

62. H. Schnecko, K.A. Jung and L. Grosse, Makromol. Chem.,

148, 67, 1971

63. E. Kohn, H.J.L. Schuurmans, J.V. Cavender and R.A. Mendelson, J. Polymer Sci., **58**, 681, 1962.
64. H.W. Coover, J.E. Guillet, R.L. Combs and F.B. Joyner, J. Polymer Sci., **A-4**, 2583, 1966.
65. G. Bier, W. Hoffman, G. Lehmann and G. Seydel, Makromol. Chem., **58**, 1, 1962.
66. W. Cooper, D.E. Eaves, G.D.T. Owen and G. Vaughan, J. Polymer Sci., **C-4**, 211, 1966.
67. G.D. Bukatov, N.B. Chumaevskii, V.A.Zakharov, G.I.Zuznetsova and Yu. I. Yermakov, Makromol. Chem., **178**, 953, 1977.
68. Yu. I. Yermakov and V.A. Zakharov in "Coordination Polymerization", Ed. J.C.W. Chien, Academic Press, New York, 1975, p.91.
69. A.A. Baulin, V.N. Sokolov, A.S. Semenova, N.M. Chirkov and L.F. Shalayeva, Vysokomol soyed., **A-17**, **1**, 46, 1975.
70. D.R. Burfield (Private communication).
71. G. Ayrey and R.J. Massa, Makromol.Chem., **175**, 3353, 1975.
72. J. Kendlin, K. Taylor and D. Tompson, "Reactions of Coordination Compounds of Transition Metals", Elsevier, Amsterdam, 1968.
73. V.A. Zakharov, G.D. Bukatov, E.A.Demin and Yu. I.Yermakov, Symposium on the Mechanism of Hydrocarbon Reactions, Siofok, 1973, p.487.
74. V.A. Zakharov and Yu. I. Yermakov, Reaction Kinetics and Catalysis Letters, **1**, (2), 247, 1974.
75. N.B. Chumaevskii, V.A. Zakharov, G.D. Bukatov, G.I. Kuznetzova and Yu. I. Yermakov, Makromol Chem., **177**, 747, 1976.
76. V.A. Zakharov, N.B.Chumaevskii, G.D.Bukatov and Yu. I. Yermakov, Makromol Chem., 177, 763, 1976.
77. G.D. Bukatov, V.A.Zakharov and Yu. I. Yermakov, Makromol Chem., **179**, 2097, 1978.
78. J.C.W. Chien and J.T.T. Hsieh, J. Polymer Sci., **14**, 1915, 1976.
79. J. Mejzlik and M. Lesna, Makromol Chem., **178**, 261, 1977.
80. J.C.W. Chien, J.Am.Chem.Soc., **93**, 4675, 1971.

NEW SUPPORTED CATALYSTS FOR STEREOREGULAR POLYMERIZATION OF α-OLEFINS

James C.W. Chien

Department of Chemistry, Department of Polymer Science
and Engineering, Materials Research Laboratory, University of Massachusetts, Amherst, MA 01003, U.S.A.

INTRODUCTION

Shortly after the polymer industry had adopted one version
or another of the Ziegler-Natta Catalysts for manufacturing poly-
(propylene) and in a few instances, poly(butene-1) and poly(4-
methylpentene-1), there were efforts to develop new catalysts.
The motives are based on the realization that the Ziegler-Natta
catalyst is relatively inefficient in the utilization of the
metal ions, only a few tenths of a percent of the titanium being
thought to participate in a polymerization. Therefore, polymer
yields are low and the metal and chlorine residue contents are
high. Various deashing procedures have been developed and patented
in polyolefin manufacturing processes. The principal research
objective is to develop high activity catalysts without loss of
stereospecificity. Other goals are better control of molecular
weight, polydispersity, particle size and distribution, and bulk
density.

In the last few years, several new catalysts have been reported or patented having either improved yield or stereospecificity, or both. They may be grouped into three classes: 1) modified $TiCl_3$ having increased activity and/or stereospecificity;
2) matrix impregnated $TiCl_3$ systems which have very high activity;
and 3) surface supported catalysts, some of which have shown very
high activity for ethylene polymerization, but none so far is
good for propylene polymerization.

Since discussion of new catalysts places special emphasis on
their activities, it would be desirable to define catalytic activity. Two definitions are useful; they are

R. W. Lenz and F. Ciardelli (eds.), Preparation and Properties of Stereoregular Polymers, 113-130.

     R (m mole Ti)=grams of polymer produced/m mole of Ti/hr./
        atmosphere of monomer
and R (gm. catalyst)=grams of polymer produced/gram of catalyst/
        hr./atmosphere of monomer

The descriptions of catalyst systems either in journals or in patents often give only the yield of polymer obtained after a given time of reaction, which is usually two to six hours. Then only average rates of polymerization, R_{av}, can be calculated. For a catalyst with constant activity, this is equal to its instantaneous rate of polymerization. If a catalyst has activity which decays rapidly, then the two quantities may be very different. Knowledge of the activity profile will afford a maximum rate of polymerization, R_{max}.

In order to have a proper perspective for the performance of the new catalysts, we give here the activity of Natta's original catalyst (1). Propylene was polymerized by 5.2 to 6.5 mM of α-TiCl$_3$, 30 to 44.5 mM of triethylaluminum with Al/Ti ratios of 4.5 to 8.6 in 250 ml of heptane at 70° and 2 atm. of monomer to produce 16 g. polymer/hr./g. TiCl3 and the rate of polymerization was constant for 30 hrs. The instantaneous activity which is the same as the average activity is: R(m mole Ti)=4.8 and R(g. catalyst)=32. For catalyst systems having TiCl$_3$ as the major component, and if one ignores the minor components such as AlCl$_3$, R(g. catalyst)≈6.7 R(m mole Ti). The polymers obtained by the above catalysts, if not subjected to a deashing procedure, can have as much as 650 ppm Ti, 2,500 ppm Al, and 1,500 ppm Cl. In general, the actual inorganic residue contents are less because the polymer is always subjected to some purification procedures.

The purpose of this paper is to discuss a few typical examples of new olefin polymerization catalysts, to critically examine physical and chemical processes limiting catalytic activity, and to describe new methods for characterization of catalysts.

MODIFIED TiCl$_3$

Pure TiCl$_3$ of any crystallographic form is not useful for olefin polymerization because of its low activities. The most common catalyst is AA-TiCl$_3$ obtained by aluminum alkyl reduction of TiCl$_4$ (2-4), having an approximate composition of 1TiCl$_3$/0.33 AlCl$_3$. Efforts to improve its activity and stereospecificity continue. AA-TiCl$_3$ is obtained from the reaction of TiCl$_4$ with Et$_2$AlCl at 0°. The product, β-TiCl$_3$·0.15 AlCl$_3$·0.35 EtAlCl$_2$, is itself a poor catalyst. However, heat treatment converts it to δ-TiCl$_3$·0.15 AlCl$_3$·0.15 EtAlCl$_2$, which is the common industrial catalyst having a surface area of 20 m^2 g^{-1} and when activated with Et_2AlCl polymerizes propylene with R_{av} (m mole Ti)≈5.

The best known recent example of modified $TiCl_3$ is that of Solvay (British Patent 1,291,068/9). The β-$TiCl_3 \cdot 0.15$ $AlCl_3 \cdot 0.35$ $EtAlCl_2$ is extracted with isoamyl ether at 35° to give β-$TiCl_3 \cdot 0.05$ $AlCl_3 \cdot 0.1$ amylether, a catalyst with can be activated but is not highly stereospecific. However, heating it to 65° with $TiCl_4$ produces α-$TiCl_3 \cdot 0.02$ $AlCl_3 \cdot 0.05$ isoamyl ether. This Solvay catalyst has R(m mole Ti)=25-30 with $AlEt_2Cl$ activation and polymer yield of greater than 98% isotactic product.

Another example is a system patented in the U.S. (4,007,132/3) which is quite similar to the Solvay catalyst. To prepare this catalyst, Hoechst reacted 5 mole of $TiCl_4$ and 4.5 mole of $AlEt_2Cl$ in 4l hydrocarbon at 0° for 8 hrs. without stirring and for 2 more hrs. with stirring. The product is then heat treated at 25° for 12 hrs., 90° for 4 hrs. and finally 110° for 6 hrs. This heat treated $TiCl_3$ (2M in 500 ml hydrocarbon) is reacted with less than stoichiometric amount of ether (0.95 mole) for 2 hrs. at 85° followed by thorough washing. The effect of heat treatment and ether on the activity and stereospecificity in propylene polymerization are illustrated in Table I. The same heat and ether treated catalyst when used in bulk polymerization for 3 hrs. has a slightly reduced R_{av} (m mole Ti)≈8 and gives 95.5% isotactic poly(propylene) as compared with the best examples in Table I.

In the same patents, Hoechst claimed that cyclopolyenes are more effective electron-donor modifiers than ether for improvement of activity and stereospecificity (Table II).

Table I: Slurry Polymerization of Propylene with Hoechst
          Modified $TiCl_3$ Catalyst*

Method of preparation of $TiCl_3$	$AlEt_2Cl$ m mole	% insoluble poly(propylene)	R_{av} (m mole Ti)
Heat and Ether	10	98.8	15.6
Heat and Ether	5	98.0	15.3
Heat alone	10	97.8	7.9
Ether alone	10	90.0	15.0

*1 m mole of $TiCl_3$ in 0.5 l hydrocarbon, propylene 6 atm., hydrogen 0.25 atm., 70°, and 2 hrs.

Table II: Effect of Cycloheptatriene-1,3,5 on the Perform-
           ance of Hoechst Heat Treated $TiCl_3$*

Cycloheptatriene, m mole	Temp. °C	% soluble polymer	R_{av} (m mole Ti)
0.2	60	0.8	16
0.3	60	0.6	15
0.2	80	1.4	19
0.3	80	1.2	18.5
0.4	80	1.0	18.5
0	80	6.0	19.5

*Heat treated $TiCl_3$ 1m mole, Al/Ti=10, propylene 6 atm., reaction time 2 hrs.

Table III: AA-Cl_3 Modified with α,β-Unsaturated Esters*

Ester	% wt. ester	% insoluble polymer	R_{av} (m mole Ti)
Methyl methacrylate	0	89	3.3
Methyl methacrylate	1.71	90	3.8
Methyl methacrylate	6.84	94	4.2
Methyl methacrylate	10	97	3.45
Methyl methacrylate	18	92	3.0
Methyl cinnamate	1	91	3.0
Methyl crotonate	3	97	3.8
Butyl acrylate	5	97	2.55
2-Ethyl hexyl acrylate	5	98	3.45

*AA-$TiCl_3$ 0.3 g., $AlEt_2Cl$ 0.5 g., 70°C, propylene 6 atm., poly-
merization time 6 hrs.

Mitsubishi (U.S. Patent 4,020,264) modified AA-$TiCl_3$ by ball-
milling 0.3 g. with 3-5 ml of α,β-unsaturated esters for 8 to 16
hrs. to increase the stereospecificity but at the expense of
some loss of activity as shown in Table III.

MATRIX IMPREGNATED $TiCl_3$

In matrix impregnated $TiCl_3$ systems, $TiCl_3$ itself is the
minor component. Numerous catalysts of this type have been
patented; the most useful matrices appear to be those based on
magnesium. Different magnesium compounds and different impregnation methods have been described. For instance, Gulf (U.S.
Patent 4,016,344) reacted 20 g. of MgO (325 mesh) with 20 ml of
methanol and 0.54 g. of dicumyl peroxide for 45°, dried and refluxed with 50 ml of $TiCl_4$ for one hr. followed by repeated

washing and drying. The catalyst contains 5-10 wt. % Ti and 11-18 wt. % Cl, and estimated mole ratios of Mg:Ti of 10 to 20:1. It is, however, not very active, having R_{av} (m mole Ti)=7.

Some of the most active catalysts are based on $MgCl_2$ reacted with titanium compounds. Patents began appearing for these systems between 1968 and 1969. Examples are: Mitsui Petrochemicals, Netherland Patent 6,911,791 and U.S. Patent 3,642,746; Hoechst, Belgium Patent 755,185; and Montecatini, Belgium Patents 744,221 and 747,846.

To illustrate this technology, we cite South African Patent 75/7382 issued to Mitsui Petrochemicals recently. The catalyst preparation was described as ball-milling 20 g. of $MgCl_2$ (dehydrated) and 6-8 ml of ethyl benzoate with or without 2-3 ml of low molecular weight hydrocarbon for 72 hrs. at 25° or 2 hrs. at 150°. The product has an approximate analysis of $1MgCl_2 \cdot 0.2 \cdot \phi COOEt$. This support was washed and dried and further ball-milled with 150 ml of $TiCl_4$ at 80° for 2 hrs. and thoroughly washed. Variations of this procedure produce various catalysts having about Ti (4%), Cl (60%), and Mg (20%) and a Mg:Ti ratio of about 10:1. In a typical polymerization of propylene (7 atm.) at 70°, 0.023 g. of catalyst containing 0.02 m mole Ti was activated by 0.12 m mole of $AlEt_3$ has R_{av} (m mole Ti)≈406 which is also the same in units of R_{av} (gm Catalyst) as defined above. Therefore, this catalyst is at least 100-fold more active than the original Natta catalyst. The maximum residue in polymers if not removed would amount to Ti (4 ppm), Al (13 ppm), Mg (16 ppm) and Cl (50 ppm). The fraction of isotactic polypropylene obtained ranged from 90 to 96%. Many other esters were also claimed such as ethyl anisate and ethyl-p-toluate but they seem to be inferior. Whole classes of titanium alkyls, alkoxides, and halides were included in the patent but only $TiCl_4$ was cited in the examples.

In a related patent (Montedison, U.S. 3,953,414) spherical support was obtained by atomizing molten $MgCl_2 \cdot 6H_2O$ and dried at 130° which are about 53-105 μ in size. This $MgCl_2 \cdot 2H_2O$ (120 g.) was refluxed at 135.5° with 2.5 l of $TiCl_4$ for 1 hr. to remove HCl and repeatedly washed with $TiCl_4$ (2x) and heptane (5x) and dried. The catalyst analyzed Ti (3%), Cl (69%), Mg (20.5%) and H_2O (3%). X-ray diffraction showed the original support to be comprised of $MgCl_2$ and $MgCl_2 \cdot H_2O$. This catalyst is very active when used with $Al(i-Bu)_3$. The average rate of polymerization of bulk propylene at 60° are R_{av} (m mole Ti)=740 and R_{av} (gm. catalyst)=500, the time of polymerization being 5 hrs. The amount of insoluble polymer is 83%. Addition of ethyl benzoate during polymerization seems to greatly decrease the rate and stereospecificity. The patent also describes that dehydration of $MgCl_2 \cdot 6H_2O$ at various temperatures ($80-135^\circ$) for 4-8 hrs., produces supports ranging in composition analyzed from pure $MgCl_2 \cdot 6H_2O$ to

<u>Table IV</u>: Structure of Support for Ethylene Polymerization[*]

Dehydration temp. of MgCl$_2$ 6H$_2$O, °C	80	95	115	135	135
Dehydration time, hrs.	4	4	4	4	8
Analysis MgCl$_2$ nH$_2$O,n	6	4	4/2	2	1
Catalyst analysis, %					
Ti	3.5	4.15	3	3.2	1.55
Cl	69	70	69	68	68
Mg	21	21	21	21	21
Surface area, m^{2}g^{-1}	98	110	65	21	21
Pore radius	51	36	45	38	63
R$_{av}$ (m mole Ti)	1000	1500	480	240	460

[*]Impregnation by refluxing 40-80 g. of support with 2 l of TiCl$_4$ at 137°. Polymerization of ethylene (5.5 atm.) with H$_2$ (7.5 atm.) at 85° with Al (i-Bu)$_3$ activation for 4 hrs.

MgCl$_2$ • 1H$_2$O and others in between having 4 and 2 molecules of water of hydration. Table IV shows the dependence of catalytic activity on the structure of the matrix materials.

There are impregnated catalysts derived from the reaction of Mg(OR)$_2$ or MgR$_2$ with chlorination agents in the presence of co-ordinating agents subsequently refluxed with TiCl$_4$ for 2 hrs. Some of the examples (Montedison, South African Patent 76/2246) are given in Table V. Other chlorination agents reported are SnCl$_4$ and SOCl$_2$; poly(dimethylsiloxane) have also been claimed as complexing agents.

We have followed the rate of polymerization of propylene initiated by a typical MgCl$_2$ matrix impregnated TiCl$_3$ catalyst. Figure 1 shows the results obtained at 60° and 3.5 atm. of monomer pressure. From the total polymer yield after 4 hrs. one calculated R$_{av}$ (m mole Ti)=661. Therefore, the initial rate of polymerization is actually about ten-fold faster. After one hr. of polymerization the catalytic activity is only about one-fourth of the initial yielded, about 66% of the total polymer.

SURFACE SUPPORTED CATALYSTS

In this class the transition metal is present only on the surface of the support. The catalyst is usually prepared by the reaction of transition metal compound with the surface hydroxyls of the support. The support materials are typically metal oxides such as Si, Al, Zn, Ti, Mg. High activity catalysts for ethylene polymerization based on Ti have been patented by Solvay and Cie (Belgium Patent 650,679 (1963); British Patent 1,024,336 (1963))

Table V: Some Examples of Impregnated Catalysts Derived from $Mg(OR)_2$ and MgR_2

Matrix Preparation

	$Mg(OEt)_2$ $SiCl_4$ δ-methacryloxy propyl-trimethoxy silane	$Mg(OEt)_2$ $SiCl_4$ benzoyl chloride	$Mg(OEt)_2$ $Cl_3SiCH\equiv CH_2$ ethyl-benzoate	$Mg(n-Bu)_2$ $SiCl_4$ ethyl benzoate	$Mg(OEt)_2$ $EtSiCl_3$ vinyl Si-$(CH_3OCH_2CH_2O)_3$
Mg Compound					
Chlorination Agent					
Complexing Agent					
Analysis of Catalyst					
Mg	12.5	16.5	18.1	18.8	19.6
Ti	3.6	2.7	2.5	3	2.85
Cl	52.6	58.2	64.4	60	68.2
Slurry Polymerization*					
R_{av}(g. catalyst)	57	42	55		24
R_{av}(m mole Ti)	78	74.5	106		40
% isotactic	89.5	92	87.5		77.5
Bulk Polymerization					
R_{av}(g. catalyst)	31	18	42	8.8	11.3
R_{av}(m mole Ti)	42.5	33.5	51	14	19
% isotactic	88.5	91	84	79.5	85

*Polymerizations are at 60° for 5 hrs. in the presence of small amounts of ethylanisate activated by $AlEt_3$.

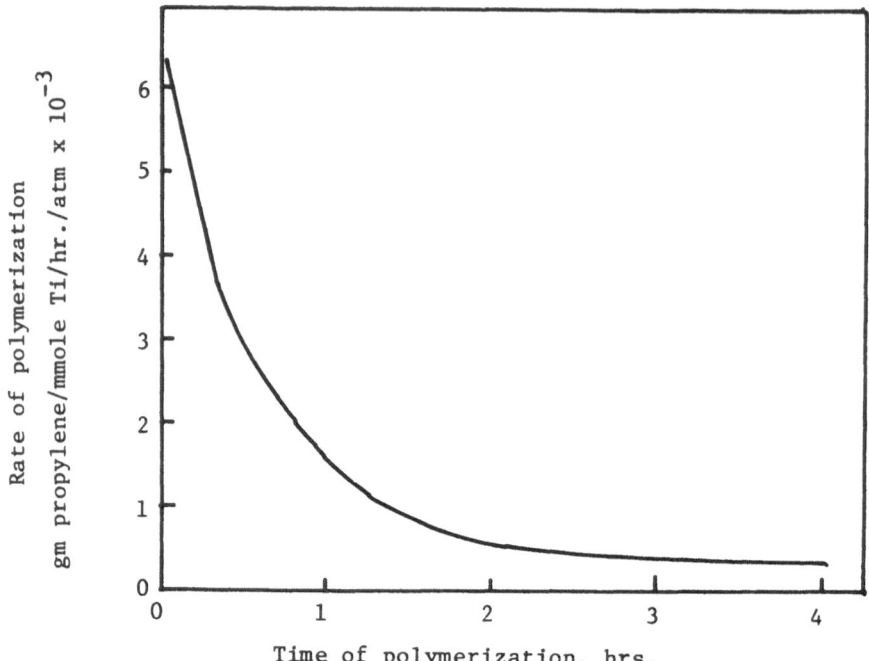

Figure 1. Profile of propylene polymerization initiated
by matrix impregnated high activity catalyst.

and chromium based catalysts (Union Carbide, U.S. Patent 3,324,095
(1967)). Some of these systems have extremely high catalytic
activity for ethylene polymerization.

    As of today there has not yet been a surface support system
which has high catalytic activity for propylene polymerization.
Worse than that, most of the catalysts give only low yields of
isotactic polymers. We have studied Mg(OH)Cl as a catalyst support (6). Depending upon the method of preparation, the support
has from 6.5 to 15 m^2/gm of surface area. The surface hydroxyls
were found by the paramagnetic probe technique (vide infra) to be
largely vicinal pairs. Tetrabenzyltitanium (B$_4$Ti), tribenzyl-
titanium chloride (B$_3$TiCl), tetra(p-methyl-benzyl) titanium (R$_4$Ti),
tri(p-methyl benzyl) titanium chloride (R$_3$TiCl), and π-cyclo-
pentadienyl-trimethyl titanium (π-C$_5$H$_5$TiMe$_3$) were reacted with
the surface hydroxyls and subsequently activated with Et$_2$AlCl.
The results of some of the polymerization experiments are given
in Table VI.

Table VI: Propylene Polymerization Catalyzed by Mg(OH)Cl
           Supported Systems*

Ti alkyls	Ti m mole/g support	electron donor	Al/Ti	R_{av} (m mole Ti)	% Isotactic
B_4Ti	0.074	-	25	1.12	77
B_3TiCl	0.073	-	12.5	2.16	67
R_4Ti	0.08	-	22	1.04	65
R_3TiCl	0.074	-	25	2.89	68
π-$C_5H_5TiMe_3$	0.072	-	26	1.56	90
B_3TiCl	0.14	ether	10	0.27	100

*Polymerization at 50° and 7.3 psig propylene.

The low activity of these catalysts is largely attributable
to the fact that only about 0.1% of the Ti is catalytically active
(vide infra). This is the direct consequence of fully covering
the support surface with titanium alkyls allowing reaction between
adjacent titanium alkyls to occur leading to deactivation. It
appears that supports of much higher surface area are needed on
which only a fraction of the hydroxyls should be reacted with
titaniumum alkyls.

One encouraging note is that, in the presence of an electron
donor, the active site can be very stereospecific.

INTRINSIC CATALYTIC ACTIVITY

To develop high activity catalysts one may aim to increase
the number of active catalytic sites, or to raise the intrinsic
activity of each site or both. The intrinsic activity is defined
by the deceptively simple equation

$$R_p = k_p [C^*] [M] \qquad\qquad (1)$$

Where k_p is the rate constant of propagation, C^* is the active
site concentration and M is the monomer concentration. Calculation of k_p requires the knowledge of R_p, $[C^*]$, and $[M]$. The uncertainty in the determinations of $[C^*]$ by various techniques
has been discussed by Tait in this volume. Depending upon the
catalyst system, the active sites may all be present initially,
or more may be produced as the catalyst agglomerates or crystals
fracture during polymerization. If there is catalyst deactivation by either chain termination, chain transfer or poison, $[C^*]$
may decrease with time. At the initial stage of reaction [M] is
the concentration of monomer dissoved in the diluent. If, during
reaction, the catalyst is completely encapsulated by the polymer

(vide infra) and there is monomer starvation due to diffusion limitation, $[M]_p$, which is now the monomer dissolved in the polymer, would vary with the distance from the particle surface to a given active site. With these complications in mind, let us compare the k_p of the high activity catalyst with those of the conventional catalyst systems based on the same active site counting technique.

The high activity $TiCl_4/MgCl_2/\phi COOEt/AlEt_2Cl$ catalyst was found to have 5 mole % of active Ti by quenching with $*CO$ after 10 min. of polymerization at $65°$ (5). The rate of polymerization during the first five min. is 6.6×10^{-3} M sec.$^{-1}$ at 6.72 atm pressure of propylene. Taking $[M]$ to be 2.69 M, then k_p=1770 M^{-1} sec^{-1}. As it was noted above, the rate of polymerization decreases rapidly with reaction (Fig. 1). Therefore, if one does not have the complete reaction profile, then the average rate, R_{av}=7.28 x 10^{-4} M sec^{-1} and the k_p would be correspondingly smaller if $[C^*]$ is assumed to remain unchanged (vide infra). The value of k_p for the low activity catalysts are summarized in Table VII (5-7). The comparison shows that the latter systems have very similar

Table VII: Comparison of Propylene Polymerization Catalysts

Catalyst	Temp. $°C$	active site mole % of Ti	k_p,M^{-1}sec^{-1}	Ref.
$TiCl_4/MgCl_2/\phi COOEt/$ AlEt$_2$Cl	50	6.5	1770	5
$(C_6H_5CH_2)_3TiCl/Mg$ (OH)Cl/AlEt$_2$Cl	50	0.09	32	6
δ-TiCl$_3$ 0.33AlCl$_3$/ AlEt$_2$Cl	70	0.14	73	7
δ-TiCl$_3$ 0.33AlCl$_3$/ AlEt$_3$	70	0.036	70	7
δ-TiCl$_3$/AlEt$_2$Cl	70	0.0025	61	7
TiCl$_2$	70	0.00013	76	7

k_p values, whereas the high activity MgCl$_2$ impregnated catalyst has distinctly larger values of k_p. Furthermore, a far greater fraction of the Ti in this catalyst is active as compared to the other systems.

FACTORS LIMITING POLYMERIZATION RATE

The activity of a catalyst may be limited by a number of causes, chemical or physical. Chemically, the change of oxidation state of the transition metal can result in deactivation. This is amply demonstrated in the soluble catalyst $(\pi-C_5H_5)_2TiCl_2/$ AlMe$_2$Cl (8,9). Similar deactivation reactions were also possible for surface supported catalysts (10). Hydrid elimination reaction can be another cause if the resulting titanium hydride species is catalytically dormant. This process has often been suggested as important for chain transfer. The rate of polymerization initiated by high activity catalysts (Fig. 1) decreases rapidly with time, which could be due to catalyst deactivation. However, first order kinetic plot of R_p vs. t is not linear. Plots of $(R_p)^{-1} - (R_p)^{-1}$ vs. t are also non-linear with the slope during early stages of polymerization smaller than the latter. This would mean a slower bimolecular deactivation occuring when the active site concentration is high. The discussion which follows would suggest that the rate of decay may be due to diffusion control. However, until the active site concentration is counted during the entire course of a polymerization, chemical contributions toward rate limitation cannot be entirely discounted.

The most probable physical cause for limiting polymerization rate is diffusion control. In Ziegler-Natta polymerization both the catalyst and the polymer product are insoluble. Therefore, the catalyst may be encapsulated by the polymer, erecting diffusion barrier to the monomer. Depending upon the uniformity, porosity, thickness, and crystallinity of the polymer coating and the reactivity of the catalytic site, there will be conditions under which the polymerization may become diffusion limited. Diffusion limitation has two important consequences: it can cause decreased polymerization rate and increase the polydispersity of the polymer (11,12). These behaviors can arise from other causes. For instance, activation of the transition metal associated with termination can reduce the polymerization. Heterogeneity of active sites can produce polymers of broad molecular weight distribution. Therefore, there has been a continuing discussion regarding the importance of diffusion limitation in Ziegler-Natta polymerizations. We have proposed three criteria for diffusion control (13): I. if the number-average molecular weight and polydispersity of the polymer calculated from kinetic rate constants as a function of time agree with the experimental values, the polymerization is not diffusion controlled; II. the polymerization may be diffusion controlled if the Thieles modulus, the ratio of the characteristic diffusion time to the characteristic reaction time, is much greater than unity; if it is much smaller than unity the polymerization is reaction controlled; III. if an initial linear dependence of rate of polymerization on catalyst concentration changes over to a square root dependence, the polymerization may be diffusion limited. Whereas criterion I is a necessary and sufficient condition as stated, its converse is not true; the other two criteria are merely necessary but not sufficient conditions.

Many examples of Ziegler-Natta polymerizations of ethylene as well as of propylene had been examined (13). The kinetic

data were tested according to the above criteria and their reactions were found not to be diffusion controlled. In particular,
in the AA-TiCl$_3$/AlEt$_2$Cl polymerization of propylene the $\overline{M}_n$ calculated from kinetic rate constants as a function of time agrees
with the experimental value according to criterion I (14, 15);
the measured polydispersity at the end of a polymerization is also
in agreement with theory. The Thiele modulus calculated from
the active site counting data (16) is only 0.04 (13) which is far
too small for diffusion control by criterion II. Finally, the
rate of polymerization is independent of either reaction time or
polymer production not satisfying either criterion II or III.

The rate constants of initiation and transfer are not known
for the high activity catalysts, thus precluding the test by
criterion I. But one can calculate the Thiele modulus α,

$$\alpha = (R_p/D_m[M]_p)^{\frac{1}{2}}S \qquad\qquad (2)$$

where D_m is the diffusion coefficient of monomer in polymer and
$[M]_p$ is the concentration of monomer dissolved in polymer. Taking
the initial value of R_p=6.6 x 10^{-3} M sec^{-1}, D_m=2 x 10^{-9} cm^2 sec^{-1}
(17), M_p =0.2M at 50° and 6.72 atm (18), and S, the initial
particle radius to be 50 μ, we calculate α=20. Therefore, according to criterion II, this polymerization may be diffusion controlled.

Several models leading to diffusion control have been discussed (11,12). In the first one, the active sites are assumed
to be fixed and a polymer core builds up, encapsulating the
catalyst. This is the fixed-site or polymer core model. In the
uniform site concentration model, polymerization breaks up the
catalyst so that $[C^*]$ may be considered nearly uniform and constant. In the third flow model, the active sites are assumed to
be convected outward with the velocity of the growing polymer
consistent with the conservation of mass. These different models
are distinguished by the dependence of rates of polymerization
on polymer production. Figure 2 is a log-log plot of the rate of
polymerization versus the quantity of polymer produced. It
shows that R_p at first decreases slowly with polymer yield then
finally to a slope of about -3. The results suggest that for
this catalyst R_p decreases eventually with (volume of polymer)$^{-1/3}$
or (radius of polymer particle)$^{-1}$.

Perhaps the most positive evidence for diffusion control in
propylene polymerization with high activity catalysts is the comparison of reactions performed at different monomer pressure.
For some catalysts, the yield in a bulk polymerization (~30 atm)
is about 50% greater than at 7 atm. Furthermore, the polymerization is more or less complete in liquid propylene in time much

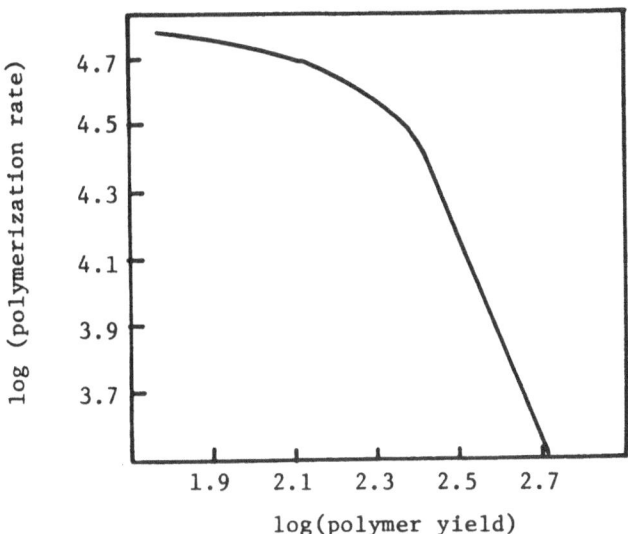

Figure 2. Log-log plot of polymerization rate vs. yield (data from Figure 1).

shorter than that at lower monomer pressure in diluent. These behaviors are consistent with diffusion rather than reaction control. For other catalysts (Table V), R_{av}'s are actually quite a bit smaller in bulk polymerization.

PREPARATION OF HIGH ACTIVITY CATALYSTS

There are numerous considerations in the designing of a high activity olefin polymerization catalyst. For stereospecific polymerization of α-olefins, the catalyst should be highly porous with $TiCl_3$ impregnated in a structure similar to those found in the α and δ crystalline forms. To accomplish this, the following procedures probably have general validity. A coordination agent, L, should be chosen which will form weak Lewis acid-base complexes with both the matrix material M and $TiCl_4$. If M is neutral or weakly acidic, L should be weakly basic. It would also be advisable to have L a molecule of moderate dimension. The proper procedure would be first to form ML_x where x<<1 by ball-milling or other means. Excesses of L should be avoided so as not to have too many L molecules occupying adjacent interstitial sites. Treatment with $TiCl_4$ should result in the extraction of L because $TiCl_4$ is a strong Lewis acid,

$$ML_x + TiCl_4 \longrightarrow M + TiCl_4L_x \qquad (3)$$

impregnation of $TiCl_4$

$$M + xTiCl_4 \longrightarrow M \cdot xTiCl_4 \qquad\qquad (4)$$

If the dimension of L is much larger than $TiCl_4$, its likely that the impregnated Ti would be coordinatively unsaturated after reduction by aluminum alkyls. For the same reason, the catalyst would be porous yet have structural integrity.

$MgCl_2$ based systems seem to be the best high activity catalysts so far disclosed. It certainly meets all the above requirements. It has a layered crystal structure like that of α- and δ-$TiCl_3$. $MgCl_2$ forms a weak complex with $\phi COOEt$ whereas $TiCl_4$ forms much stronger 1:1 and 2:2 complexes with $\phi COOEt$. Ball-milling of $MgCl_2$ produces a solid analyzed as $MgCl_2 \cdot 0.2\phi COOEt$.

Similar reasonings as above seem to apply to the modified $TiCl_3$ systems. Here, an ether is used to extract $AlCl_3$ and $EtAlCl_2$ from AA-$TiCl_3$.

PARAMAGNETIC TECHNIQUES FOR THE CHARACTERIZATION OF SUPPORTED CATALYSTS

Earlier we have developed two paramagnetic probe techniques (10,17) for the characterization of catalyst supports which should be valuable for the study of supported catalysts as will be shown in this section. The two techniques are complimentary to each other. In the first method we make use of the following principles:
1) VCl_4 is a 3d' molecule with electronically doubly degenerate ground state. It has extremely short spin-lattice relaxation times and its EPR spectra is not observable except at liquid He temperatures.
2) Reaction of VCl_4 with surface hydroxyl groups of a catalyst support removes this degeneracy and its EPR becomes readily detectable at all temperatures.
3) If the V is singly attached, i.e. via a single -O-V bond, its EPR spectrum is nearly isotropic,

$$\omega_0 = g_{iso}\, \beta_0\, H/\hbar + a_{iso}\, M$$

where ω_0 is the resonance frequency, g_{iso} is the isotropic g value, a_{iso} is the isotropic hyperfine splitting constant, H is the external field, β_0 is the Bohr magneton and M is the nuclear quantum number.
4) If the V is attached to the surface through two bonds (-O-V-O-) resulting from a reaction with a pair of vicinal hydroxyls, the EPR is axially asymmetric

$$\omega_0 = g\,\beta_0 H/\hbar + AM + \frac{A_\perp^2(A_\parallel^2 + A^2)}{4A^2\, g\beta_0 H/\hbar}\, [^-I(I + 1) - M^2] \qquad (5)$$

where $g^2 = g_\parallel^2 \cos^2\theta + g_\perp^2 \sin^2\theta$, I is the nuclear angular momentum, and $A^2 = A_{\shortparallel}^2 \cos^2\theta + A_{\perp}^2 \sin^2\theta$

5) When two V atoms are separated by a distance of 12Å or less, dipolar interaction results in spectral broadening. This line broadening is temperature independent if the two atoms are both not free to rotate, i.e. multiply attached. It is temperature dependent if one or both of the V atoms are singly attached, because at -195° only a small fraction of the frozen orientation has the two V atoms in closest approach.

6) If two V atoms are very close to each other and particularly with two VCl_4 reacted with two geminal hydroxyl groups of the same metal atom support, then their EPR will not be observable at any temperature because of excessive line broadening.

Studies have been made with two types of fumed silica, several forms of alumina (17) and Mg(OH)Cl (6). In these experiments VCl_4 is reacted with the surface hydroxyls of the material to various degrees of coverage. EPR spectra were recorded at -195° and 25° and their integrated intensities were compared with analysis by atomic absorption. The results permit estimates of the type of hydroxyls present on the support surfaces, i.e. isolated, vicinal and geminal hydroxyls and their approximate spatial distributions.

The second paramagnetic probe technique is based on another set of principles:
1) $TiCl_4$ is a $3d^0$ diamagnetic molecule.
2) $TiCl_4$ can react with surface hydroxyls of various types described above in the same manner as VCl_4.
3) Addition of alkylating agents at low temperature results in reduction only if two alkylated Ti(IV)'s can undergo the bimolecular processes depicted below.

These experiments were performed with two types of fumed silica (17) and the results are in entire agreement with the VCl_4 probe experiments. It is particularly striking when $EtAlCl_2$ was used; it was able to readily convert all geminally attached Ti(IV) to Ti(III) even at room temperature.

It seems that the above techniques can be readily adapted to study matrix impregnated catalysts though we have not yet done so. For instance, $MgCl_2$ can be ball-milled with varying amounts of $\phi COOEt$ to produce $MgCl_2 \cdot n\phi COOEt$ where n is 0.02 to 0.2. This can be refluxed with VCl_4 followed by washing. Isolated VCl_4 will give temperature independent narrow line-width EPR spectra whereas clusters of VCl_4 will have temperature dependent broadened EPR. Comparison with atomic absorption results should give the fraction of VCl_4 too closely situated with another one for EPR observation. Similar experiments using $TiCl_4$ followed by addition of $AlEtCl_2$ should reveal those $TiCl_4$ molecules which can undergo bimolecular reduction after alkylation. Also, reaction with

AlEtCl or AlEt$_3$ can show whether reduction by dialkylation of
isolated TiCl$_4$ can occur.

FURTHER IMPROVEMENTS ON HIGH ACTIVITY CATALYSTS

The present high activity catalyst systems such as TiCl$_4$/
MgCl$_2$/ϕCOOEt/AlEt$_2$Cl can be further improved. Firstly, the activity may still be increased since it has only about 5% active
Ti. This could be due to incompleteness in the counting technique. For instance, a chain could have been terminated before
*CO is introduced, even though this was done after only 10-15
min. of reaction. Alternatively, the low molecular weight polymers containing the label may have escaped isolation. If most
or all of the propagating chains were counted, then other explanations must be sought for the low amount of active sites.
The matrix MgCl$_2 \cdot \phi$COOEt may not have a stoichiometric structure
in the sense that the ϕCOOEt does not occupy regular lattice sites.
Then there can be two or more adjacent ϕCOOEt. The consequence
is that there will be, after refluxing with TiCl$_4$, interstitial
Ti's which have none, one or more adjacent Ti's. The mechanism
of reduction and the active oxidation state of Ti become important. For isolated TiCl$_4$ reduction can only occur via

$$(6)$$

If the divalent Ti is catalytically active, then the best catalyst should have only isolated TiCl$_4$ in the matrix. This is by
no means certain. In Table VI a $k_p = 76$ M^{-1}sec^{-1} was given for
TiCl$_2$, but only 1.3 x 10^{-6} of the Ti was reported to be active (7).
This active Ti can very well be trace amounts of Ti (III) in the
material. When there are adjacent TiCl$_4$ impregnated then the
more probable reduction pathway is a bimolecular one of

$$(7)$$

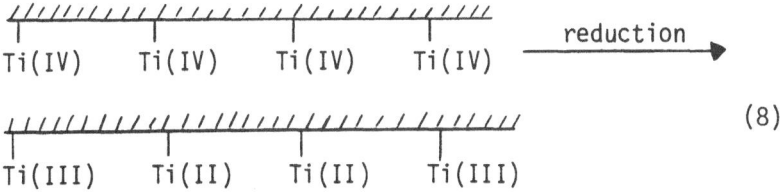

The resulting Ti(III) alkyls will be catalytically active but
they may undergo further reduction either prior to or during
polymerization. But this process could leave other Ti(III)
species separated to remain active. Just for the sake of illus-
tration, consider four adjacent Ti's

$$
\begin{array}{cccc}
\text{Ti(IV)} & \text{Ti(IV)} & \text{Ti(IV)} & \text{Ti(IV)}
\end{array}
\xrightarrow{\text{reduction}}
$$

$$
\begin{array}{cccc}
\text{Ti(III)} & \text{Ti(II)} & \text{Ti(II)} & \text{Ti(III)}
\end{array}
\tag{8}
$$

where now the outer Ti(III) species will remain catalytically
active.

The rapid decay of activity is another drawback of the present systems. This may be due to deactivation processes as described above. It may also be due to diffusion control (vide
supra). Of the three models, R_p decreases most markedly with
reaction for the fixed site or polymer core model. Figure 2 suggests that this applies to the catalyst under discussion. The
flow model would have the smallest change of R_p with polymerization. Therefore, a catalyst which is more readily fractured or
fragmented may well have nearly constant activity.

Finally, there is also room for improvement with regard to
stereospecificity of the high activity catalysts.

REFERENCES

1. G. Natta, J. Polym. Sci., 34, 21 (1959).
2. E. Tornquist, C.W. Seelback and A.W. Langer, U.S. Patent,
    3,128,352 (1964).
3. E. Tornquist, J.T. Richardson, Z.W. Wilchinsky, and R.W.
    Looney, J. Catalysis, 8, 189 (1967).
4. E. Tornquist and A.W. Langer, U.S. Patent, 3,032,510 (1962).
5. J.C.W. Chien, unpublished results.
6. J.C.W. Chien and J.T.T. Hsieh, J. Polym Sci. Polym. Chem. Ed.,

$\underline{14}$, 1915 (1976).
7. V.A. Zakharov, N.B. Chumayerskii, G.D. Bukator, and Yu. Yermakov, Reaction Kinetics and Catalysis Letters, $\underline{2}$, 329 (1975).
8. J.C.W. Chien, J. Am. CHem. Soc., $\underline{81}$, 86 (1959).
9. G. Henrici-Olive and S. Olive, Angew. Chem. Internat. Ed., $\underline{6}$, 790 (1967).
10. J.C.W. Chien, J. Catalysis, $\underline{23}$, 71 (1971).
11. W.E. Schmeal and J.R. Street, AIChE J., $\underline{17}$, 1188 (1971).
12. D. Singh and R. P. Merrill, Macromolecules, $\underline{4}$, 599 (1971).
13. J.C.W. Chien, J. Polym. Sci. Polym. Chem. Ed., in press.
14. J.C.W. Chien, J. Polym. Sci. A, $\underline{1}$, 1839 (1963).
15. J.C.W. Chien, J. Polym. Sci. A, $\underline{1}$, 425 (1963).
16. Yu. I. Yermakov and V.A. Zakharov, in "Coordination Polymerization", ed. J.C.W. Chien, Academic Press, New York, N.Y. (1975) p. 91.
17. J.C.W. Chien, J. Am. Chem. Soc., $\underline{93}$, 4675 (1971).

THE CONTROL OF DIOLEFINS SELECTIVE POLYMERIZATION ·

Ph.Teyssié[+], A. Devaux, P.Hadjiandreou, M. Julémont,
J.M. Thomassin, E. Walckiers and R. Warin.
Laboratory of Macromolecular Chemistry and Organic
Catalysis, University of Liège, Sart-Tilman, 4000 Liège,
Belgium.

INTRODUCTION

After about twenty years of sometimes frantic research activity
in the field of conjugated dienes coordination polymerization, using
catalytic selective transition metal complexes, it seems worthwhile
to make a tentative evaluation of the achievements and of the challenges ahead. It is the goal of the present review, which does not
intend to be exhaustive at all, to emphasize the most significant
advances which have been accomplished, as well as some of the key
problems that we are still facing.

At the industrial level, the achievements have been bright, both
in quantity and quality.
Suffice it to emphasize the most important ones, and first the production of high cis-(98 to 99 %) 1,4-polydienes : cis-1,4-polybuta-
diene, maybe the most interesting synthetic elastomer known to date,
is produced in amounts of about 750.000 Tons per year, while cis
1.4 polyisoprene, the practical equivalent of natural rubber from
Hevea tree, is around 500.000 Tons. Trans-1.4-polyisoprene has also
been produced in smaller quantities as a substitute for gutta-percha
(originating from Balata tree). Furthermore, cis- and trans-1.4-poly-
butadienes of variable compositions have been used as totally amorphous elastomers enjoying interesting properties ($f.i.$ better fati-
gue properties).
More recent developments have also to be pointed out, in particular
that of tactic crystalline 1,2-polybutadienes having high melting
points; the use of the syndiotactic product has been strongly promoted
in Japan for films production, and even the amorphous 1.2 polymer
seems to have desirable properties as a component in elastomers blends.
Finally, very recent reports, also from Japan, claim improved phy-

131

R. W. Lenz and F. Ciardelli (eds.), Preparation and Properties of Stereoregular Polymers, 131-150.

sical properties in block copolymers including a long sequence of
cis 1,4 units followed by a short one of trans-1,4 or 1.2 units [1].

    Our mechanistic knowledge of these reactions has also progres-
sed accordingly, and we begin to understand many of the basic factors by which they are governed. However the numerous informations
provided by a refined structural analysis of the polymerization
history, has revealed an increasingly complex picture of these catalysts behaviour.
In fact, it becomes obvious that we have underestimated the refi-
nements of these controls, and in particuler the determinant
role of rather small modifications in the overall geometry of the
complexes in the reaction mixture itself. The following chapters
will be devoted to the analysis of these controls at different
levels, after a critical discussion of the main basic mechanisms
which have been proposed to account for the main features of olefins and diolefins polymerization by coordination complexes.

I. BASIC MECHANISMS PROPOSED FOR UNSATURATED HYDROCARBONS POLYME-
   RIZATION BY COORDINATION CATALYSIS.

A. General Considerations

    It should be stated first that, in spite, of the use of simple
η^3-allyl model complexes discussed below for mechanistic studies,
most of the synthetic work effected has called upon the now classical Ziegler-Natta type catalytic systems, which can be represented by the general formula :
$$M_T X_n + M_s R_m Y_p + L$$
(where usually M_T is a transition metal, M_s a metal from groups
I to III, X and Y counter-anions, R an hydrido or alkyl group,
and L a donor ligand). These systems often behave as heterogeneous
catalysts, but in the case of diolefins polymerization, practical
reasons (solution viscosity, product processing, ...) have led to
the choice of soluble compounds.
However, in terms of the mechanisms at the molecular level, it
seems well stablished that heterogeneous catalysts do not behave
differently from the homogeneous ones (except maybe the existence
of additional adsorption and diffusion parameters).

    Approaching these molecular coordination mechanisms, one has
to consider first the bonding situation between the growing chain
and the active metal complex center It is indeed obvious that the
main prerequisite in these selective polymerizations is the perma
nent bonding of the chain to a metal of the complex, and the deter-

minant role of the transition metal in this respect is clearly
documented (although apparently this role can also be played by
lithium in some coordinated-anionic polymerizations).
If a coordination vacancy is available in cis-position to the bonded polydiene chain, the σ-allyl end group will tend to put itself
into the more stable but temporarily inactive π or η^3-allyl
(Cotton's notation[2]) situation (fig.1).

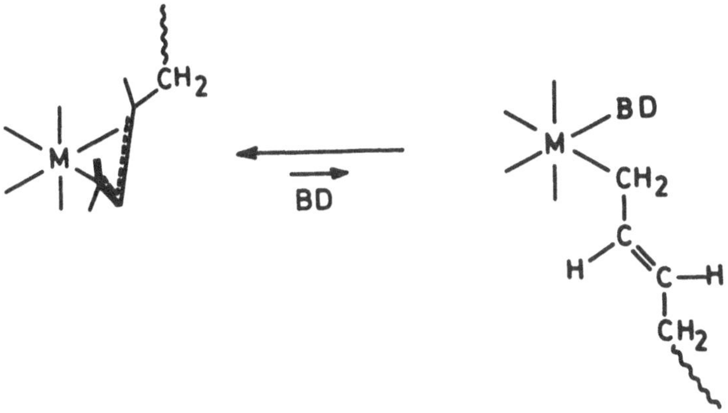

Fig.1 : η^1(or σ) - η^3(or π) isomeric equilibrium
in allylic complexes of transition metals

Three different mechanistic proposals dominate the scene of
coordinatively-controlled mechanisms.

B. The Cossee-Arlman scheme, and its refinement by Rodriguez-Van Looy.

This proposal might be viewed as a crystallization of many
attempts to establish a rationalizing picture of the available
data in the field, and is somehow related to previous suggestions
by Natta and coworkers [3].(Fig.2).
This cis-rearrangement (or cis-insertion) scheme, has been proposed more than 15 years ago by P. Cossee [4] : it involves essentially a π-bonding of the monomer (mono- or bidentate depending on the number of positions available) on a vacancy located in a cis-position versus the η^1-bonded polymer chain, followed
by a purely electronic rearrangement (typical of the pericyclic
type, and involving a small migration of the η^1-bonded chain,
to the monomer). This liberates a new coordination position,
which can be used further, in a repetitive catalytic process.
The other steps of this chain process, namely the initiation
and termination reactions, involve respectively the fixation on
the complex of a H or R group foreshadowing the growing chain,

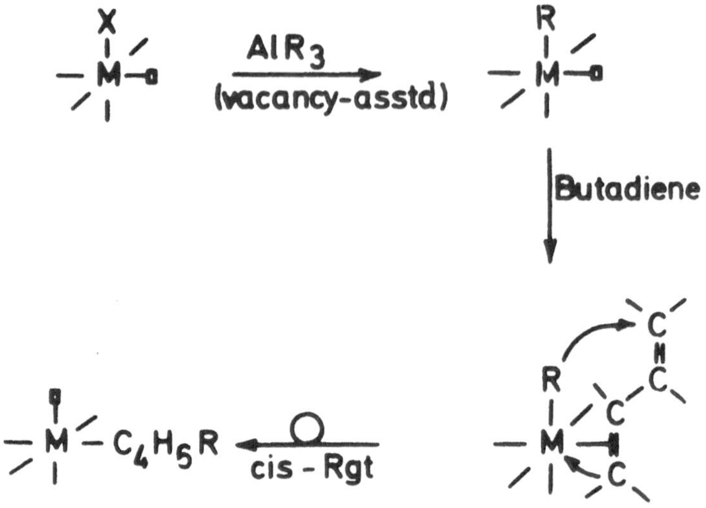

Fig.2 : Cis-rearrangement mechanism in coordination polymerisation.

and β- or δ-elimination transfer reactions.

To accomodate many intriguing results, Rodriguez and Van Looy, in a series of very elegant and rather decisive papers [5], have developed a refinement of the Cossee's proposal for,olefins poly-merizations) including the M_S derivative (most often an alumium alkyl) as one of the ligands of the M_T coordination sphere, possi-bly bridged to it by a μ-bond involving the first carbon of the growing chain.

C. The "exosphere" mechanism.

On the other hand, several authors (6,7,8,9) have considered a completely different type of electronic rearrangement,closely related to the allylic transposition involved *f.i.* in the addition of metal-alkyl derivatives to ketones [10]. (Fig.3).
It has been supported by the interesting work of Powell (8) on palladium-catalyzed 1,2 addition reactions, where the new C-C bond has been shown indeed to be formed outside of the coordination sphere.Like in the Cossee's scheme, it implies first the reversi-ble bonding of the monomer to the η^1-allyl form of the active

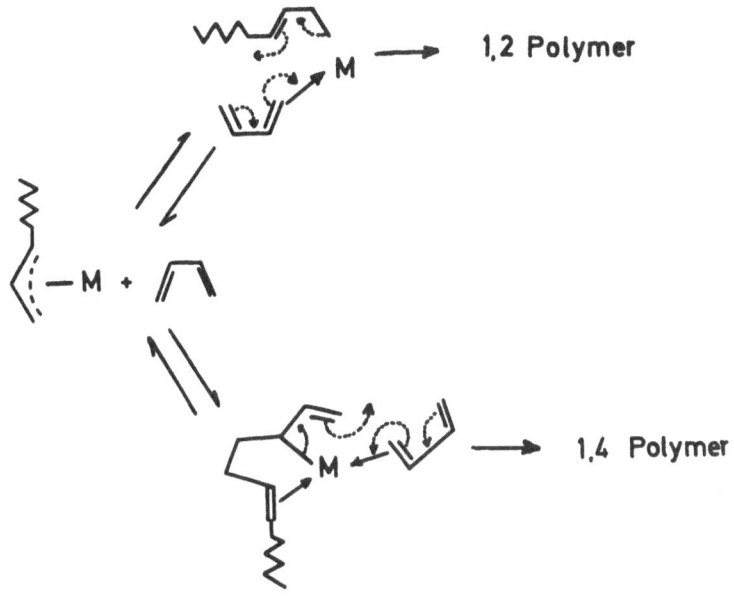

1,2 Polymer

1,4 Polymer

*Fig.3 : Outer-sphere pericyclic rearrangement in coordina-
tion polymerization.*

complex, but involves a σ-bond between the metal and the <u>least
substituted</u> carbon of the allyl group. The subsequent linking
of the C_3 atom of this allyl group to the free C^4-carbon of the
coordinated diene proceeds probably through a concerted process.
(Fig.3)

This mechanism has been tentatively extended to 1,4 polymerization by the other transition metals, like nickel, even though
the stereochemistry involved is completely different. However,
its application raises several difficulties which may cast doubt
on its validity in that case : it does not allow indeed the interpretation of the cis-trans controls discussed below; on the other
hand, it would imply unlikely situations, like the coordination
of the diene by the most substituted double bond [11], and σ-bon-
ding of the η^1-allyl group to the metal by the carbon atom carrying the growing chain : in the case of nickel, this last assumption is inconsistent with the chemical shifts of the C^{13} atoms in
η^3-allyl-, η^3-crotyl- and η^3-crotyl-mono(triphenylphosphite)-
nickel-trifluoracetates.(Respectively : δC_1 : 52.1, 46.2 and
41.0 ppm; δC_3 (substituted) 52.1, 69.8, and 96.1 ppm, versus TMS).
Further investigations should determine if cis insertion predominates in 1,4 addition,while allylic transposition is the main process to obtain branched addition products.

D. Other proposals.

Very recently, Green [12] has challenged the importance of the cis-rearrangement reactions, and has proposed an alternative mechanism based on the formation by α-elimination, of a carbene linking the chain to the metal : this carbene would then form a metalla-cyclobutane with the incoming monomer, and a reductive elimination would finally restore the initial alkyl complex. In addition to some problems raised by the many molecular motions in the complex for every propagation step, this mechanism would imply in the case of diolefins unlikely situations from a steric point of view. Although this pathway is probably valid for reactions related to olefins metathesis processes, it necessitates further experimental evidence to be applied to olefins and diolefins usual coordination polymerizations.

Finally, several attempts are now being made [13] to interpretate these polymerizations on more theoretical grounds : in spite of the great interest of such ultimate answers, they have not yet reached a point where they can be used as a general working tool for rationalizing and predicting diolefins experimental behaviour.

II. MODEL COMPLEXES FOR THE STUDY OF DIOLEFINS COORDINATION POLYMERIZATION.

A. Structure.

On support of the ideas developed above (particularly under I.A.), many stable η^3-allyl complexes of transition metals, isolated under a pure form, were found to be more or less active catalysts for this type of polymerization.

Interestingly, it has been shown that the overall rate of the reaction depends on the electronic density in the coordination sphere, and very high rates indeed have been promoted by increasing the electron withdrawing character of the counter-anion [14] : in this way, monometallic (although binuclear) nickel complexes have been synthesized which have activities comparable [15] to the standard bimetallic Ziegler-Natta systems (wherein the aluminum derivative is probably acting as a ligand controlling the right electronic distribution). Bis η^3-allylnickel trifluoroacetates (ANiTFA, prepared from both 1,5-COD or $Ni(CO)_4$ [16]) is probably the most efficient of these complexes, and has been choosen for most of the mechanistic studies reported below, the more as it has the ability to produce a very high cis-polybutadiene (up to 98 % under suitable conditions).

Moreover, the general properties of the polybutadienes obtained compare very favorable with those of Ziegler-type polymers (M.W. and M.W. distributions, % unsaturation, % gel, physico-mechanical properties after vulcanization).

B . General kinetic behaviour

The detailed course of these chain reactions could be investigated thanks to the use of the specially labelled monomer $D_2C=CH-CH=CD_2$, reacting on a non-deuterated ANiTFA. Its main features, pertaining to the reaction scheme given below, can be summarized as follow :

initiation $\quad$ 1 $C_4H_2D_4$

propagation $\quad$ n $C_4H_2D_4$

1. As shown also by detailed NMR studies (18), the allyl complex in solution is binuclear (N.M.R. and cryoscopic data), and undergoes a dynamic rearrangement between the symmetric (tail-to-tail) and 2 disymmetric (head-to-tail) isomers, under the temperature conditions where polymerization takes place.
Contrarily to the case of palladium, this dynamic behaviour does not promote a rearrangement isomerization of the allyl group (no coalescence of syn and anti proton) : in other words, it has to be ascribed to the dissociation and recombination of Ni-O bonds. Moreover, added monomer is always seen under its free, uncomplexed, form, in the NMR time-scale.

2. Under specific conditions (see below) the polymerization process is a "living" one (no termination nor transfer) wherein all the Ni atoms are active. This situation allows to observe accurately (19-20) the kinetic behaviour of the catalyst in the polymerizing medium, again by NMR spectroscopy.

     The allyl group of the complex initiates the polymer chain and becomes the free end of this chain, formed by successive insertions of monomer molecules.
During all this process, the complex remains, in the NMR time-scale, under the form of a syn - binuclear entity (carrying a chain on each nickel atom), producing high-cis or mixed cis.trans polybutadiene (depending on the solvent). In other words, its behaviour is typical of a "dormant" polymerization where most of the time no propagation occurs, unless the complex puts itself temporarily under the reactive η^1-isomeric form.

3. Relative initiation and propagation rates have been determined (ratio of CH_2^e and CH_2^i resonance peaks) by the same NMR technique on the polymerizing mixture (19,21) : although k_p is greater than k_i, they are of the same order of magnitude, which is expected for complexes having rather similar structures.
If there is no determinant dissociation step of the binuclear catalyst into a mononuclear one, the overall rate obeys the simple relationship $R_p = k(BD)(Ni)$, with usual $\Delta S^{\neq}$ and $\Delta H^{\neq}$ values for this type of polymerization, $i.e.$ around - 20 e.u. (22) and 10 to 15 Kcal/Mole (22,23), respectively.

4. The fact that every Ni atom is active and initiates one growing chain suggests the existence of a living system, which has been confirmed by different experiments (19,24).
However, transfer reactions have been observed at high conversions and monomer concentrations, which probably arise from hydride shifts reactions, already suggested in the literature (25) and confirmed by NMR spectroscopy (19,24) : they may be substantially decreased by controlling the temperature and modifying the nature of the solvent ($f.e.$ o-dichlorobenzene).

Under these conditions, and by using vacuum-type techniques, good "living" behaviours can be controlled and applied to the synthesis of different products.

III. THE SELECTIVE CONTROL OF THE DIFFERENT ISOMERIC UNITS AND
     OF THEIR DISTRIBUTION IN THE POLYDIENE CHAIN.

This problem is certainly one where coordination catalysis has proven best its nearly incredible and sometimes completely unexpected potentialities in controlling the selectivity of organic reactions, $i.e.$ at four levels : selectivity versus substrate, regioselectivity, stereo (and enantio) selectivity, and more recently, "chronoselectivity" (in other words the "coding" ability). Although the basic reasons governing these abilities are not yet fully understood, this chapter will present the actual exploratory and tentative views which might represent, in spite of their crude character, a working tool for further more refined investigations.

A. Control of the regioselectivity (structural isomers 1,2/1,4).

It is well documented that this control depends essentially on the nature of the transition metal involved, although in some cases some specific ligands can influence it to some extent. For example, 1,2 polybutadiene can be obtained in the presence of Pd, Cr and Mo-derivatives, while 1,4 polymers are preferentially produced by Ti, Co, Ni and Rh complexes (see $e.g.$ (26)); in both cases, steric purities as high as 98 to 99 % have been often obtained.

The lack of experimental methods allowing a progressive variation of this regioselectivity (from pure 1.2 to pure 1.4) by a systematic modification of the catalyst structure, has prevented up to now a thorough investigation of this kind of control. To date, the only simple proposal remains that of Arlman (27), based on the geometry of the centre, $i.e.$ the respective distances between the C_2 or the C_4 atom of the coordinated olefin, and the first C_α of the growing chain, undergoing the rearrangement. In that respect, the key controlling factor would obviously be the radius of the metal ion in the complex. Alternatively, the change in regioselectivity might be coupled with a change in mechanism from cis-rearrangement to allylic transposition (8), although there is no experimental evidence for such a possibility.

On the other hand , the electronic density modifications due to the coordination of the monomer, as well as substitution (in chain

and monomer), must also have a significant influence : a direct
consequence could be a stabilization of the h'-bond between the
metal atom and the most substituted carbon atom of the allylic
chain (C_γ), leading after further monomer insertion to the formation of a 1,2 unit.

Obviously, more experiments e.g. with dienes substituted in
specific positions, are required to solve these oroblems.

Anyhow, in the case of a 1,2 selectivity, the control of the
iso- or syndio-tactic placement will in turn depend on the steric
and kinetic parameters operative for monoolefins, with some possible assistance from the π-electrons of the vinyl bond.

B. Control of the stereoselectivity (geometrical isomers cis 1,4/
   trans 1,4).

1°) Influence of the ligands .

As already mentioned   , (bis (η^3-allylnickel trifluoroacetate),
ANiTFA, promotes in paraffinic solvents the rapid formation of a
very high cis-1,4-polybutadiene. Moreover, in the presence of
different additional ligands, the stereochemistry of the process
is drastically modified as indicated in fig.4,, while the regio-

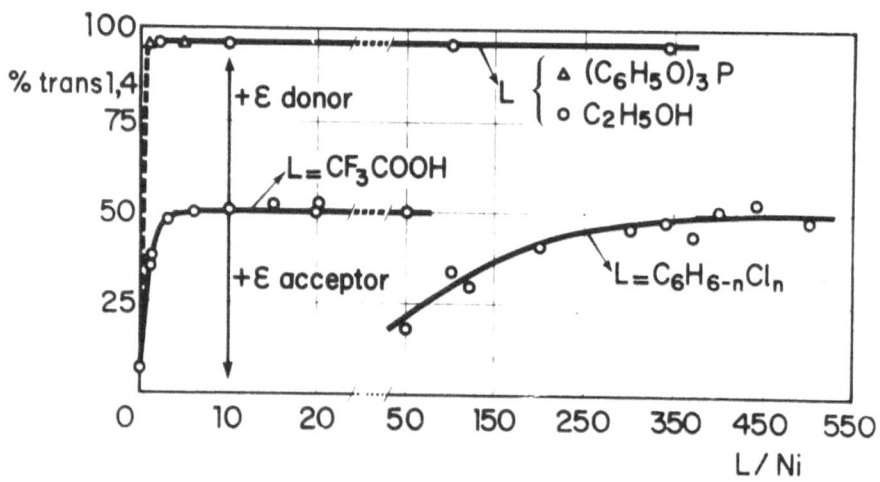

Fig.4 : Stereoselective controls by coordination equili-
        bria in 1,4 polymerizations of butadiene.

selectivity remains unchanged (99 % 1,4).

When one tries to interpretate this behaviour, common to most catalytic systems, it appears that all of the results reported up to now in the literature may confirm again the hypothesis presented by Cossee and Arlman (4,27). It is based either on a monodentate s-trans, or a bidentate-cis coordination of the diolefin monomer, leading respectively to the formation of trans or cis configurations in the polymer. This hypothesis is strongly supported by the fact that α-TiCl$_3$, which offers only one coordination vacancy at the active centers (both in Cossee and Rodriguez models), promotes the formation of a trans 1,4 polymer (27), while β-TiCl$_3$ which offers also centers with 2 and more vacancies favors the formation of a mixture of homo-cis and homo-trans polymers (28).

In the same manner, one can explain the influence of small (stoechiometric) amounts of strongly donating ligands (P(OR)$_3$,R$_2$O, ROH) on soluble systems, i.e. ANiTFA.The 2 coordination vacancies liberated a) by the temporary conversion of the η^3-allyl complex into a transient h'-allyl situation, and b) by temporary dynamic opening of a Ni-O μ-bridge (see above and below),can ensure the formation of the cis-isomer; blocking one of them by the strong ligand leads to a monodentate monomer coordination and the trans-isomer.

The overall situation is depicted in fig.5, which calls for a

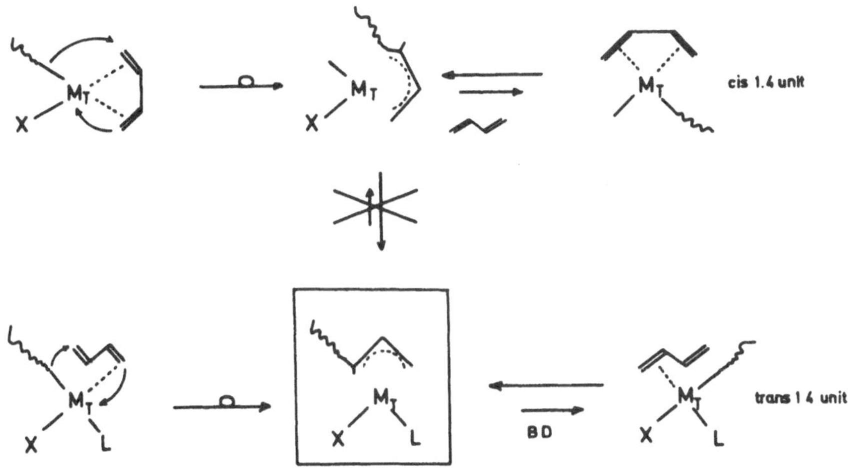

Fig.5 : Mechanism of the cis or trans stereoselection in 1,4 polymerization of butadiene.

few remarks.
In the course of the polymerization, the monomer insertion may
lead either to a η^1 or to an η^3-allyl complex. In the first case,
several monomer insertions may take place before going back to the
stable "dormant" _syn_ η^3-allyl isomer of the growing chain-end (17).
In the latter case, only the formation of a very reactive _anti_
η^3-allyl structure will ensure the formation of the cis-1,4-poly-
butadiene often observed experimentally : such a stereoselectivity
would imply the absence of any rapid anti-syn isomerization, in
accordance with our NMR studies on ANiTFA (see above), and with
the Su-Collette diene-olefin dimerization mechanism (29). It is
difficult to decide which one of these two possible pathways is
actually followed, but whatever it can be, the general scheme will
not be changed.

In other words, this convincing scheme excludes, as it was
sometimes claimed (30), a necessarily determinant role of the
permanent isomeric form (syn or anti) of the catalytic complex in
controlling the stereoselectivity (trans or cis); that is ruled
out first by several NMR studies where syn-complexes in situ have
been shown to produce high cis-polymer, as well as by the difficulties inherent to such a proposal (no "isomeric memory" of the η^3-
allyl group when it goes to the η^1- form, or alternatively a very
unlikely reaction between two π-bonded entities (monomer and η^3 -
allyl group). Again, as it was just reminded above, this statement
does not preclude the formation of a cis-unit from a transient anti-
isomer, the existence of which is not a prerequisite however.

Very recently, these views have been definitively supported
by the very elegant work of Porri (31), using cis-cis-1,4-dideute-
rio-butadiene as monomer, and correlating the different stereore-
gularities of the polymers obtained (trans-1,4-threo-diisotactic
and cis-1,4-threo-disyndiotactic structures).

Finally, one has to mention also alternative interpretations,
in particular a recent one by Furukawa (32), suggesting a control of
the cis-isomerism by the coordination to the nickel atom, of the
first free double-bond in the chain ("backbiting"). However, these
interesting proposals, which have not been ascertained up to now
by experimental data, would also raise serious problems in terms
of intramolecular movements and electrons displacements.

2°) Influence of the counter-anions.

Using a series of bis (η^3-allylnickel-X) complexes, one can
also obtain pure (99 %) 1,4-polubutadienes, with a stereoselectivity
going from 98 % cis (X = CF_3-COO^-) to 99 % trans (X = I^-), with
a monotonous variation through Cl^- and Br^- (15,14). This type of
control depends more on the anion group-electronegativity than on
its bulkiness, as indicated by the formation of very high cis-

polymers from X⁻ = 1.3.5-trinitrophenate (picrate salt).

The profound influence of the anion electronegativity on the electronic distribution in the coordination sphere, and in particular on the η^3-allyl group (higher dissimmetry), has been demonstrated by NMR spectroscopy (17,18); this influence might account for an easier liberation (by bridge-splitting) of the necessary vacancies, and bidentate coordination of the monomer.

3°) Obviously a double control may occur in the case of the 4-substituted dienes, where again an isotactic or syndiotactic placement of the substituent can take place, mostly for steric reasons.

C. Control of specific binary isomeric compositions : the equibinary polydienes.

Fig.4 demonstrates the possibility to control, by the addition of specific ligand, the formation of polymers containing exactly equal amounts of cis and trans isomers. It was shown later that this type of behaviour is quite general in coordination catalysis, and the name "equibinary polydienes" has been coined for these polymers containing equal amounts of 2 different isomeric units. Other typical examples include the equibinary (1,4 cis or trans -1,2) polybutadienes (33), (cis 1,4-3,4) polyisoprene (34), (1,2-3,4) polyisoprene, obtained in the presence of various complexes of nickel, molybdenum and cobalt.

Such a phenomemon exhibits all the characteristic features of a selective and competitive coordination reaction , as it can be seen from fig. 4. Indeed :
- this equibinary composition is reached asymptotically, and with amounts of ligands depending on their nature (apparent interaction constants can be calculated from the curvature of the plots in the figure);
- the phenomenon is reversible : elimination of ligand (*i.e.* by distillation), or addition of stronger ligands (donor or acceptor), lead to higher cis or higher trans contents respectively;
- the control is highly selective : the amount of 1,2 isomer remains very low (ca. 1 %), the equibinary composition can be obtained over a rather wide range of concentrations and temperatures, and the polymer cannot be separated into fractions of different cis/trans contents. Furthermore, intermediate compositions *i.e.* 75 % cis, can be resolved (24) into a rather high cis fraction and a 50/50 one.

These results show clearly that there are two different and specific catalytic sites in equilibrium (separated, or on the same binuclear complex), which yield respectively the high cis and the equibinary polymer.

It should also be emphasized that these cis/trans isomerisms are directly and irreversibly controlled by the catalyst during the polymerization; all the attempts to isomerize a preformed polybutadiene by these catalytic systems have been so far unsuccessfull.

An interpretation of this intriguing behaviour was very difficult at first sight, and had to wait for additional results before being tentatively approached; a possible overall view will be discussed in the next section.

D. The "coding" control of the isomers distribution in equibinary polydienes : the concept of "chronoselectivity".

A priori, an equibinary composition could result either from a mixture of equal amounts of homo-cis and homo-trans polybutadienes, or alternatively from a random, or an alternate, or even a stereoblock placement of cis and trans units in the same chains. The first possibility could be easily discarded, since the polymers obtained cannot be separated into different compositions by classical fractionnation methods, and the product has none of the characteristic physical properties of the pure cis- and trans-polybutadienes.
The remaining hypotheses have been approached either by chemical methods (ozonolysis) (33) for (1.4-1.2) polybutadiene, or by high-resolution H-NMR spectroscopy at high temperature with careful spin decoupling (35) which allows a quantitative analysis of the isomeric triads in the (1,4-cis-1,4 trans) polymers.

The results obtained (Table I) are absolutely astonishing, and demonstrate again how structural controls by coordination complexes are extremely sensitive to reaction conditions. They indicate clearly that by changing the solvent, or the ligands in the complex (anions or additives), one can control a continuum of situations, going from a purely random distribution of cis and trans units (in benzene) through a more or less alternating placement ($\leqslant$ 80 %, in CH_2Cl_2), up to a stereoblock situation (with various mean block lenghts, depending on the conditions).

Besides the mechanistic interest of the problem, it can be realized immediately that these controls might have important consequences in terms of applications. The random product is a perfectly amorphous (no melting point) 1.4 polydiene of very low Tg (-100°C), while the stereoblock polymer is in fact, when the mean D.P.of the blocks is high enough, a thermoplastic elastomer made in one step from one very usual monomer.

A. $(ANiTFA)_2$: C_6H_6 : 49 cis/49 trans

Triads	CTC	TTC	TTT	CCC	CCT	TCT
Observed	0.13	0.25	0.12	0.13	0.26	0.11
Bernoullian	0.125	0.25	0.125	0.125	0.25	0.125

B. $(ANiTFA)_2$: CH_2Cl_2 : 43 cis/56 trans

Triads	CTC	TTC	TTT	CCC	CCT	TCT
Observed	0.24	0.25	0.07	0.03	0.19	0.22
Markov 1st o.	0.20	0.25	0.08	0.04	0.20	0.23
Bernoullian	0.11	0.27	0.18	0.08	0.22	0.14

C. $(ANiCl)_2$; o-dichlorobenzene; 47 cis/49 trans

CCC and TTT > 95 %; $\bar{n}$ = 15 units

cis stereoblocks		Trans stereoblocks	
T_g	− 109°C	$T_{\alpha \to \beta}$	55° C
T_m^o	− 7,5°C	T_m^o	120° C

Table I. : Various isomers distributions in equibinary 1,4-polybutadienes obtained with $(ANiX)_2$ complexes.

In terms of selective control, it is still very hazardous to propose a mechanistic scheme accounting for this unexpected situation, which incidentally exists probably in Ziegler-Natta type polymerizations but has never been recognized as such.
Very tentatively, one could approach the problem starting from NMR investigations which have demonstrated (17,18) the binuclear nature of the complex even in the polymerization mixtures, as well as its dynamic rearrangement by scission and reformation of the nickel-oxygen bonds. If one admits that a partial opening (one Ni-O bond) of the binuclear complex provides a nickel atom with two vacancies (in the η^1-allyl form) and another one with only 1 vacancy, one has a system where one site can produce a cis unit, and the other site a trans one (statistically ofcourse)

Now, depending on the relative rates of the monomer insertion (k_i) versus the complex dynamic rearrangement (k_r), one might receive either a perfectly random product (if $k_r >> k_i$), or a stereoblock one (if $k_i >> k_r$). In this latter case, there should be no more thermodynamic control of the 50/50 composition, and that is experimentally observed; and in between the two extreme situations, one should have a zone of "statistically" alternating placements, as it is indeed the case. This purely exploratory hypothesis is schematized in fig.6.

Obviously, the kinetic ratio k_i/k_r must be dependent on structural factors in the complex, but also on collision complexes with solvent molecules, all features which could explain the experimental observations related above.

This "dual-site" mechanism (which might also arise from bimolecular interactions between 2 allyl-complexes), has received preliminary support from several types of experiments : suffice it to remind the polymerization (36) of isoprene by cobalt complexes which yields a (cis-1,4-3,4) equibinary polymer, but wherein the vinyl-type 3,4 units (and not the cis-1,4 ones) are selectively replaced by methylmethacrylate units when this vinyl comonomer is present.

Anyhow, whatever the mechanism, one faces here a small informational machine which can develop a thermodynamic control of an equibinary composition, and a kinetic control of the two isomers distribution : in other words, it represents the first synthetic example of a very simple, but characteristic kinetic "code". The terms of "chronoselectivity", or "troposteric control" as suggested by Professor T.H. Esch   , might be considered to characterize such bahaviours.

In any event, it is also worthwhile to emphasize that such a bis (η^3-allyl) nickel complex is able, under minor structural or kinetic modifications, to produce from the sole monomer butadiene, 4 completely different products displaying very different properties : a pure cis-1,4- (an excellent elastomer with a m.p. around 0°C), a pure trans-1,4- (a crystalline plastic, m.p. around 140°C), a random equibinary (cis 1,4-trans 1,4)- (a totally amorphous elastomer), and a stereoblock (cis 1,4-trans 1,4)-polybutadiene (a thermoplastic elastomer).

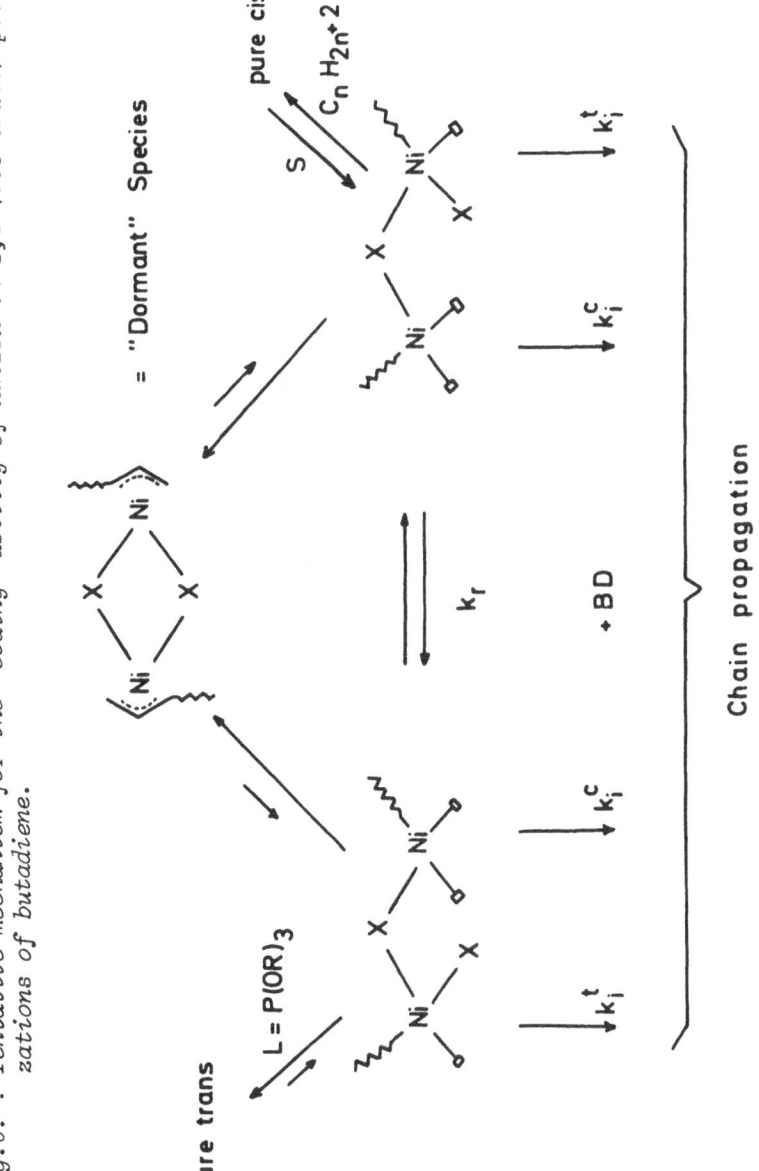

Fig.6. : Tentative mechanism for the "coding" ability of ANiTFA in 1,4 (cis-trans) polymerizations of butadiene.

pure trans

L = P(OR)$_3$

= "Dormant" Species

pure cis

S ⇌ C$_n$H$_{2n+2}$

k_r + BD

Chain propagation

If k_i^x = rates of butadiene insertion,

statistics of placement = f (k_i / k_r)

CONCLUSIONS

Although our knowledge of these polymerizations has progressed considerably in the last 15 years, one is still astonished by new discoveries in the field revealing the unexpected refinement and versatility of the elaborate selectivity controls involved.

There is no doubt however that much remains to be done in many respects, from the mechanistic point of view as well as in terms of new synthetic advances; to mention only two examples among numerous ones : a better clarification of the nature of the electronic rearrangements involved, and a good control of the chain termination allowing the synthetic use of the "living" polymers obtained would be higly desirable.

From a more general point of view, one might also wish more advances in two specific directions.
The first one pertains to some unification of concepts between these coordination systems and the reactions initiated by lithium derivatives. The second one is related to the obvious observation that these elaborate selective controls depend strongly on the overall geometry of the active complex in situ; it would be most helpful to obtain, e.g., by the simultaneous application of different powerful physical methods, more experimental information about these actual structures.

One might conclude by saying that even without taking into account possible unexpected new breakthroughs in the field, there are still a lot of exciting challenges in this fascinating area of research, where chemists have reached a level of selectivity control unequalled till now in synthetic organic chemistry.

ACKNOWLEGDMENTS

One of the authors (Ph.T.) wants to express his appreciation tho his former coworkers at the Institut Français du Pétrole, namely, F. Dawans, J.C. Maréchal and J.P. Durand, for their devoted and enthusiastic collaboration. The support of B.F. Goodrich Chemical C°, Cleveland (U.S.A.), and the Service de la Programmation de la Politique Scientifique, are also gratfully acknowledged.

REFERENCES

1) M. Takayanagi et al., U.S. Patent 3.595.850 (1971);
   S. Sugiura et al., U.S. Patent 3.935.180 (1976).
2) F.A. Cotton, J. Am. Chem. Soc., 90, 6230 (1968).
3) G. Natta and G. Mazzanti, Tetrahedron, 8, 86 (1960).
4) See : P. Cossee, in : Stereochemistry of Macromolecules, Vol.1,
   p. 145, A.D. Ketley Editors, M. Dekker, New York 1967.
5) L. Rodriguez and H. Van Looy, J. Polym. Sci., A1, 4, 1905,
   1971 (1966).
6) Y. Sokolov and I.Y. Poddubnyi, Adv. Chem. Ser., 91, 250 (1969).
7) Y. Takahashi, S. Sakai, and Y. Ishii, J. Organometal. Chem.,
   16, 177 (1969).
8) R.P. Hughes and J. Powell, J. Am. Chem. Soc., 94, 7723 (1972).
9) P. Heimbach, Angew. Chem. Int. Ed., 12, 975 (1973).
10) M. Julémont and Ph. Teyssié, in preparation for"Aspects of
    homogeneous catalysis", Vol.4, R. Ugo Editor, Reidel, Dordrecht
    Holland.
11) C.A. Tolman, J. Am. Chem. Soc., 96, 2780, (1974).
12) K.J. Ivin et al., and M.L.H. Green et al., J.C.S., Chem. Com.,
    604 (1978).
13) See f.i. P. Heimbach, J. Mol. Cat., in press, 1979; also
    J. Furukawa, plenary lecture, IUPAC Symposium on Macromole-
    cules; Madrid 1974.
14) J.P. Durand, F. Dawans and Ph. Teyssié, J. Polym. Sci., A1,
    979 (1970); F. Dawans and Ph. Teyssié, ibid , B7, 111 (1969).
15) J.P. Durand, F. Dawans and Ph. Teyssié, J. Polym. Sci. B6 ,
    760 (1968).
16) F. Dawans, J.C. Maréchal and Ph. Teyssié, J. Organometal.
    Chem., 21, 259 (1970) : U.S. Pat. 3.542.695 (1967), 3.660.445
    (1972), 3.719.653 (1973), and French Pat. 1.556.962 (1967).
17) R. Warin et al., J. Polym. Sci., B11, 177 (1973).
18) R. Warin, M. Julémont and Ph. Teyssié, J. Organometal. Chem.
    in preparation.
19) J.M. Thomassin et al., J. Polym. Sci., A1, 13, 1147 (1975).
20) Ph. Teyssié et al., in "Coordination polymerization (A memo-
    rial to K. Ziegler", J.C.W. Chien Editor, Acad. Press,
    New York (1975).
21) M.I. Lobach et al., J. Polym. Sci., B9, 71 (1971).
22) J.F. Harrod and L.R. Wallace, Macromolecules, 2, 449 (1969);
    ibid, 5, 682, (1972).
23) J.C. Maréchal, F. Dawans and Ph. Teyssié, J. Polym. Sci., A1,
    8, 1993 (1970).
24) J.M. Thomassin, Ph.D. Thesis, Univ. of Liège, 1975.
25) J.F. Harrod and R. Wallace, Macromolecules, 5, 685, (1972).
26) Ph. Teyssié and F. Dawans, in "Stereo Rubbers", p. 79 sq.,
    W.M. Saltman Editor, J. Wiley and Sons, New York (1977).
27) E.J. Arlman, J. Catalysis, 5, 178, (1966), See also ref.4
28) G. Natta, Actes 2è Congr. Inter. Catalyse, Paris, 1960, 67.
29) J. Collette and L.Su, Adv. Organometal. Chem., 17, in press, 1979.

30) S. Otsuka and M. Kawakami, Kogyo Kagaku Zasshi, 68, 874 (1965).
31) L. Porri and M. Aglietto, Makromolekulare Chem., 177, 1465
    (1976); Colloque Section Sud G.F.P., La Grande Motte, Editor
    F. Schue, Octobre 1978.
32) J. Furukawa, Plenary Lecture, IUPAC Symposium on Macromolecules
    Madrid 1974.
33) Ph. Teyssié, F. Dawans, and J.P. Durand, J. Polym. Sci., C22
    221 (1968); J.P. Durand, and Ph. Teyssié, J. Polym. Sci., B6,
    299 (1968); J.P. Durand, F. Dawans, and Ph. Teyssié, J. Polym.
    Sci., A1, 987 (1970).
    J. Furukawa, K. Haga, E. Kobayashi, Y. Iseda, T. Youshimoto,
    and K. Sakamoto, Polymer J., 2, 371 (1971); J. Furukawa,
    E. Kobayashi, and T. Kawagoe, ibid, 5, 231 (1973
34) F. Dawans and Ph. Teyssié, Makromolekulare Chem., 109, 68
    (1967)
35) E.R. Santee, V.D. Mockel and M. Morton, J. Polym. Sci., B11
    453 (1973); M. Julémont et al., Makromolekulare Chem., 175,
    1673 (1974).
36) F. Dawans and Ph. Teyssié, Europ. Polym. J., 5, 541 (1969).

o

o   o

STEREOCHEMISTRY OF THE METATHESIS REACTION OF
OLEFINS

H. Höcker

Institut für Organische Chemie der
Universität München, Karlstraße 23
D-8000 München 2, W-Germany

I. Introduction

The disproportionation of olefins first reported by
Banks and Bailey[1]) and - as a special case - the
ring-opening polymerization of cycloalkenes[2,3])
comprise a re-arrangement of the carbon skeleton
of the olefins by opening and closure of double
bonds. Thus the reaction - generally referred to
as metathesis reaction - is a trans-alkylidenation
reaction.

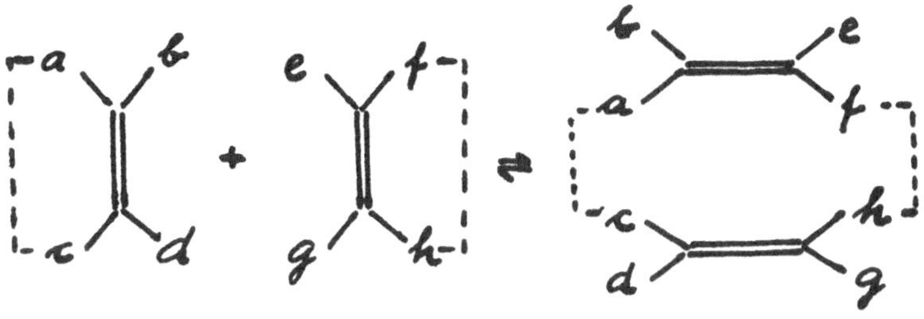

$$(1)$$

Using cycloolefins - as indicated by the dashed
lines in eq. 1- the reaction continues to yield
polymers (polyalkenylenes) which are in equilibrium
with a homologeous series of oligomers (Fig. 1)[3])
and - in the case of cyclopentene - with the monomer
as well (ceiling temperature ca. 150°C)[4]). While

R. W. Lenz and F. Ciardelli (eds.), Preparation and Properties of Stereoregular Polymers, 151-162.

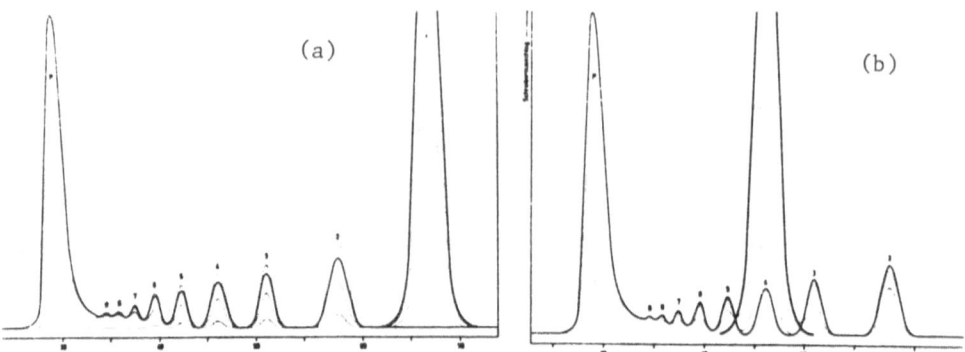

Fig. 1. Oligomer distribution in the metathetical
         polymerization of cyclododecene (a) and
         of the tetramer of cyclododecene, cyclo-
         octatetracontatetraen (b).

the polymer can be assumed to consist of acyclic
species the oligomers are proven to be cyclic. Since
the polymer/oligomer-equilibrium is independent of
temperature it can be described by a ring chain
equilibrium[5]). In fact, the plot of the logarithm
of the equilibrium concentration of the oligomers
(which to a first approximation is equal to the
equilibrium constants of the corresponding oligomers)
vs. the logarithm of the degree of polymerization
results a straight line with a slope of -2.5 as
predictable according to the theory of Jacobson and
Stockmayer[5]) (Fig.2).

The catalysts promoting the reaction are Ziegler-
Natta type catalysts, the most active ones comprising Mo, W, Re as transition metals, e.g. the
classical Calderon system $WCl_6/C_2H_5OH/C_2H_5AlCl_2$ [6]),
$WCl_6/Sn(CH_3)_4$ [7]), and $Mo[P(C_6H_5)_3]_2Cl_2(NO)_2(CH_3)_3$
Al_2Cl_3 [8]). A new generation of catalysts are transition metal-carbene complexes, e.g. $(CO)_5W=C(C_6H_5)_2$ [9])
Those catalysts are usually referred to as soluble
in suitable solvents such as aromatic hydrocarbons.
On the other hand a wide spectrum of heterogeneous
catalysts is known, e.g. molybdenum oxide supported
on alumina and combined with $(i-bu)_3Al$ [10]).

The mechanism of the reaction is still being discussed. Initially a pairwise mechanism was proposed[11])
with a cyclobutane derivative (or alternatively, a
tetramethylene complex[12])) as the key intermediate.
Since recently, however, the so-called carbene-
mechanism first proposed by Hérisson and Chauvin [13])
is strongly favored by most the investigators:

$$L_x M = C < \quad \rightleftharpoons \quad L_x M - C < \atop {\overline{C - C <}} \quad \rightleftharpoons \quad L_x M \quad C < \atop {\overline{C \quad C <}} \quad (2)$$

The key intermediate is a transition metal-carbene
which forms a metallacyclobutane derivative with
an olefinic double bond. According to this mechanism
the reaction is <u>not</u> a "pairwise" reaction. There are
three alkylidene groups involved rather than four
and the polymerization is a chain reaction rather
than a stepwise reaction.

The stereochemistry of the reaction has been of interest for a long time already in order to pre-
determine the structure of the products. In recent
years, however, the stereochemistry has been investigated in order to gain additional support for
one or the other mechanism. Most investigations
have been performed with acyclic olefins. But there
are some involving cycloolefines as well.

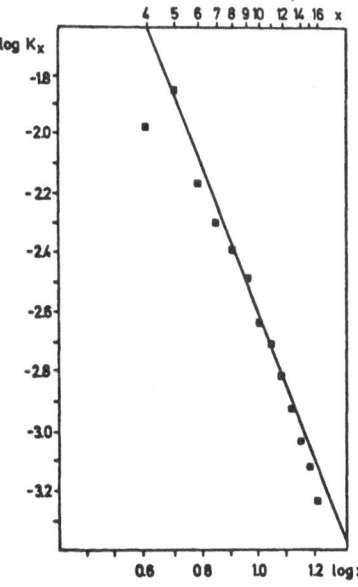

Fig. 2.   Jacobson-Stockmayer[5]) plot of the oligomers
obtained by metathesis of cyclooctadiene.

These investigations are complicated by isomerization reactions which have to be considered. It is
to be decided whether the stereochemistry observed
is a result of the metathesis reaction or of isomerization reactions and whether the isomerization
occurs via metathesis.

II. Survey of the Results

In the following a brief summary of the observations
with respect to the stereochemistry of the metathesis reaction will be given.

1. Generally it has been found that the trans-isomers
react slowlier than cis-isomers. As an example,
Fig.3 shows the time-conversion for cyclododecene.
                              curves
Moreover, Basset et al.[14]) report that no metathesis
takes place when starting from the pure trans-2-
pentene with the catalytic system $W(CO)_5[P(C_6H_5)_3]/$
$C_2H_5AlCl_2/O_2$ and that the metathesis occurs -
though with varying induction periods - when the
cis-isomer is added in minute concentration ($<0.2\%$).

2. According to early results of Calderon and
others the cis/trans ratio of the reaction products
approaches the thermodynamically controlled equilibrium value. This, however, is a consequence of
isomerization reactions, since the equilibrium
values derivate from those found upon extrapolation
to 0% conversion.

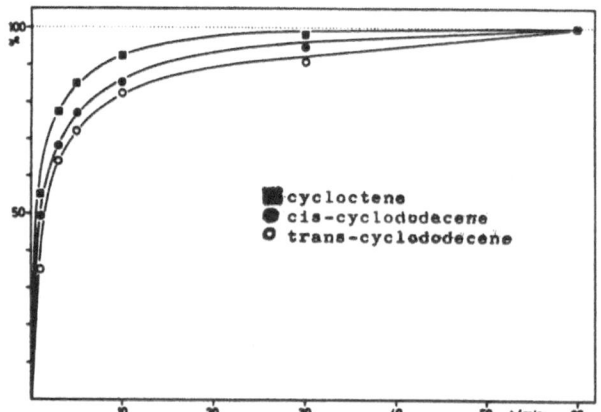

Fig. 3.   Time-conversion plot of the metathetical
          polymerization of cyclooctene (COE), cis-
          cyclododecene, and trans-cyclododecene (CD).

3. Cis-trans isomerization has been proven by Bilhou
et. al.[14]) to occur with a rate comparable to that
of the productive metathesis reaction; this has
been taken as an indication that the cis-trans isomerization can be regarded as a regenerative or
non-productive metathesis reaction.

4. At an extrapolated conversion of 0% it is observed that the cis-products are slightly favored
when starting from a cis-isomer. Thus, according to
Bilhou et al.[14a]) cis-2-pentene yields a trans/cis-
ratio of 0.73 for 2-butene and 0.88 for 3-hexene.
The effect is even stronger when heterogeneous
catalysts are used, i.e. 0.37 for 2-butene[14b]).
According to Katz et al.[15]) carbene catalysts yield
95.3% cis-2-butene and 92.5% cis-3-hexene besides
96% cis-2-pentene when starting from pure cis-2-
pentene.

5. Starting from trans-isomers, e.g. trans-2-
pentene, the trans-products are favored. Generally
a trans/cis-ratio of 2 is observed. The stereospecificity of the carbene catalyst is even higher,
though it is lower than with respect to the cis-
isomer (27% cis-2-butene, 16.6% cis-3-hexene and no
cis-2-pentene).

6. The polymerization of cis-cyclooctene yields a
homologeous series of oligomers with increasing
trans-content (Fig.4). The dimer consists of two

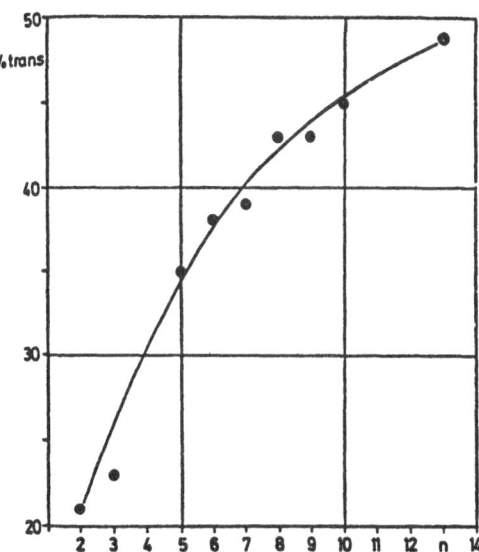

Fig. 4.   Trans-content of COE-oligomers as a
          function of the degree of oligomerization.

isomers, the cis-cis and the cis-trans isomer, both
initially produced with equal rate (Fig.5). The
concentration of the cis-trans isomer, however,
decreases again as the mixture approaches the
thermodynamic equilibrium. Almost independent of
the initial monomer concentration the fraction of
the cis-cis isomer in the dimer amounts to about
45% in the beginning and 75% at the end of the
reaction (Fig.6).

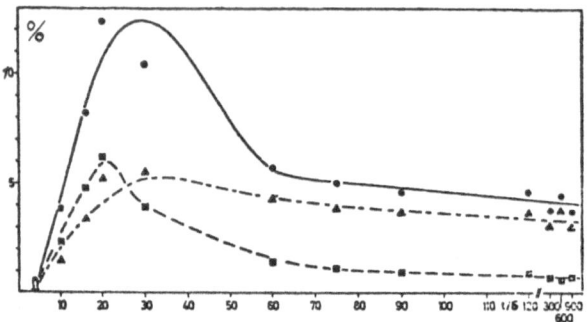

Fig. 5. Yield of the dimer of COE (●) as well as
         yield of the cis-cis (▲) and of the cis-
         trans-isomer (■) as a function of time of
         polymerization at $[WCl_6]=2.5.20^{-4}$ M with
         $WCl_6/C_2H_5OH/C_2H_5AlCl_2$ as catalyst.

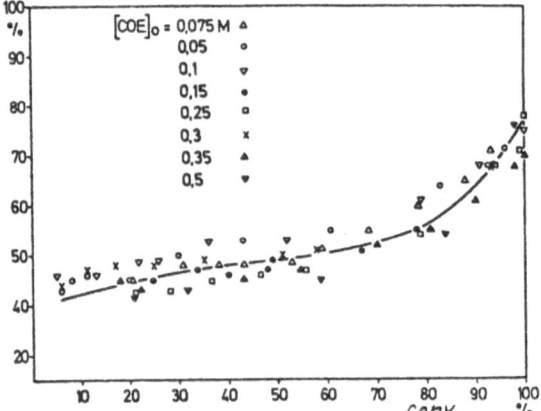

Fig. 6. Fraction of the cis-cis-isomer of the
         dimer of COE as a function of conversion
         at different initial [COE] with $WCl_6/$
         $Sn(CH_3)_4$ as catalyst.

7. Polyalkenylenes obtained with most catalysts
under normal conditions exhibit a trans-content of
double bonds >50%. There are catalysts, however,
which allow the formation of highly cis-configurated
polyalkenylenes. Pampus et al.[16]) have reported a
catalyst consisting of WCl_6, $Sn(C_2H_5)_4$, $(C_2H_5)_2O$
(1:2:2) which at -30°C converts cyclopentene to a
polymer with more than 92% cis-configurated double
bonds while above 0°C the polymer shows ≈80%
trans-configurated double bonds.

According to Katz et al.[17]) $(C_6H_5)_2C=W(CO)_5$ produces polyalkenylenes with more than 90% cis-
configurated double bonds.

8. With molybdenum-catalysts, the formation of
trans-products from trans-isomers is particularly
favored.As well, cyclopentene and norbornene are
converted to highly cis-configurated polymers.

III. Diskussion

The interpretation of these experimental results is
by no means unambiguous. All approaches start from
the metallacyclobutane species introduced above as
the hypothetical key intermediate of the reaction.
Following the arguments of Katz et al.[15]) this key
intermediate is a non-planar species the substituents of which being in axial and equatioral
position. The decision which one of the possible
products is formed depends on the highest stability
or the lowest steric interactions between juxtaposed axial groups and ring atoms. From Fig. 7a it
is seen that the least steric interactions occur
when a cis-olefin is converted to a cis-product via
a metallacyclobutane in which the axial substituent
is juxtaposed to the metal since the carbon-tungsten
bonds (ca. 2.39 Å [19])) are longer than C-C-bonds.

In the case of the trans-olefin (Fig. 7b) the
situation is not that evident which is in agreement
with the lower stereospecificity observed with
trans-olefins and the tungsten-carben as a catalyst.
However, the formation of a trans-product via a
metallacyclobutane with all substituents in
equatorial position obviously should be the most
favored one.

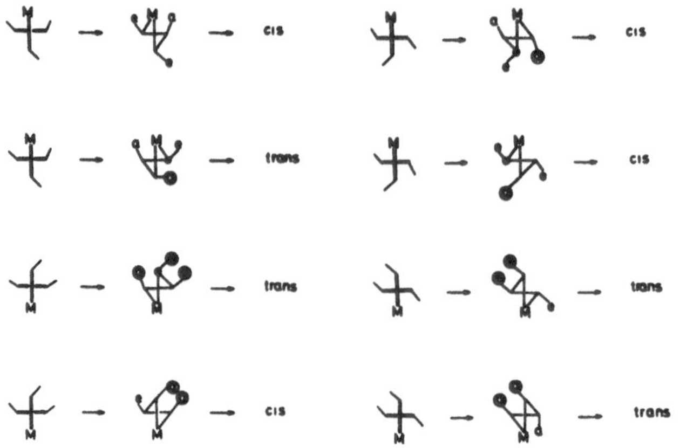

(a)   (b)

Fig. 7.   Conformations of the hypothetical metalla-
           cyclobutane intermediate after Katz et al.

Katz et al.[20] have shown in a rather complicated
but straightforward way that the same scheme is
able to account for the high fraction of Z-con-
figuration (85%) of polyisoprene formed from
1-methylcyclobutene (as an example for the metathesis reaction of a trisubstituted ethylene.
Still the picture seems to be over-simplified. With
increasing bulkiness of the substituent R the stereospecificity decreases as observed by Basset et al.[21].
This is in disagreement with the scheme proposed,
however, it can be explained assuming an interaction between the substituents and the ligands of
the metal.

In fact, Basset et al.[21] suggest that catalysts of
low stereoselectivity might exhibit an empty site
of coordination at the metal while those with high
stereoselectivity coordinatively might be more
saturated. This would also account for the observation of Ivin et at.[22] who found a higher cis-
contend when some coordinating species such as
ethyl acrylate is added to the catalyst.

It has been observed that - as a rule - the
tungsten-carben catalysts exhibit much higher
stereospecifidty than any WCl6-containing catalyst.
Exceptions are made by certain molybdenum con-

taining catalysts[18b]) and by a tungsten halide containing catalyst at low temperatures[16]).
Low stereospecificity according to Katz et al.[15])
may be attributed to a heterolytical cleavage of
the tungsten-carbon bond in the tungstena-
cyclobutane system in the presence of Lewis acids
yielding a 3-metallopropyl cation. Now the restriction to rotation is suspended until new ring
formation. Another explanation might be given by
assuming the following equilibrium:

$$L_xW=C\langle^{R_1}_{R_2} + C_2H_5AlCl_2 \rightleftharpoons L_x\overset{\oplus}{W}\equiv C-R_2 + [C_2H_5AlCl_2R_1]^{\ominus} \quad (3)$$

This assumption is based on experiments of Fischer
et al.[23]) who reported the following reactions:

$$(CO)_5W=C\langle^{OCH_3}_{C_6H_5} + BCl_3 \xrightleftharpoons[-78°C]{-15°C} \begin{array}{c} Cl(CO)_4W\equiv C-C_6H_5 \\ + \\ BCl_2(OCH_3) \end{array} \quad (4)$$

$$(CO)_4[t\text{-}P(CH_3)_3]M=C\langle^{OR}_R + BCl_3 \longrightarrow \begin{array}{c} (CO)_4[P(CH_3)_3]\overset{\oplus}{M}=C-R \\ + \\ [ROBCl_3]^{\ominus} \end{array} \quad \mathbf{(5)}$$

In fact, it was observed that the stereospecificity
of certain systems diminishes (i) when the Al/W ratio
is increased $(WF_6/(C_2H_5)_3Al_2Cl_3)$ [24]) and (ii) when
the temperature is raised[25]).

Molybdenum seems to behave differently from
tungsten. With trans-olefins high stereospecificity
is observed[18]) when Mo-catalysts are used. As well,
the polymerizations of cyclopentene and norbornene[27])
result the highest stereospecificity with Mo-con-
taining catalysts. This might be explained by a lower
cleavage probability of the metallacyclobutane ring
or by a smaller equilibrium constant for eq. 3.

Finally, a special effect briefly should be discussed. Ivin et al.[28]) proved the non-randomness of
cis- and trans-configurated double bonds in polypentenamer by ^{13}C-NMR spectroscopy. They conclude
the presence of two different kinds of active
species in one of which the last formed double bond
remains coordinated to the metal atom of the
catalyst.

We have arrived at a similar conclusion concerning
an additional coordination of a double bond on the
grounds of the molecular weight distribution of the
oligomers in the initial stage of the reaction
which corresponds to a Schulz-Flory distribution
controlled by addition and transfer reactions. As
well the formation of oligomers of cyclooctene and
cyclododecene and the monomer-polymer equilibrium
in the case of the metathetical polymerization of
cyclopentene can be made plausible by this concept
(Fig. 8).

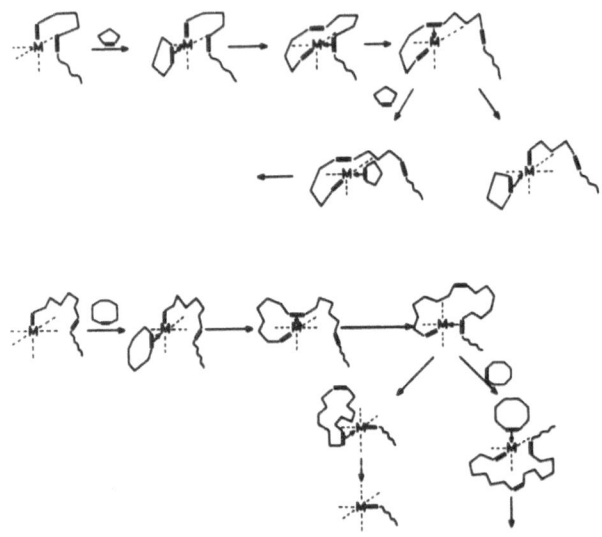

Fig. 8.   Metathesis of cyclopentene and cyclo-
          octene via carbene-mechanism with co-
          ordination of an additional double bond.

References :

1.   R.L. Banks, G.C. Bailey, Ind.Eng., Chem.Prod.
      Res.Dev. 3, 170(1964)
2.a) N. Calderon, M. Morris, J.Polym.Sci., Part A-2
      5, 1283(1967)
  b) G. Dall'Asta, Rubber Chem.Technol. 47 511(1974)
3.a) H. Höcker, R. Musch, Makromol.Chem. 175, 1395
      (1974)
  b) H. Höcker, W. Reimann, K. Riebel, Z. Szenti-
      vanyi, Makromol.Chem. 177, 1707(1976)

4.   E.A. Ofstead, N. Calderon, Makrom.Chem. 154,
      21(1972)
5.   H. Jacobson, W.H. Stockmayer, J.Chem.Phys. 18,
      1600(1950)
6.a) N. Calderon, H.Y. Chen, K.W. Scott, Tetrahedr.
      Lett. 3327(1967)
  b) N. Calderon, E.A. Ofstead, J.P. Ward,
      W.A. Judy, K.W. Scott, J.Am.Chem.Soc. 90,
      4133(1968)
7.   P.B. Van Dam, M.C. Mittelmeijer, C. Boelhouwer,
      J.C.S.Chem.Comm. (1972) 1221
8.a) E.A. Zuech, J.Chem.Soc., Chem.Commun. 1182
      (1968)
  b) E.A. Zuech, W.B. Hughes, D.H. Kubicek, E.T.
      Kittleman, J.Am.Chem.Soc. 92, 528(1970)
9.a) C.P. Casey, T.J. Burkhardt, J.Am.Chem.Soc.
      95, 5833(1973)
  b) T.J. Katz, N. Acton, Tetrahedron Lett. 4251
      (1976)
10.   E.F. Peters, B.L. Evering, U.S.Patent
      2.963.447(1960)
11.a) C.P.C. Bradshaw, E.J. Howman, L. Turner, Catal.
      7, 269(1967)
  b) N. Calderon, E.A. Ofstead, J.P. Ward,
      W.A. Judy, K.W. Scott, J.Am.Chem.Soc. 90, 4133
      (1968)
  c) C.T. Adams, S.G. Brandenberger, J.Catal. 13,
      360(1969)
12.   G.S. Lewandos, R. Pettit, Tetrahedron Lett.
      789(1971)
13.   J.L. Hérisson, Y. Chauvin, Makromol.Chem. 141,
      161(1970)
14.a) J.L. Bilhou, J.M. Basset, R. Mutin, W.F.
      Graydon, J.Am.Soc. 99, 4083(1977)
  b) J.M. Basset, J.L. Bilhou, R. Mutin,
      A. Theolier, J.Am.Chem.Soc. 97, 7376(1975)
15.   T.J. Katz, W.H. Hersh, Tetrahedron Lett. 585
      (1977)
16.   G. Pampus, G. Lehnert, Makromol.Chem. 175,
      2605(1974)
17.   T.J. Katz, S.J. Lee, N. Acton, Tetrahedron
      Lett. 4247(1976)
18.a) W.B. Hughes, J.Chem.Soc., Chem.Commun. 431
      (1969)
  b) G. Doyle, J.Catal. 30, 118(1973)
19.   M.R. Churchill in "Perspectives in Structural
      Chemistry", Vol.III, J.D. Dunitz and
      J.A. Ibers, eds., John Wiley and Sons, N.Y.
      1970, p. 91 ff.

20.   T.J. Katz, J. McGinnis, C. Altus, J.Am.Chem.
       Soc. 98, 606(1976)
21.   M. Leconte, J.L. Bilhou, W. Reimann,
       J.M. Basset, J.C.S.Chem.Comm. (1978) 341
22.   K.J. Ivin, D.T. Laverty, J.J. Rooney, P. Watt,
       Rec.Trav.Chim.Pays-Bas 96, M 54(1976); Pro-
       ceedings International Symposium on Metathesis
23.   E.O. Fischer, S. Walz, W.R. Wagner, J.Organo-
       metallic Chem. 134, C 37(1977)
24.   P. Günther, F. Haas, G. Marwede, K. Nützel,
       W. Oberkirch, G. Pampus, N. Schön, J. Witte,
       Angew. Makromol.Chem. 14, 87(1970); 16/17,
       27(1971)
25.a) R.J. Minchak, H. Tucker, Polym.Prepr.Am.Chem.
       Soc.Div.Polym.Chem. 13, 885(1970)
   b) G. Pampus, G. Lehnart, Makromol.Chem. 175,
       2605(1974)
26.   G. Dall'Asta, G. Motroni, Angew.Makromol.Chem.
       16/17, 51(1971)
27.   G. Sartori, F. Ciampelli, N. Cameli, Chim.Ind.
       (Milan) 45, 1478(1963)
28.   K.J. Ivin, D.T. Laverty, J.J. Rooney,
       Makromol.Chem. 179, 253(1978)

STEREOREGULATION OF HOMOGENEOUS FREE RADICAL, CATIONIC AND ANIONIC POLYMERIZATION REACTIONS

Robert W. Lenz

Materials Research Laboratory,
Chemical Engineering Department,
University of Massachusetts,
Amherst, Massachusetts   01003 USA

R. W. Lenz and F. Ciardelli (eds.), Preparation and Properties of Stereoregular Polymers, 163–184.

    The preceding discussions have been concerned almost
entirely with the sterochemical control and the formation of
stereoregular polymers by heterogeneous catalysts, presumably
at chiral active sites on the catalyst surface. This chapter
is concerned with the formation of stereoregular polymers by
homogeneous reactions, primarily ionic polymerization reactions,
in which the active endgroup is freely accessible in a matrix
of solvent molecules. The solvent matrix is not necessarily
passive, and indeed except for free radical polymerization
reactions (for which specific solvent effects on the stereochemistry of polymerization reactions has not yet been
convincingly demonstrated) and for polymerization reactions
of free carbanions (but probably not for free carbenium
ions), solvation can play a very important role in the
stereochemistry of propagation.

    It is generally assumed that the stereosequence distribu-
tions, or tacticity, obtained in polymerization reactions of
vinyl and related monomers is primarily a kinetically-controlled
process (1).   That is, as in other important aspects of chain-
growth polymerization reactions (e.g., molecular weight and
copolymer composition), tacticity is determined by the relative
rates of competitive reactions, in this case the rates of
isotatic, k_I , and syndiotactic, k_S , additions, as shown
in Equation 1.

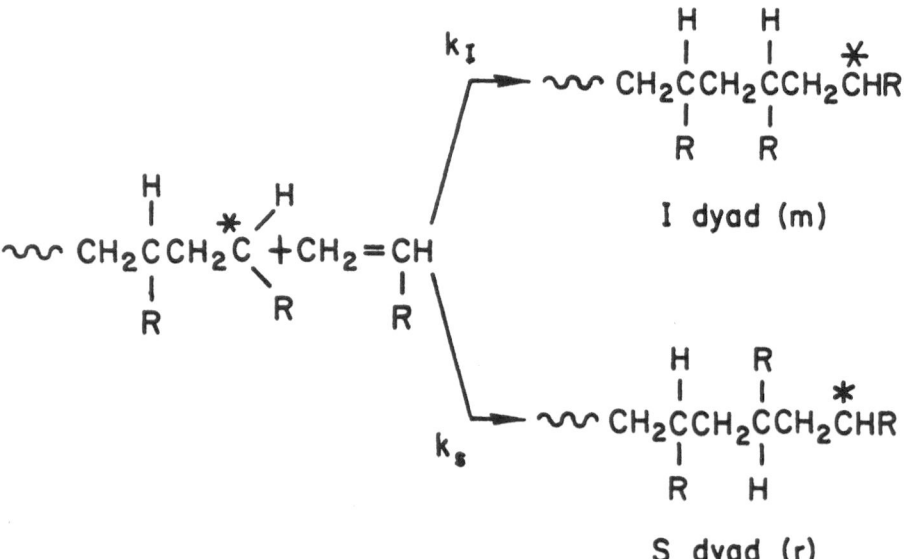

Equation 1

Note that in these equations the designations I or m (for meso) are used synonamously to designate isotactic dyads, and S or r (for racemic) for syndiotactic dyads. Accordingly to the m-r designation, an isotactic triad would be an mm sequence, a syndiotactic triad an rr sequence and a heterotactic triad an mr (or rm) sequence (2).

The differnces in activation free energies which account for tacticity differences through kinetically-controlled processes are illustrated schematically on a reaction coordinate diagram in Figure 1, in which the activation energy for isotactic placement is shown to be larger by the amount $\Delta\Delta G^*$ for a given penultimate unit configuration (3). Determination of tacticity as a function of polymerization temperature, therefore, permits calculation of the differences in enthalpies, $\Delta\Delta H^*$ and entropies, $\Delta\Delta S^*$, of activation by the Arrhenius equation.

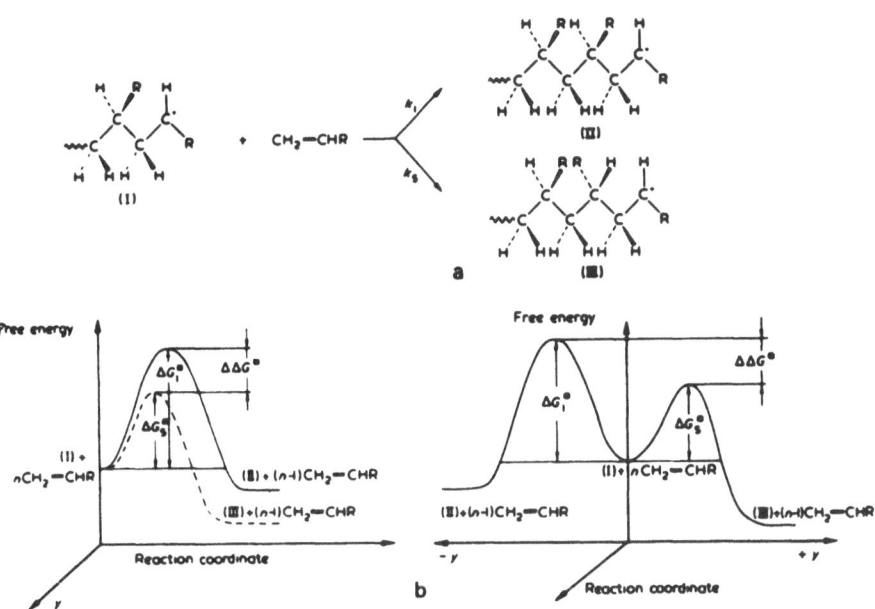

Figure 1.   Reaction coordinate diagram for the addition of a vinyl monomer to a free radical group (taken from Reference 3).

Free Radical Polymerization Reactions

In Table 1 are the tacticity data for the polymerization of methyl methacrylate with free radical initiators at various temperatures (4). From the dyad ratios, which are assumed to be a direct measure of the ratio of the rate constants k_I and k_S, the values of $\Delta\Delta H^*$ and $\Delta\Delta S^*$ were calculated to be 1 kcal/mole and 1 eu/mole, respectively, in favor of the formation of r dyads. In the case of methly methacrylate polymerization, this degree of variation in tacticity can make a substantial difference in the physical properties of the polymers obtained, particularly in the glass temperatures as shown by the data in Table 2 (4,5). Indeed, the highly syndiotactic polymer of Table 1, which was prepared at -78°C was capable of crystallizing.

Table 1. Effect of Temperature on Tacticity in the Free Radical Polymerization of Methly Methacrylate (4)

Polymerization Temp., °C	Syndiotactic Dyad Content (r) from NMR	Calculated Rate Constant Ratio: k_S/k_I
- 78	0.88	7.4
0	0.79	3.8
50	0.77	3.3
100	0.73	2.7

Table 2. Physical Properties of Stereoregular Poly-(methyl methacrylate) (4)

Principal Configuration	Isotactic Dyad Fraction, m	T_g, °C	T_m, °C
Syndiotactic	0.0*	160*	---
Syndiotactic	0.18	126	---
Syndiotactic	0.25	114	200
Isotactic	0.72	62	---
Isotactic	0.81	54	---
Isotactic	1.0	48	160

*Extrapolated value

The mechanism for this type of steroregulation is most likely based on the steric and polar interactions between the substituents on the penultimate unit quaternary carbon atom and those on the sp^2 carbon radical at the active end of the growing polymer chain. That is, the differences in activation energies can be attributed to the steric or polar repulsions between the R groups on the last two units, as shown in Equation 2, as the active endgroup changes from a planar sp^2 to a tetrahedral sp^3 configuration in the process of forming the new carbon-carbon bond (6).

As can be seen from this equation, where the last two units are in a trans conformation, formation of an r unit would have the lesser activation free energy because the two R groups on the terminal and penultimate units are the farthest apart when these units are in an anti position. The geometry of this placement can also be shown by the use of the Neumann projection for the last two units in either a trans (left) or gauche (right) conformation, as indicated in Equation 3.

Tacticity data for the free radical polymerizations of a number of mono and disubstituted ethylenes are collected in Table 3 (3). As seen from this data, the predominant triad compositions in most cases are either the heterotactic or syndiotactic triads, but, in general, the r dyad placements are the predominant mode of addition.

A similar analysis can be made for the stereochemistry of the free radical polymerization of conjugated deines, which can form at least three different types of repeating units as shown in Equation 4 for the simplest case, 1,3-butadiene.

Again, the relative rates of the three competing reacitons which can generate either a 1,2- (ignoring tacticity), a trans-1,4, or a cis-1,4 unit will determine the structure of the polymer, and again, the differences in activiation free energies of these reactions should determine these relative rates (7).

In Table 4 is the data for polymer composition versus reaction temperature in the free radical polymerization of 1,3-butadiene (8A). An Arrhenius plot for this data is given in Figure 2 (8B), which shows that a reaction temperature of approximately $105^{\circ}K$ would be needed to form a polymer containing 95% trans-1,4 units, which are the units formed with the lowest activation energy presumably for steric reasons.

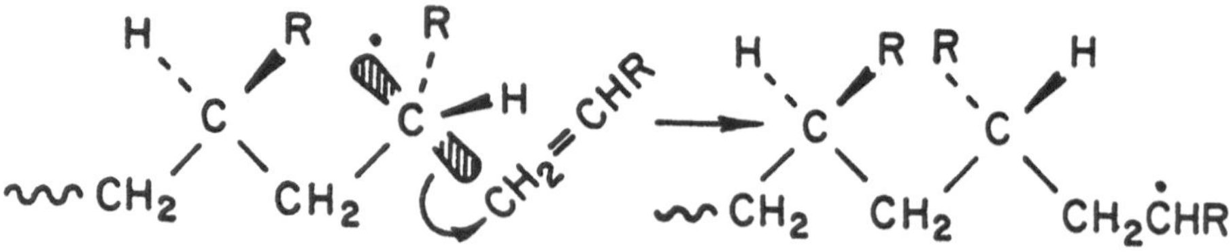

r

Equation 2

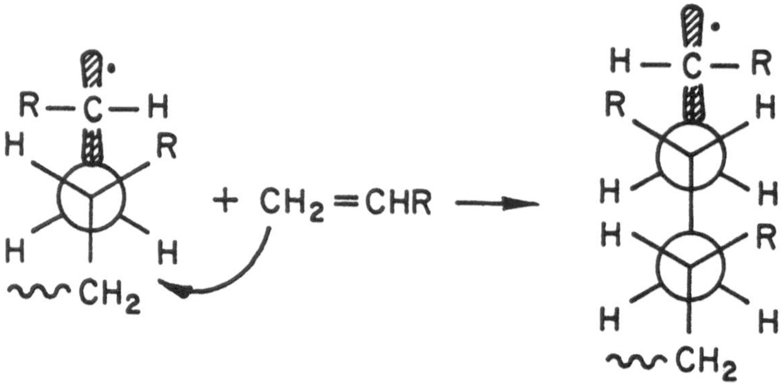

trans - anti

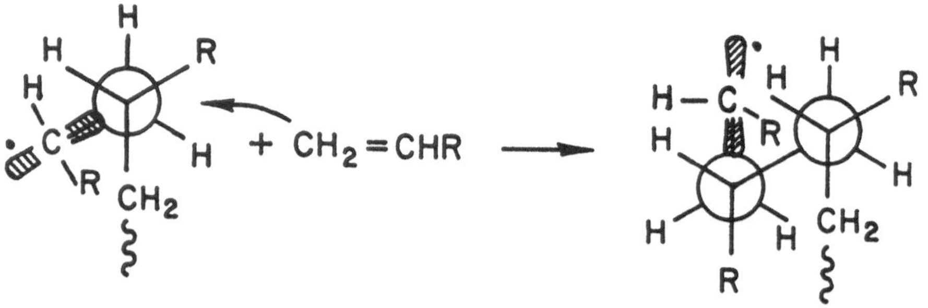

gauche - anti

Equation 3

Table 3.   Polymer Tacticity in Free Radical
           Polymerizaiton Reactions

| Monomer | Temp, °C | Triad Tacticity | | | r Dyads |
		I	H	S	
Monosubstituted Elhylenes					
styrene	100	0.06	0.40	0.55	0.74
vinyl chloride	25	0.21	0.45	0.34	0.56
	40	0.19	0.49	0.32	0.57
vinyl formate	30	0.30	0.40	0.30	0.50
	40	0.33	0.39	0.32	0.47
vinyl acetate	30	0.29	0.39	0.32	0.58
4-vinylpyridine	60	0.06	0.45	0.49	0.71
Disubstituted Ethylenes					
α-methylstyrene	100	0.06	0.40	0.55	0.74
benzyl methacrylate	60	0.07	0.37	0.56	0.74
triphenylmethyl methacrylate	60	0.64	0.22	0.14	0.25
iso-propenyl acrylate	60	0.20	0.47	0.33	0.56
methacrylonitrile	80	0.20	0.50	0.30	0.55
methyl α-chloro- acrylate	70	0.12	0.35	0.53	0.74

Equation 4

Table 4. Effect of Temperature on Repeating Unit
          Composition in the Free Radical Poly-
          merization of 1,3-Butadiene (8)

Polymerization Temp., °C	Repeating Unit Mole fractions 1,2-	trans-1,4	cis-1,4	Polymer Melting Temp., °C
-20	----	0.78-0.84	----	37
-10	0.16	0.78	0.07	33
0	0.18	0.73	0.09	23
25	0.19	0.68	0.13	23
50	0.21	0.64	0.15	0
75	0.20	0.56	0.24	--
100	0.20	0.55	0.25	--

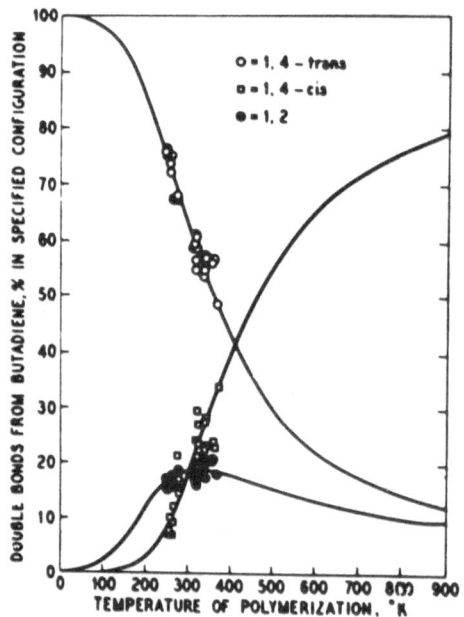

Figure 2. Arrhenius plot for types of units formed as a
function reaction temperature in the free radical polymerization of 1,3-butadiene; top curve is for 1,4-<u>trans</u>, middle curve
for 1,4-<u>cis</u>, and lower curve for 1,2-units (taken from
Reference 8A).

Ionic Polymerization Reactions

The direct relationships between activation free energies and tacticity observed for free radical polymerization reactions are seldom encountered in ionic polymerization reactions In these systems, temperature not only affects the propagation rate constant, but it also controls the ion pair equilibrium of the active endgroups. That is, in most ionic polymerization reactions, the active endgroup is in the form of some type of ion pair, and each type of ion pair will most likely have its own ratio of rate constants, k_I to k_S which controls tacticity. As a result, two or more different kinds of active endgroups can exist in dynamic equilibrium with each other and grow concurrently, as indicated below: (9)

$$
\begin{array}{ccc}
 & K_{cs} & K_{sf} \\
P_n^*Y \rightleftharpoons & P_n^* |S|Y \rightleftharpoons & P_n^* |S+S|Y
\end{array}
$$

contact   solvated   free ion
ion pair   ion pair

$$
+ M \downarrow (k_I/k_S)_c \quad + M \downarrow (k_I/k_S)_s \quad + M \downarrow (k_I/k_S)_f
$$

$$
P_{n+1}^*Y \rightleftharpoons P_{n+1}^* |S|Y \rightleftharpoons P_{n+1}^* |S+S|Y
$$

where P_n^* , Y, S and M represent active polymer chain endgroup, counterion, solvent and monomer, respectively; and c , s and f are for contact ion pair, solvated ion pair and free ion, respectivley.

This type of behavior greatly complicates the problem of analyzing relationships between relative reaction rates (k_I/k_S) and tacticities in ionic polymeriztaion reactions since a given polymer molecule can grow with different tacticities at different stages during its lifetime depending on the type of end group which exists at any particular stage. These competitive equilibria, ion pair interchange rates and stereochemical propagation rates are not only affected by temperature but also by monomer structure, solvent, and counterion. Hence anionic and cationic polymerization reactions are much more complicated, but also much more versatile, than free radical reations. Data for polymer tacticities obtained in anionic and cationic polymerization reactions of vinyl-type monomers as a function of polymerization conditions are collected in Tables 5 and 6, respectivley (3).

Table 5. Polymer Tacticity in Homogeneous Anionic Polymerization Reactions

Monomer	Polymerization Conditions Temp., °C	Solvent	Counterion	Probable Mech.*	Triad Tacticity I	H	S	r Dyads
styrene	-40	toluene	Li	c	isotactic			-
	30	THF	Li	c-s	atactic			0.65
	-20	THF	Na	c-s	syndiotactic			-
vinyl phenyl ketone	-25	toluene	Zn	c	isotactic			0.15
α-methyl-styrene	4	cyclohexane	Li	c	0	0.31	0.69	0.84
	-78	THF	Na	c-s	0.10	0.50	0.40	0.65
p-isopropyl-α-methylstrene	-25	DCE	Li-TMEDA	s	0.13	0.24	0.63	0.75
methyl methacrylate	-70	toluene	Li	c	0.67	0.21	0.13	0.23
	-78	THF	Li	c-s	0.31	0.32	0.37	0.53
	0	THF	Li	s	0.06	0.38	0.56	0.75
	-60	pyridine	Li	s	0.07	0.33	0.60	0.76
	-60	EtNH2	Li	s-f	0.02	0.28	0.70	0.86
	10	EtNH2	K	c-s	0.10	0.41	0.49	0.70

*Probable mechanism according to type of ion pair likely to be present under the conditions of the reaction: c-contact ion pair; s-solvated ion pair; f-free ions.

Table 6   Polymer Tacticity in Homogeneous Cationic Polymerization Reactions

Monomer	Polymerization Conditions			Probable Mech.*	Triad Tacticity			r Dyads
	Temp., °C	Solvent	Initator		I	H	S	
styrene	-78	toluene	BF_3-OEt_2	c-s	atactic			0.60
methyl vinyl ether	-78	toluene	BF_3-OEt_2	c-s	0.65	0.27	0.08	0.21
	-78	tol/hexane	BF_3-OEt_2	c	1	0	0	1
	-78	tol/$CHCl_3$	BF_3-OEt_2	c-s	0.41	0.37	0.22	0.40
t-butyl vinyl ether	-78	CH_2Cl_2	BF_3-OEt_2	c-s	0.19	0.50	0.31	0.66
α-methyl-styrene	-78	CH_2Cl_2	$SnCl_4$	s-f	0.01	0.07	0.92	0.96
	-78	CH_2Cl_2	$TiCl_4$	s-f	0.01	0.13	0.86	0.93
p-methoxy-α-methylstrene	-78	CH_2Cl_2	$TiCl_4$	s-f	0.00	0.08	0.92	0.96
p-fluoro-α-methylstrene	-78	CH_2Cl_2	$TiCl_4$	s	0.02	0.25	0.73	0.86

* See footnote Table 4.

A generalized mechanism has been proposed to rationalize the effect of ion pair structure on the stereochemistry of addition (6). The principal difference between this mechanism and that suggested above for free radical polymerization reactions is that the reactive carbon atom at the end of the growing chain may or may not be shielded from the approach of the monomer by the counterion (or in some cases, the counterion may play a positive role by associating or complexing with the incoming monomer).

According to this mechanism as illustrated below, if the active endgroup exists as a contact ion pair, the presence of the counterion forces the monomer to approach the reactive center from the opposite side (backside, bs, attack). The preferred conformation of the penultimate and terminal units is still assumed to be <u>anti</u>, with the R substituents on opposite sides of the planar zig-zag chain (r conformation), and the counterion, Y , is assumed to be as far removed from the penultimate unit as possible. In this case the new dyad formed would now be isotactic with a central m methylene group as shown in Equation 5.

Alternatively, a solvated ion pair or a free ion in which the counterion is too far away to interfere with the approaching monomer, would as in a free radical propagation reaction, prefer a frontside, fs, attack for the conformation shown, so that the syndiotactic or r dyad would again be formed. Hence, the four possibilities for direction of approach of the monomer relative to the placement of the counterion (bs or fs) in combination with endgroup conformations (<u>anti</u> or <u>syn</u>) are listed below with the dyad tacticities which would be formed from each combination:

Conformation		Approach		Stereochemistry
syn	+	bs	⟶	m dyad
syn	+	fs	⟶	r dyad
anti	+	bs	⟶	r dyad
anti	+	fs	⟶	m dyad

The <u>syn</u> conformations, in which the R groups in both the penultimate and ultimate units are on the same side of the plane in a zig-zag conformation, may not exist as such, but instead this apparent conformation may exist as the result of the formation of a reversible ring structure between the active engroup and a preceding repeating unit, as shown in Equation 6.

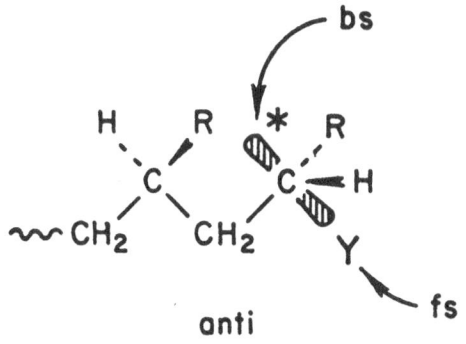

anti

Equation 5

Equation 6

This type of cyclic structure has been proposed to account for the formation of an isotactic polymer in the anionic polymerization of methyl methacrylate, (10) under conditions favoring the existence of tight ion pair endgroups, according to the reaction sequence in Equation 7.

According to this mechanism, formation of the cyclic endgroup structure, with chelation of the lithium cation by the ring and coordination of the chelated cation to the incoming monomer, forces the formation of an isotactic dyad. A similar type of active endgroup cyclic structure has also been proposed to account for the formation of the isotactic polymers in the anionic polymerization of 2-vinly pyridine (11) and in the cationic polymerization of vinyl ethers.(12)

The mode of incorporation of a monomer into the polymer chain, as to whether the addition reaction to the double bond is ultimately either cis or trans, is also of interest in describing the complete mechanism of the propagation reaction as illustrated in Equation 8.

The stereochemistry of this reaction generally would have little effect on polymer tacticity in polymerization reactions involving solvated ion pairs or free ions because either the active endgroup would be free to rotate, 1, or the counterion could move to a new location, 2, or both as shown in Equation 9.

However, in reactions involving tight ion pairs, after the addition reaction the counterion would presumably be fixed on one side of the plane of the incoming monomer to essentially form a specific configuration for the new active center, and the mode of addition of the ion pair, therefore, would control the stereochemistry of the reaction as indicated shown in Equation 10.

According to this scheme, two successive backside (bs) attacks by the monomer on the ion pair with concommitant cation transfer would result in a net trans addition, and vice versa for frontside (fs) attack. Tacticity would then be determined by the relative positioning or orientation of the R group at the active center and the R group in the incoming monomer. This orientation is shown to be anti in the equation above, and for this orientation, two successive bs attacks which result in trans addition would lead to an r or syndiotactic placement. Conversely, two successive fs attacks which result in cis addition would generate an m or isotactic placement.

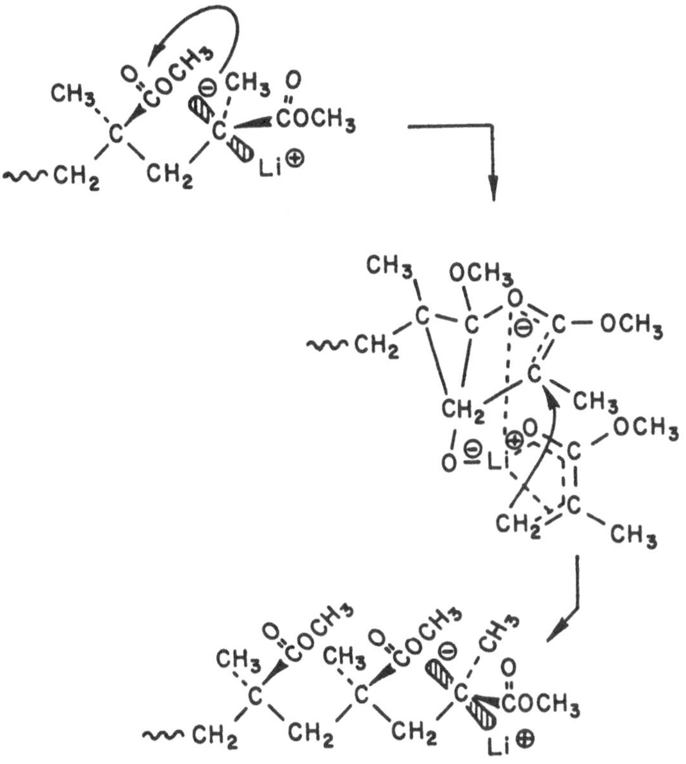

Equation 7

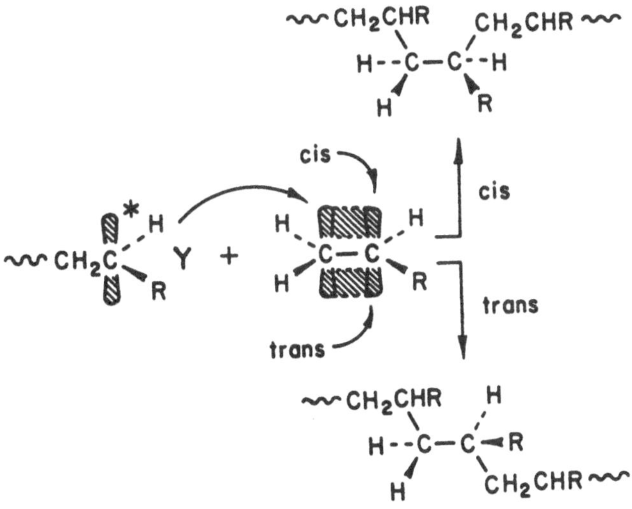

Equation 8

Equation 9

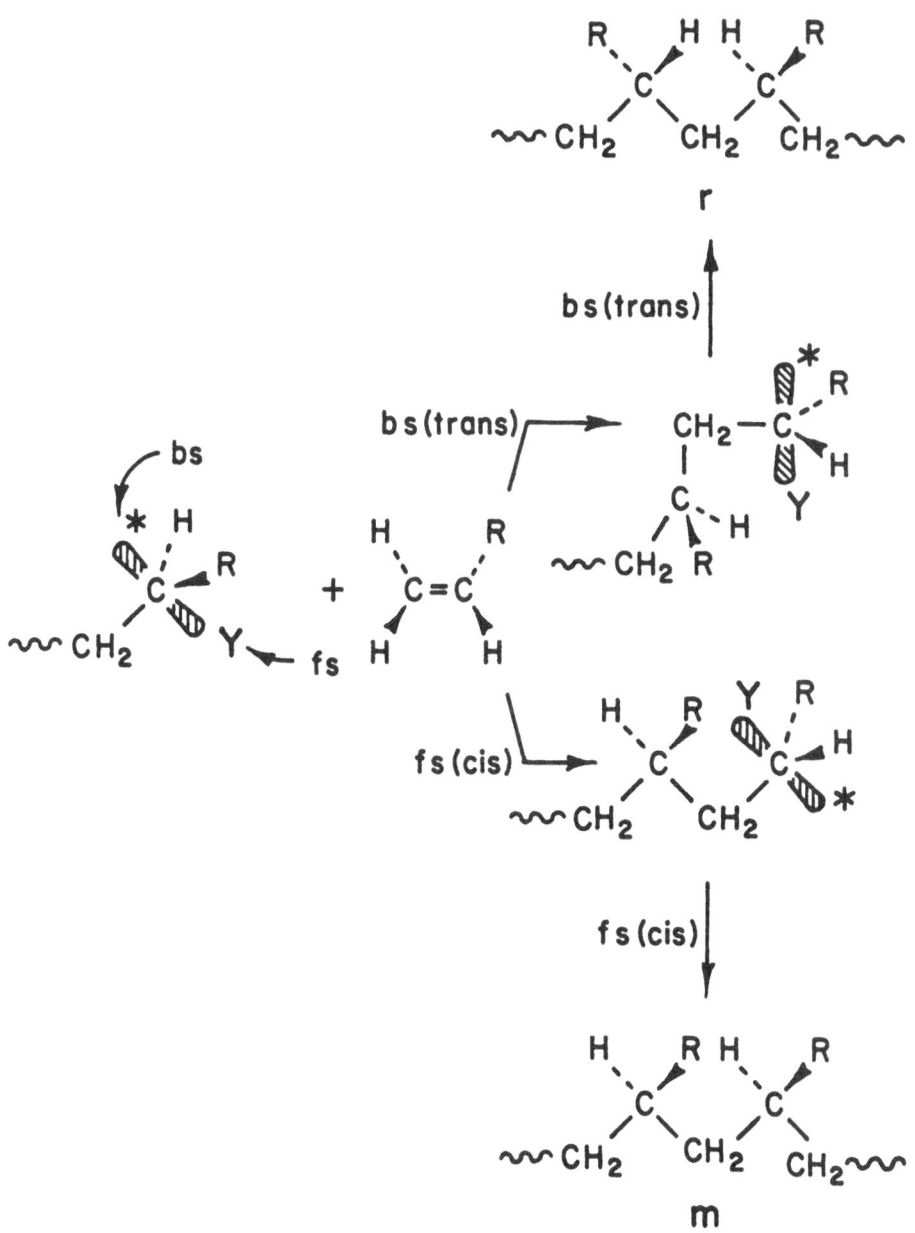

Equation 10

The use of monomers deuterated on the methylene position permits the assignment of these addition mechanisms in stereospecific polymerization reactions from NMR identification of the configuration of the backbone methylene group formed (13). For example, in the isotactic polymerization of cis-deuterated ethyl methacrylate under reaction conditions which form predominantly contact ion pairs, cis addition was demonstrated from the predominantly threo configuration of the m methylene units formed (14), as shown in Equation 11.

The same type of polymerization reaction carried out under conditions where solvated ion pairs were expected yielded random arrangements of methylene configurations in polymers, which were either atactic or largely syndiotactic.

These considerations concerning the mechanism of stereochemical control in ionic polymerization reactions, therefore, can account for the tacticities obtained in most homogeneous anionic and cationic polymerizations as shown in Tables 5 and 6, respectively (3).

Statistical Analysis of Stereochemistry

The determination of polymerization mechanisms, according to the stereochemical control of the propagation reaction, has been greatly aided (or indeed made possible) by high resolution NMR spectroscopy. By this analytical method it is possible to determine the types of configurational sequences which exist along the polymer backbone, and if relative amounts of sequences in three to five successive repeating units can be measured, much can be said about the nature of the active center and its interaction with the incoming monomer as indicated in the discussion above.

Statistical analysis of triad, tetrad and pentad configurations has been used extensively to determine the probable configuration of the active center. If the carbon atom of the active endgroup is in either a planar sp^2 configuration (as in the case of free radical or carbenium ion polymerization reactions) or a rapidly inverting sp^3 configuration (as is likely the case in an anionic polymerization reaction involving free ions), the endgroup has no specific configuration, and only two possible rate constants or probabilities need be considered: (1) the relative rate or probability of forming an isotactic or m dyad, P_m , and (2) the relative rate or probability of forming a syndiotactic or r dyad, P_r, which would be equal to $1-P_m$. Hence the probability of generating an isotactic triad would be $P_m \cdot P_m$ or P_m^2 and, similarly,

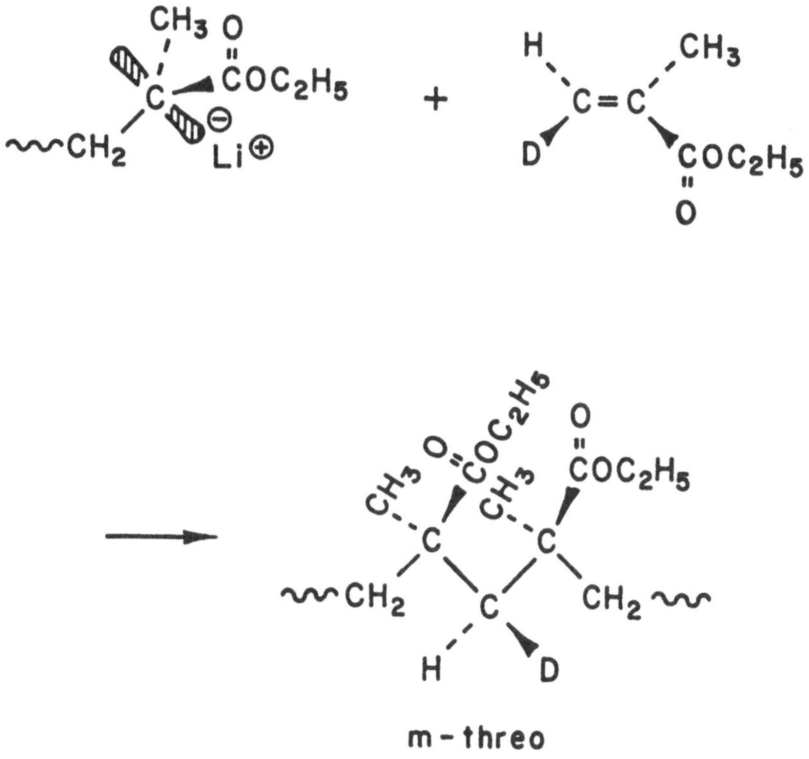
CH3 O
COC2H5
C
CH2
Li⊕
H
CH3
C=C
D
COC2H5
O
CH3
O
COC2H5
CH3
O
COC2H5
C
C
CH2
C
CH2
H
D
m-threo

Equation 11

the probabilities of forming heterotactic and syndiotactic triads would be $P_m(1-P_m)$ and $(1-P_m)^2$, respectively.(15) These statistical relationships are refereed to as Bernoullian, and if the triad composition, as determined by NMR analysis, corresponds to such a distribution, then the nature of the active center as indicated above is established.

If, however, the active center has a fixed sp^3 configuration, as would be expected in a contact ion pair or in heterogeneous catalysis, then four different reactions are possible.with four different relative rate constants or probabilities, as shown in Equation 12 (15).

In fact, there are really only two independent probabilities to consider because $P_{m/m} + P_{m/r} = 1$, $P_{r/r} + P_{r/m} = 1$, and such a statistical distribution is referred to as a first-order Markov sequence.

To prove the existence of a Bernoullian sequence, only triad information is needed, but tetrad information is required to verify a first-order Markov sequence. More complicated stereochemical mechanisms are possible of course (such as reactions controlled by penultimate configurations), and these would have to be fitted by more complex statistical analyses requiring knowledge of more than two probabilities. However, most detailed analyses to date on the tacticities of polymers obtained in homogeneous free radical or ionic polymerization reactions have been found to conform to either Bernoullian or first-order Markovian statistics.

Conclusions

Stereochemical control of polymerization reactions has resulted in, and continues to offer great promise for, the synthesis of new materials with highly desireable properties. A knowledge of the underlying mechanisms of such stereochemical control is important in determining the important factors which lead to the synthesis of stereoregular polymers and in predicting future directions of this field. Perhaps more so than in any other branch of organic chemistry, the opportunity for creative control of product structure offers a tremendous aesthetic and intellectual challenge to the research worker in polymer chemistry.

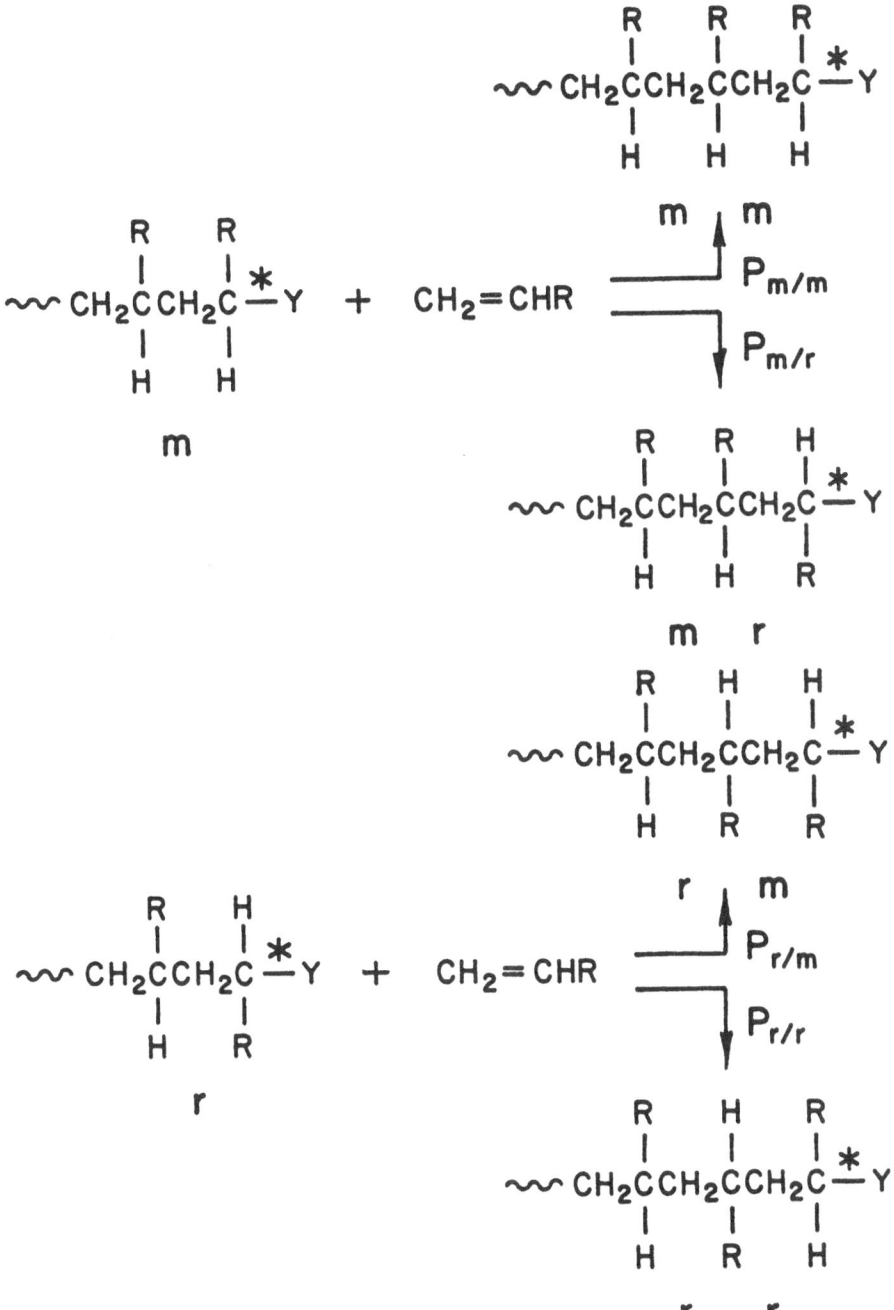

Equation 12

Literature Cited

1. R. W. Lenz, "Organic Chemistry of Synthetic High Polymers", Interscience Publishers, New York, 1967; p. 348 ff.
2. F. A. Bovey, "High Resolution NMR of Macromolecules", Academic Press, New York, 1972; p. 68 ff.
3. P. Pino and V. W. Suter, Polymer, $\underline{17}$, 977 (1976).
4. E. V. Thompson, J. Polymer Sci.: Part A-2, $\underline{4}$, 199 (1966); E. W. Goode, F. W. Owens, R. P. Fellman, W. H. Snyder and J. E. Moore, J. Polymer Sci., $\underline{46}$, 317 (1960).
5. G. Dever, F. E. Karasz, W. J. MacKnight and R. W. Lenz, J. Polymer Sci., Polymer Chem. Ed., $\underline{13}$, 2151 (1975); F. A. Karasz and W. J. MacKnight, Macromolecules, $\underline{1}$, 537 (1968).
6. T. Kunitake and C. Aso, J. Polymer Sci., Part A-1, $\underline{8}$, 665 (1970).
7. R. W. Lenz, Reference 1, p. 345 ff.
8. (A) L. Mandelkern, M. Tryon and F. A. Quinn, J. Polymer Sci., $\underline{19}$, 77 (1956); (B) F. E. Condon, J. Polymer Sci., $\underline{11}$, 139 (1953).
9. R. W. Lenz, Reference 1, p. 499 ff.
10. W. Fowells, C. Schuerch, F. A. Bovey and F. P. Hood, J. Am. Chem. Soc., $\underline{89}$, 1396 (1967); T. J. Leitereg and D. J. Cram, J. Am. Chem. Soc., $\underline{90}$, 4019 (1968).
11. C. F. Tien and T. E. Hogen - Esch, Polymer Letters, $\underline{16}$, 297 (1978).
12. D. J. Cram and K. R. Kopecky, J. Am. Chem. Soc., $\underline{81}$, 2748 (1959).
13. F. A. Bovey, Reference 2, p. 171 ff.
14. W. Fowells, C. Schuerch, F. A. Bovey and F. P. Hood, J. Am. Chem. Soc., $\underline{89}$, 1396 (1967).
15. F. A. Bovey, Reference 2; p. 72 ff. p. 151 ff.

COEXISTENCE OF ACTIVE SPECIES IN ANIONIC POLYMERIZATION OF ETHYL METHACRYLATE

Koichi Hatada, Tatsuki Kitayama, Hikaru Sugino,
Mitsuru Furomoto, and Heimei Yuki
Department of Chemistry, Faculty of Engineering Science
Osaka University, Toyonaka, Osaka, Japan

INTRODUCTION

It has been suggested that the broad molecular weight distribution of poly(methyl methacrylate) prepared in toluene with
anionic initiator could be explained in terms of a combination of
two narrower distributions produced simultaneously occurring
mechanisms(1). Cottam et al.(2) suggested the existence of two
active species uncomplexed and complexed with lithium methoxide,
respectively, in the polymerization of MMA with BuLi. Recently
it was reported that in the polymerization of MMA in a mixture
of toluene and THF with Grignard reagent the polymers obtained
consist of two components, methanol-soluble and methanol-insoluble polymers, which are considered to form at two distinct and
independent active centers(3). In the polymerization of methyl
α-ethylacrylate by BuLi in toluene there exist two types of
active species which produce isotactic and syndiotactic polymers,
respectively(4).

In our previous paper we reported that two kinds of active
species exist in the anionic polymerizations of ethyl methacry-
late(EMA)(5). In the present work the poly(ethyl methacrylate)s
obtained were tested by gel permeation chromatography(GPC) in
order to check the validity of the hypothesis of multiple active
species.

EXPERIMENTAL

The EMA was obtained from a commercial source and purified
by fractional distillation under nitrogen pressure. The monomer

R. W. Lenz and F. Ciardelli (eds.), Preparation and Properties of Stereoregular Polymers, 185-190.
Copyright © 1979 by D. Reidel Publishing Company.

thus purified was distilled over calcium hydride under high vacuum.

Polymerization was carried out in a glass ampoule under dry nitrogen. The reaction was stopped by adding a small amount of methanol, and the mixture was poured into a large amount of methanol. After standing overnight, the precipitated polymer was collected by filtration, washed with methanol and dried in vacuo at room temperature. The combined filtrate was evaporated under reduced nitrogen pressure. The residue was dissolved in benzene and a small amount of insoluble material was removed by filtration. The fraction soluble in methanol was recovered by freeze-drying from the benzene solution.

The triad tacticities of poly(ethyl methacrylate)s were measured by peak eliminated Fourier transform NMR method(6) in toluene-d_8 at 100°C. The ^{1}H-NMR spectra were taken on a JNM-FX-100(JEOL) spectrometer at 100MHz.

GPC was performed by using a JASCO model FLC-A10 with Shodex GPC column A-80M using tetrahydrofuran as a solvent.

RESULTS AND DISCUSSION

The polymerizations of EMA were carried out in toluene with BuLi as initiator at various temperatures. The results are shown in Table 1. The polymer obtained at -78°C consists of a methanol-soluble and a methanol-insoluble fraction, which were proved to be isotactic and syndiotactic, respectively, by NMR spectroscopy. The gel permeation chromatogram of the isotactic fraction showed the high and low molecular weight peaks as shown in Figure 1 and the isotacticity of high molecular weight part (I 89, H 6, S 5%) was found to be higher than that of low molecular weight one (I 74, H 14, S 12%). The syndiotactic fraction was unimodal in

Table 1   Polymerization of EMA in Toluene by BuLi at Various Temperatures for 24hr[a]

Temp. (°C)	Insoluble in MeOH				Soluble in MeOH			
	Yield (%)	Tacticity(%)			Yield (%)	Tacticity(%)		
		I	H	S		I	H	S
-78	18	19	33	48	81	80	14	6
-40	6	44	25	31	88	79	14	7
-20	0	—	—	—	94	81	13	6
0	0	—	—	—	94	78	17	5
-78[b]	2	75	14	11	98	77	16	7

[a] EMA 10mmol, toluene 10ml, BuLi 0.3mmol.
[b] Initiated with 1,1-diphenylhexyllithium.

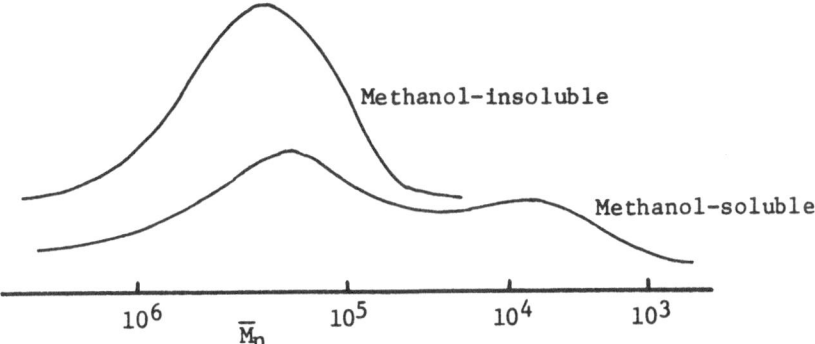

Figure 1   Gel permeation chromatogram of poly(ethyl methacrylate) prepared in toluene with BuLi at -78°C for 24hr.

molecular weight distribution. The results indicate the coexistence of isotactic, syndiotactic and oligomeric active species in
butyllithium polymerization of EMA in toluene at -78°C.

Time dependence of the polymerization of EMA in toluene with
BuLi at -78°C is given in Table 2. The yield and the molecular
weight of isotactic and oligomeric fractions were estimated from
the bimodal chromatogram of the methanol-soluble fraction. With
an increase in the polymerization time the yield and the molecular weight of isotactic and syndiotactic polymers increased and
the molecular weight distribution for both fractions became
gradually wide. The oligomeric product appears to form rapidly
at the early stage of the polymerization and the $\bar{M}_w/\bar{M}_n$ ratio
stays relatively constant in the course of the polymerization.
The numbers of polymer molecules formed were calculated and
listed in Table 3. The numbers for isotactic and syndiotactic
polymers gradually increased during the polymerization, while the
number of the oligomeric fraction remained almost constant or
slightly decreased.

With increasing the polymerization temperature the amount of
the syndiotactic fraction decreased and only an isotactic polymer

Table 2   Polymerization of EMA in Toluene with BuLi
for Various Polymerization Times[a]

Time (hr)	Syndiotactic			Isotactic			Oligomeric		
	Yield (%)	$\bar{M}_n \times 10^{-3}$	$\bar{M}_w/\bar{M}_n$	Yield (%)	$\bar{M}_n \times 10^{-3}$	$\bar{M}_w/\bar{M}_n$	Yield (%)	$\bar{M}_n \times 10^{-3}$	$\bar{M}_w/\bar{M}_n$
0.17	0.1	—		0.7	46.7	1.39	7.6	2.45	2.44
1	1.8	143.2	1.23	3.6	49.4	1.34	20.5	2.57	2.29
4	8.5	179.2	1.66	24.8	63.4	2.33	24.8	3.34	2.24
24	18.0	201.5	1.86	51.8	113.0	2.38	29.2	4.13	2.79

[a] EMA 10mmol, toluene 10ml, BuLi 0.3mmol.

Table 3   Number of Polymer Molecules Formed During the
Polymerization in Toluene with BuLi at -78°C[a]

Polymn Time (hr)	Syndiotactic		Isotactic		Oligomeric	
	Yield (%)	N^b (mmol)	Yield (%)	N^b (mmol)	Yield (%)	N^b (mmol)
0.17	0.1	——	0.7	0.0002	7.6	0.035
1	1.8	0.0001	3.6	0.0008	20.5	0.091
4	8.5	0.0005	24.8	0.0045	24.8	0.085
24	18.0	0.0010	51.8	0.0052	29.2	0.081

[a] EMA 10mmol, toluene 10ml, BuLi 0.3mmol.
[b] Number of polymer chains formed during the polymerization.

was obtained above -20°C (Table 1). The gel permeation chromatogram of the methanol-soluble fraction revealed that the amount
of the oligomeric product also decreased with an increase in the
polymerization temperature. The effect of the time on the polymerization at -20°C is shown in Table 4. The reaction was very
rapid at this temperature and almost finished in 90 seconds.
The polymer obtained was methanol-soluble and isotactic, and did
not include syndiotactic fraction. The polymer formed for the
first 15 seconds was low in the molecular weight and broad in the
molecular weight distribution. With increasing the polymerization time the molecular weight increased and the distribution
became narrower. The result is contrary to that in the polymerization at -78°C.

In the polymerization of methyl methacrylate(MMA) with BuLi
the initiator disappears almost instantaneously on mixing the
reactants(2,7) and reacts first with many more carbonyl groups
to produce lithium methoxide than with the monomer vinyl double
bond at lower temperature(8). When 1,1-diphenylhexyllithium was

Table 4   Polymerization of EMA in Toluene with BuLi
at -20°C for Various Polymerization Times[a]

Time (sec)	Soluble in MeOH[b]					
	Yield (%)	Tacticity(%) I H S			$\overline{M}_n \times 10^{-3}$	$\overline{M}_w/\overline{M}_n$	N^c (mmol)
15	19	70	19	11	3.45	3.32	0.063
30	39	73	18	9	10.29	3.04	0.043
90	90	76	18	6	25.98	1.84	0.039
210	92	76	17	7	32.93	1.70	0.032

[a] EMA 10mmol, toluene 10ml, BuLi 0.3mmol.
[b] No methanol-insoluble fraction was obtained in
the polymerization at -20°C.
[c] Number of polymer chains formed during the polymerization.

used as initiator, the amount of lithium methoxide formed was much smaller than with BuLi. The isotacticity was lower for butyllithium-initiated polymer than for diphenylhexyllithium-initiated one. From these results it was suggested that lithium methoxide in the reaction mixture was related to the formation of syndiotactic blocks in the polymer(8).

When the polymerization of EMA was carried out in toluene with 1,1-diphenylhexyllithium as initiator at -78°C, a large amount of methanol-soluble, isotactic polymer formed with a very small amount of methanol-insoluble fraction. The latter fraction was also isotactic in this case. The results indicate the possibility that in the polymerization of EMA in toluene with BuLi the propagating species for syndiotactic polymer exist in the form of ion pairs complexed with one or more molecules of lithium ethoxide, and the isotactic active species are less connected with the ethoxide or free from it. At higher polymerization temperatures the coordination of ethoxide weakens and the rate of propagation at the chain ends becomes very rapid, which results in a decrease in the formation of methanol-insoluble, syndiotactic fraction. In the polymerization of MMA with alkyllithium the amount of the lithium methoxide formed in the first few seconds decreased with an increase in the polymerization temperature(8). This may be also the case in the polymerization of EMA, which is related with a decrease in the amount of syndiotactic fraction at higher polymerization temperature. The methanol-insoluble fraction prepared at -40°C had rather stereoblock characteristics. The result may be due to the existence of more and less complexed active species with lithium ethoxide which slowly interconvert.

As shown in Table 2 the $\overline{M_w}/\overline{M_n}$ ratios of isotactic and syndiotactic polymers formed in the polymerization in toluene at -78°C with BuLi increased with increasing the polymerization time. In the polymerization of methyl methacrylate in toluene with alkyllithium no termination reaction occurred at the growing chain ends for high molecular weight polymer(8,9), and the initiator disappeared instantaneously on mixing the reactants (2,7). So the broadening molecular weight distribution mentioned above should be due to the transformation of a part of the oligomeric living chains formed at the initial stage into the active species for isotactic or syndiotactic polymer during the polymerization.

The poly(ethyl methacrylate) prepared in toluene with C_6H_5MgBr at -78°C could be also separated into the methanol-soluble and the methanol-insoluble fractions, which were isotactic and syndiotactic, respectively(5). Gel permeation chromatogram of these fractions (Fig. 2) revealed that the molecular weight of syndiotactic fraction was much higher than that of isotactic fraction. The results should be the strong indication

for the coexistence of isotactic-favouring and syndiotactic-
favouring active species in the polymerization with C_6H_5MgBr.
In Figure 2 each peak appears to be bimodal, which may show the
complicated structure of the propagating chain ends.

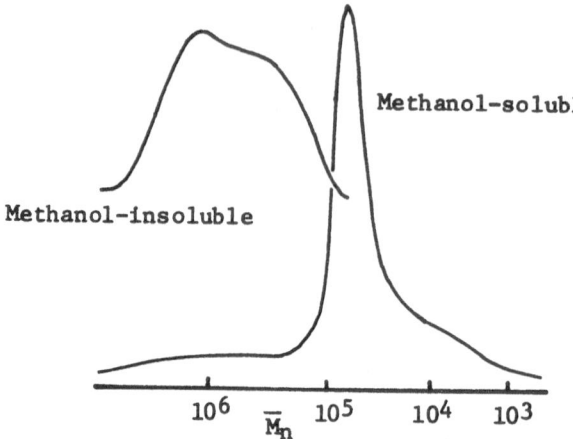

Figure 2   Gel permeation chromatogram of poly(ethyl
    methacrylate) prepared in toluene with C_6H_5MgBr
    at -78°C.
    EMA 10mmol, toluene 10ml, C_6H_5MgBr 0.03mmol,
    Tacticity: methanol-insoluble fraction I 8,
    H 21, S 71%; methanol-soluble fraction I 75,
    H 12, S 13%.

REFERENCES

1) T. J. R. Weakley, R. J. P. Williams, J. D. Wilson, J. Chem.
   Soc., 3963 (1960)
2) B. J. Cottam, D. M. Wiles, S. Bywater, Can. J. Chem., 41,
   1905 (1963)
3) B. O. Bateup, P. E. M. Allen, European Polymer J., 13, 761
   (1977)
4) K. Hatada, S. Kokan, T. Niinomi, K. Miyaji, H. Yuki, J. Polym.
   Sci., Polym. Chem. Ed., 13, 2117 (1975)
5) K. Hatada, Y. Umemura, H. Yuki, Preprint of the IUPAC symposium on macromolecules in Dublin, July, 1977, p 63.
6) K. Hatada, K. Ohta, Y. Okamoto, T. Kitayama, Y. Umemura,
   H. Yuki, J. Polym. Sci., Polym. Lett. Ed., 14, 531 (1976)
7) D. L. Glusker, E. Stiles, B. Yoncoskie, J. Polym. Sci., 49,
   297 (1961)
8) D. M. Wiles, S. Bywater, Trans. Faraday Soc., 61, 150 (1965)
9) K. Hatada, T. Kitayama, K. Fujikawa, K. Ohta, H. Yuki,
   Polymer Bulletin, 1, (1978) in press.

ION PAIR EFFECTS ON THE ANIONIC OLIGOMERIZATION STEREOCHEMISTRY OF VINYL PYRIDINES AND RELATED MONOMERS

W.L. Jenkins, R.A. Smith, C.F. Tien, and T.E. Hogen-Esch

Center for Macromolecular Science, Department of Chemistry, University of Florida, Gainesville, Florida 32611 U.S.A.

INTRODUCTION

Reaction (1) carried out by slow in vacuo distillation of 2-vinylpyridine onto Li or Na salts [1] at $-78°$ in THF followed by reaction with CH_3I yields isotactic (or meso) products [7]-[11].[1-3] However, for larger or more extensively coordinated

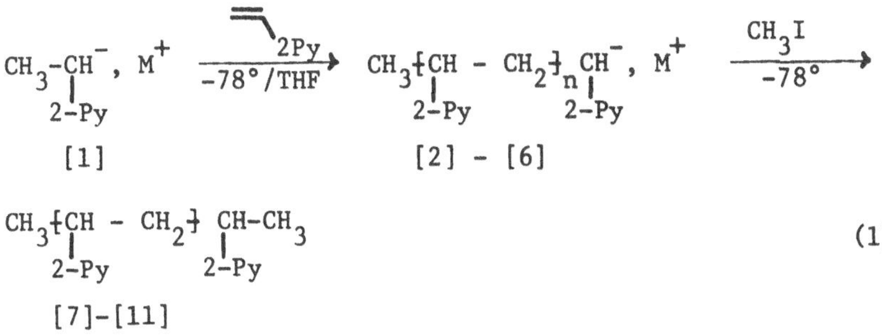

$$CH_3-\overset{-}{\underset{\underset{2-Py}{|}}{CH}}, M^+ \xrightarrow[-78°/THF]{2Py} CH_3 \underset{\underset{2-Py}{|}}{\overset{}{\{}CH} - CH_2 \}_n \overset{-}{\underset{\underset{2-Py}{|}}{CH}}, M^+ \xrightarrow[-78°]{CH_3I}$$

[1]   [2] - [6]

$$CH_3 \underset{\underset{2-Py}{|}}{\{CH} - CH_2 \} \underset{\underset{2-Py}{|}}{CH-CH_3} \tag{1}$$

[7]-[11]

[2], [7] n=1; [3], [8] n=2; [4], [9] n=3; [5], [10] n=4-5; [6], [11] n≈10-12

cations the formation of "dimers" [7] is much less stereoselective. Epimerization shows the various stereoisomers of [7], [8], and [9] to be of similar stability so that the stereochemistry is kinetically determined. These results may be explained by considering diastereomeric ion pairs [12] and [13] in their respective chelated forms [12a] and [13a]. Beta carbon stereochemistry of oligomerization of $E-d_1$-2-vinylpyridine [14] and $Z-d_2$-2-vinylpyridine [15] was shown to be related to the cation

191

R. W. Lenz and F. Ciardelli (eds.), Preparation and Properties of Stereoregular Polymers, 191-194.
Copyright © 1979 by D. Reidel Publishing Company.

and solvent dependent equilibrium (2), (Table 1).

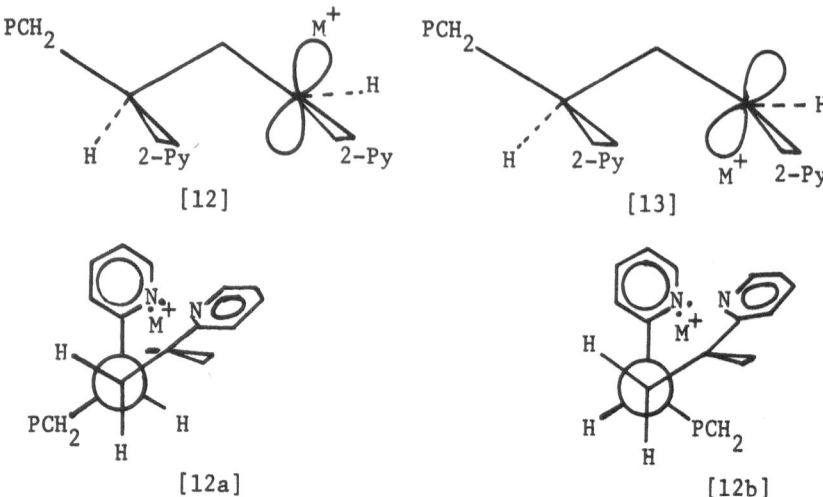

TABLE 1:   EFFECT OF CATION AND SOLVENT ON EQUILIBRIUM (2) AND ON
           BETA CARBON OLIGOMERIZATION STEREOCHEMISTRY OF [14]-[16]

Cation/ Anion	Solvent	% (E)	Monomer/ Product	Selectivity %	Addition Stereochemistry
Li/[1]	THF	96	[14]/[7]	91	trans
Li/[2]	THF	--	[14]/[8]	67	trans
Li/[3]	THF	--	[14]/[9]	90	trans
Li/[1]	THF	96	[16]^c/b	51	--
Li/[1]	THF^a	38	--	--	--
K/[1]	THF	80	[15]/b	75	--

^{a}One equivalent of cryptand (2,2,1) added. ^{b}Deuterated product.
^{c}Z-d$_1$-4-vinylpyridine.

$$H_3C\diagdown \diagup H \;,\; M^+ \qquad\qquad\qquad H\diagdown \diagup CH_3,\; M^+$$

(E)   (Z)   (2)

The results of Table 1 are consistent with a reaction
mechanism (eq. 3) showing the monomer approach to the (E) isomer

of [1]. This approach results in the threo placement of deuterium whereas a similar reaction of the (Z) isomer would result in the erythro-deuterio compound.

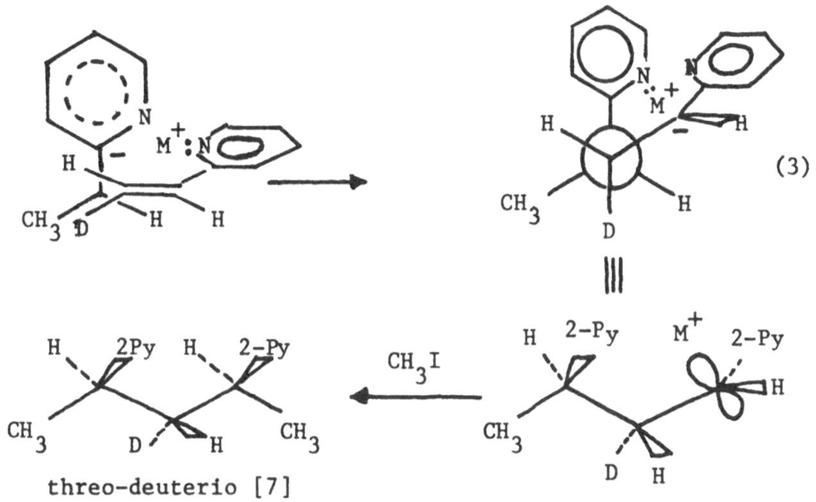

(3)

threo-deuterio [7]

RECENT RESULTS

Although the lower molecular weight products [7]-[9] prepared with Li or Na as counterion in THF are highly isotactic, the corresponding polymers ($M \sim 10^4$) give a rather broad ^{1}H nmr. This is in marked contrast to the proton nmr spectra[4] of P_2VP initiated by Et_2Mg in toluene.[5] It is, however, possible that the broad bands are caused by side products that are sometimes found during preparation of oligomers. In order to explore this question further, a low molecular weight polymer [11] was prepared at -100°C to prevent side products and was capped with ^{13}C enriched CH_3I. The methyl signal in the ^{13}C nmr spectrum indicated that the polymer is highly stereoregular. This is confirmed by the ^{13}C nmr spectrum of the K-tBuO/DMSO treated polymer that shows a complex $^{13}CH_3$ signal indicating extensive racemization of the chain to "atactic" material. The proton spectrum of [11] like that of [10] shows a sharp methyl doublet that is quite different from that of the racemized material. It is thus likely that the broadened methylene and methine regions of the proton spectrum of these polymers are due to side products. Several possibilities will be outlined.

We have also examined the oligomerization of alpha-methyl-2-vinylpyridine according to equation (4). Oligomers [18] are crystalline products with sharp melting points and the tetramer

$$(CH_3)_2 \underset{2-Py}{\overset{-}{C}}, \ M^+ \quad \xrightarrow[-78°/THF]{\overset{CH_3}{\underset{2-Py}{\diagup}}} \quad \xrightarrow{CH_3I} \quad CH_3 \underset{2-Py}{\overset{CH_3}{-}} \underset{n}{\overset{|}{C}} - CH_2 \underset{}{\overset{}{\}}_n - \underset{2-Py}{\overset{CH_3}{\underset{|}{C}}} - CH_3 \quad (4)$$

[17]   [18]

(n=3) appears to be a meso product. We are currently in the process of isolating and characterizing the higher oligomers. Interestingly, there is no evidence for the formation of side products as found for 2-vinylpyridine.

REFERENCES

1. C.F. Tien and T.E. Hogen-Esch, Macromolecules, 9, 871 (1976).

2. C.F. Tien and T.E. Hogen-Esch, J. Am. Chem. Soc., 98, 7109 (1976).

3. C.F. Tien and T.E. Hogen-Esch, Polymer Letters, 16, 297 (1978).

4. K. Matsuzaki and T. Sugimoto, J. Polym. Sci., A-2, 5, 1320 (1967); K. Matsuzaki, T. Matsubara, and T. Kanai, J. Polym. Sci. (Chem. Ed.), 15, 1573 (1977) and references therein.

5. G. Natta, G. Mazzanti, P. Longi, G.D. Asta, and F. Bernardi, J. Polym. Sci., A-1, 51, 487 (1961).

AUTO- AND INDUCED STEREOREGULATION OF POLYISOCYANIDES

Frank Millich

Polymer Chemistry Division, Department of Chemistry
University of Missouri-Kansas City, K.C., Mo.,
U.S.A.   64110

Polyisocyanides, $\left\{ \substack{C \\ \| \\ N-R} \right\}_n$, represent a class of polymers with unique primary structure that sets them apart from conventional polymers. Their polymeric properties in several ways are unique in kind or degree, and their exploration has already, and further promises, to broaden our experiences and understanding of macro-molecular chemistry and physics. These are the subjects of some recent reviews (1-5). Poly(α-phenylethyl isocyanide), α-PPEI, is the progenitor member of the polymer class, being the first and most characterized of the few currently known soluble poly-isocyanides of high molecular weight. It assumes a rigid-rod con-formation, and it yields liquid crystals in solution. Most poly-isocyanides are strongly aggregated solids and are insoluble in solvents other than strong acids, wherein they are protonated and show time-dependent viscosity changes. This phenomenon has led to a new viscosity concept of isoviscosity and isohydrodynamic volume (6), which has recently received some independent support (7,8). Polyisocyanides form racemic mixtures of complete stereo-regular helices, and give evidence of dipole coupling of the vicinal imine groups. Hence, polyisocyanides are proposed as models of fixed dimensional coordinates for future investigation of optical, electrical and solution thermodynamic properties.

Millich and Sinclair (9) reasoned that polyisocyanides natu-rally form cylindrical helices of a special type of asymmetry. The syntheses of oligo(vicinal imines) is currently one subject of my research (10). The onset of helicity may be expected even among the early members of the homologous series of oligo(vicinal imines) and also their oxygen analogs, the oligo(vicinal ketones) (11). Calvin and Wood (12), considering the conformation of pentan-2,3-4-trione (Fig. 1), concluded that a planar all s-cis

R. W. Lenz and F. Ciardelli (eds.), Preparation and Properties of Stereoregular Polymers, 195-199.

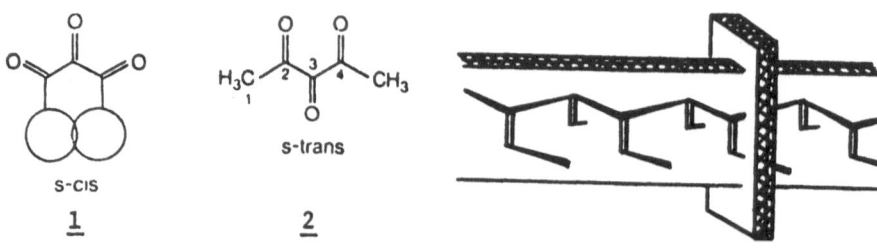

Fig. 1.   Planar conformations   Fig. 2.   Sketch of extended
of pentan-2,3-4-trione (12).   chiral polyisocyanides (16).

conformation of the triketone, 1, is impossible owing to steric
repulsion between the terminal methyl groups; and, a planar
s-trans conformation, 2, (preferred for aliphatic α-diketones because it allows maximal overlap of π-orbitals and minimizes unfavorable interactions between adjacent dipoles) was also rejec-
ted on the bases of uv spectra and of unfavorable electrostatic
interactions between the oxygens of C-2 and C-4. In oligo(vicinal
imines) the dipole and electrostatic repulsions must be less, but
the steric repulsions of N-substituents are very prominent.

    α-PPEI thus forms a cylindrical 4/1 helix of a tight 4.1 Å
axial length/turn, and 15 Å cylinder diameter (2). The dimensions
of the helix are consistently supported by Debye-Scherrer X-ray
data (9,13), electron micrography, small angle X-ray scattering
and solution thermodynamics (14), and circular dichroism (15).

    In addition to helical chirality, polyimines embody another
element of asymmetry owing to the syn-anti isomeric geometry of
imines, for which the term "herringbone polymer" has been applied
to polyisocyanides by S.M. Aharoni. Millich and Baker (16) first
published the following conclusions:
    "Although syn-anti geometric isomerism about the nitrogen is
    possible, in principle, in poly(α-phenylethyl isocyanide),
    only one of these forms is practically possible in a low
    energy form of the sterically compressed helix..." "There-
    fore, in the polymer model (of Fig. 2) one can expect a
    constructively additive specific rotational contribution
    from each mer,...and this structure represents an exception
    to the generalization of Frisch, et al. (17)."
    "The possible resolution of racemic polymer product into
    optically active polymeric antipodes by means of optically
    active resolving substrates...has not yet been attempted."
Efforts to synthesize an efficient resolving chromatographic
column were begun by Millich and Yau, but first success fell to
Drenth and coworkers (18), who conceived of using an optically
active poly(sec-butyl isocyanide) as the chromatographic column
packing to resolve the racemic polymer mixture of poly(tert-butyl
isocyanide).

These first two features of asymmetry in polyisocyanides--
i.e., the helical screw direction and the declivity of the twill
of pendant groups--are formed autogenically during polymerization
and during any post-polymerization "self-annealing" process. A
third element of asymmetry is introduced if the N-substituents
contain centers of asymmetry. The polymerization of a racemic
α-PEI by the heterophasic sulfuric acid catalyst system (19) led
to an optically inactive product (16); and the polymerization of
each enantiomer separately led to optically active polymers of
high molar refractivity (M_D^{27} 500 deg. cm^2/g), which is tenfold
greater than that of the monomer (16). The specific rotations of
both the levo- and dextrorotatory polymers were comparable, yet
the former showed slightly greater values of specific rotation,
molecular weight and yields (16), suggesting an easier polymerizability. There is the conceptual possibility that the polymer
propagation procedes stereoselectively for steric reasons.

Thermal racemization of the levo- and dextro-rotatory α-PPEI
polymers in refluxing toluene (i.e. 111°C) has been carried out
(16). The results clearly show at least one thermally sensitive
mode of racemization for each polymer and at least one thermally
resistance chiral structural feature. Also, the levo- and
dextro-rotatory polymers are distinguished by showing different
quantitative change by the thermal treatment.

Drenth and coworkers (20) have proposed a directive mechanism for helix formation specific to the polymerization of isocyanide monomers by their nickel (II) coordination catalysts.
They further specify rules governing a catalyst-induced stereoregulation for the incorporation of chiral isocyanide monomers
into the polymers during the chain-propagation process (21). The
controlling features are a consequence of substituent relative
size, and, where it occurs, substituent coordination to nickel
(II). This catalyst-induced stereoregulation probably does not
apply to the sulfuric acid polymerization of isocyanides.

Optical activity in polyisocyanides has been shown to arise
in yet another way. The nature of the supramolecular structure
in the solid and liquid crystalline states of the synthetic homopolymers of PPEI have been characterized, and the latter correlated to molecular ordering in concentrated solutions and solid
films (10). Oriented films gave well-defined electron diffraction patterns from selected areas, and the reflections were
indeed in terms of a pseudo-hexagonal triclinic unit cell with
a=b=14.92 A, c=10.33 A and α=93.4°, β=90.5° and γ=118.2°. Kiamco
and Hellmuth found evidence of liquid crystalline properties of
concentrated solutions of α-PPEI (13) which may serve to explain
the anomalously large, concentration-dependent, freezing point
depressions produced by this polyisocyanide in benzene (4). Kiamco
summarizes the optical properties as follows: a highly birefrin-

gent liquid crystalline phase, present only in concentrations
greater than 35% by weight and in solid films, is optically negative and uniaxial in character, whereas dilute solutions of the
same do not show any optical activity. The solid films are also
circularly dichroic for some wavelength region where one component of the circularly polarized light is transmitted without
attenuation. The character of α-PPEI is said to be derived from
the cholesteric liquid crystalline supramolecular arrangement of
PPEI helices. Further support of the existence of a helical
supramolecular structure is given by the presence of large spheru-
lites of up to 5 cm. in diameter.

The cholesteric texture of α-PPEI scatters white light and
produces dark lines. Consequently, the optical rotation has its
origins primarily in the supramolecular helix and the contribution of the asymmetric molecules are negligible since isotropic
solutions (1 mm thick or less) appear black under crossed polarized light and do not show any circular dichroic or optical
rotatory dispersion spectra. The α-PPEI samples studied were
polymerized from racemic monomer, and may be considered to be
racemic mixtures of oppositely wound helices.

The rotational spectra at room temperature of racemic α-PPEI
($M_n=10^5$) films prepared from several solvents show large CD ellip-
ticities whose sign depends on the solvent used (13). A Moffitt-
Yang plot for the film from benzene over the wavelength range
380-520 nm gives a linear plot of the ORD data (using 290 nm as
the λ_o) with a slope, b_o, equal to -670 deg-cm^2/mole (13), as do
other polymers of helical conformation. The sign and magnitude
of b_o are similar to those determined for poly(γ-benzyl L-glutamate) (22) and for poly(γ-dodecyl L-glutamate (23). The CD-spectra
of these racemic α-PPEI films differ from that obtained with
poly(D-α-phenylethyl isocyanide) in dilute solution (24), which
shows only one adsorption band. Similarly, the CD-spectrum of
(-)poly(tert-butyl isocyanide) in cyclohexane (1.875$\frac{hmole}{1}$) gives
sign of only one absorption band over the region around 280 nm
(15) for which only the n-π* transition is responsible.

The optical activity of these liquid crystals is not particularly associated with the presence of a chiral carbon atom in
the N-substituents of α-PPEI, since preliminary polarizing-micro-
scope studies on fractionated α-PPEI, a polyisocyanide of more
flexible conformation in solution and lacking a chiral carbon
atom, also indicate the spontaneous reversible formation of a
highly birefringent mesophase in concentrated solutions (13).
These observations suggest that small differences in conformations due to concentration changes give rise to the occurrence
of a phase separation and molecular ordering.

One may say in summary that of the four origins of asymmetry
in polyisocyanides discussed above, only one is predicated to be
induced, i.e., the nickel(II)-catalyzed (but not the sulfuric
acid catalyzed) polymerization of chiral monomers. The other
three are reasoned to be autogenic. One may expect that polyiso-

cyanides, wherein the N-substituents are of modest or larger size and which have not undergone rearrangements, spontaneously yield non-mesoid asymmetric products which are at the very least racemic modifications, and, even in this case also, may be optically active as a consequence of their supramolecular organization.

(Abridged text)

REFERENCES

1. F. Millich, "Polyisocyanides", in Macromolecular Reviews, A. Peterlin, Ed. (in press).
2. F. Millich, "Rigid Rods and the Characterization of Poly-isocyanides", Adv. in Polymer Sci., 19, 117 (1975).
3. F. Millich, "The Polymerization of Isocyanides", Chemical Reviews, 72, 101 (1972).
4. F. Millich, "Polyisonitriles", in Encyclopedia of Polymer Science and Technology, Suppl Vol. 15, H.F. Mark, N.G. Gaylord and N.M. Bikales, Ed., Wiley-Interscience, N.Y. (1971), p. 395-410.
5. See also, W. Drenth and R.J.M. Nolte, "Poly(iminomethylenes): Rigid Rod Helical Polymers", Accounts of Chemical Research, (in press).
6. F. Millich, E.W. Hellmuth and S.Y. Huang, J. Polymer Sci., 19, 117 (1975).
7. S.M. Aharoni, J. Appl. Polym. Sci., 21, 1323 (1977).
8. S.M. Aharoni, Polymer Letters, 12, 549 (1974).
9. F. Millich and R.G. Sinclair, Polym. Sci., C, 22, 33 (1968).
10. F. Millich and W.H. Sohn, "Schiff's Base and Other Syntheses of Oligoisocyanides", Polymer Preprints, Amer. Chem. Soc., Div. Polym. Chem., 19(2), 97 (1978).
11. M.B. Rubin, "The Chemistry of Vicinal Polyketones", Chem. Revs., 75, 177 (1975).
12. M. Calvin and C.L. Wood, J. Amer. Soc., 62, 3152 (1940).
13. E.A. Kiamco, Ph.D. Thesis, U.M.K.C., K.C., Mo. (1975).
14. S.Y. Huang, Ph.D. Thesis, U.M.K.C., K.C., Mo. (1973).
15. A.J.M. van Beijnen, R.J.M. Nolte, W. Drenth and A.M.F. Hezemans, Tetrahedron, 32, 2017 (1976).
16. F. Millich and G.K. Baker, Macromolecules, 2, 122 (1969)
17. H.L. Frisch, C.Schuerch and M.Szwarc, J.Polym.Sci., 11, 559 (1953).
18. R.J.M. Nolte, A.J.M. van Beijnen and W. Drenth, J. Amer. Chem. Soc., 96, 5932 (1974).
19. F. Millich, R.G. Sinclair and G.K. Baker in "Macromolecular Syntheses",Vol.4,W.J. Bailey, Ed., Wiley, N.Y., 1972, p.19.
20. R.J.M. Nolte, J.W. Zwikker, J. Reedijk and W. Drenth, J. Mol. Catal., 4, 423 (1978).
21. A.J.M. van Beijnen, R.J.M. Nolte, J.W. Zwikker and W. Drenth, J. Mol. Catal., 4, 427 (1978).
22. W. Moffitt and J.T. Yang, Proc. Nat. Acad. Sci. (US), 42, 596 (1956).
23. J. Smith and R. Woody, Biopolymers, 12, 2657 (1973).
24. F. Millich, unpublished data.

PRESENT UNDERSTANDING OF THE STEREOSPECIFIC POLYME-
RIZATION OF CYCLIC MONOMERS : THREE-MEMBER RINGS.

Nicolas Spassky, Ardéchir Momtaz and Maurice Sépulchre

Laboratoire de Chimie Macromoléculaire, associé au
CNRS - Université Pierre et Marie Curie
4, Place Jussieu 75230 Paris Cédex 05 (FRANCE)

I - UNDERSTANDING

I - INTRODUCTION

The polymerization of cyclic three-member rings monomers is performed using ionic type initiators. Anionic and cationic polymerization produce generally random amorphous polymers very often
of low molecular weight. On the contrary highly active coordination type initiators are able to promote ring-opening polymerization leading to high molecular weight stereoregular products.
The interest for such stereospecific polymerization started beginning of 50'S when crystalline polypropylene oxide of h.m.w.
was synthesized. Numerous researches performed and still under
investigation in the field of oxiranes were summarized in several
reviews (1-5). The study of stereospecific polymerization of thiiranes started later, early 60'S, and the main results were also
recently reviewed (6-8). The comparison of different aspects of
polymerization of oxiranes and thiiranes were examined in few pu-
blications (9-12). Practically no work have been succeeded in the
field of coordination type polymerization of aziridines, although
the cationic polymerization was extensively studied.
The aim of this paper is to present the main aspects of the stereospecific polymerization of oxiranes and thiiranes with emphasis on the more recent results obtained in this field. Unpublished data are included and new possible ways are discussed.

The ring-opening of a monosubstituted three-member ring cyclic
monomer could occur in α or β position and leads to different
monomeric configurational units in the macromolecule, as shown
below

R. W. Lenz and F. Ciardelli (eds.), Preparation and Properties of Stereoregular Polymers, 201-223.
Copyright © 1979 by D. Reidel Publishing Company.

syndiotactic dyad

isotactic dyad   irregularity

According to the definition, a stereospecific polymerization is
a polymerization in which a tactic polymer is formed.
Thus, we shall consider such initiators which lead to macromolecules with a great predominance of isotactic enchainments, since
polymers with predominant syndiotactic enchainments are not yet
known in ring-opening mechanism.
Irregularities, i.e. opening in α position, should be avoided,
unless they are systematically introduced by some specific mechanism.
It is also important to notice that in mono-substituted monomers
one of the carbon atoms is asymmetric. Thus, the usual monomer
is an equimolecular mixture of two stereoisomers, generally called (R) and (S) refering to the configuration of the asymmetric
carbon.
As seen, this introduces a further distinction in the macromolecular chain : the isotactic dyads could be of -RR- or -SS- structure, while there is only one type of syndiotactic dyad -RS- (it
would be better according to stereochemistry to call threo a
-RR- type enchainment and meso an -RS- dyad and to reserve the
terminology of iso and syndio to -RRR- and -RSR- type triads
respectively).
It is therefore necessary to distinguish between chains containing both type of stereoisomeric units and those containing only
one of them.
The latter could be produced in a stereoselective polymerization,
a process in which a polymer molecule is formed from a mixture of
stereoisomeric monomer molecules by incorporation of only one
stereoisomeric species.
Such ideal process could be represented in our case by the following scheme :

racemic monomer   selective   purely isotactic chains

(R + S)   process   poly R + poly S

In case of partial selectivity, the macromolecular chain contains
both type of units, i.e. poly $R_n S_m$, with predominance of one of
them.

2 - STEREOSPECIFIC POLYMERIZATION USING ACHIRAL INITIATORS
2-1) Initiator systems

Salts, oxides, organometallic derivatives of divalent and trivalent metals are the most usual stereospecific initiators used
both for oxiranes and thiiranes. Many of them were already described in reviews (1-6). The most studied of them result from
modification reactions of organometallic compounds such as $ZnEt_2$
the hydrolysis of which was described by Furukawa, Tsuruta (1959)
and Colclough (1964), the alcoolysis mainly studied by Tsuruta
(1962), the aminolysis by Tani (1967) and the glycolysis by
Colclough (1965) and by us (1970). Colclough (1963), Vandenberg
(1963) and Saegusa (1965) have extensively studied products of
hydrolysis and modification of AlR_3. $CdEt_2$ and $MgEt_2$ have also
been tried under modified forms as initiators. Bimetallic
μ- oxoalkoxides obtained by a different way were shown to be very efficient initiators by Teyssié and his group (1967).
We shall discuss in that follows some structural, kinetics and
stereospecific aspects of few of these initiators.
Although most of them apply to both oxiranes and thiiranes, one
must mention that cadmium derivatives are good initiators for
thiiranes but unable to polymerize oxiranes.
Lets examine shortly some properties of two very commonly used
systems which are $ZnEt_2$-alcohol and $ZnEt_2$-water.
The reaction between diethylzinc and an alcohol involve two main
steps :

$$ZnEt_2 + ROH \longrightarrow EtZnOR + EtH \nearrow \qquad (1)$$

$$EtZnOR + ROH \longrightarrow Zn(OR)_2 + EtH \nearrow \qquad (2)$$

Alkylalkoxide and dialkoxide species are produced, but generally as result of the reaction one obtains mixed species of the
type :

$$\left[EtZnOR \right]_x \left[Zn(OR)_2 \right]_y$$

The kinetics of the reaction with different classes of alcohols
was studied by Tsuruta (2) using IR spectroscopy. Products with
predominance of dialkoxide species (y $>$ x) are excellent initiators for the polymerization of oxiranes and thiiranes, while
products with predominance of alkylalkoxide species are generally not polymerizing oxiranes but polymerizing thiiranes.
Only few defined crystalline and soluble species like
$(EtZnOMe)_6$. $Zn(OMe)_2$ were isolated and studied (13). The
crystallinity is not favorable to catalytic behaviour. Thus,
crystalline $Zn(OMe)_2$ synthesized from LiOMe and $ZnCl_2$ has no catalytic activity. The latter could appear if one introduces disorder in the crystalline lattice (14).
It was shown that initiators based on long chain alcohols such

as nonanol are less active than those obtained with long poly-
ether chain like $CH_3O(CH_2CH(CH_3)O)_{11}$ H (15).
Finally it must be pointed out that the real efficiency of these
systems is very poor : only 10^{-2} of the introduced initiator is
used.
Another initiator which was extensively used in polymerization
of oxiranes and thiiranes is diethylzinc-water system.
The stereospecificity is depending on the value of components
ratio $ZnEt_2/H_2O$. Both for oxiranes (2) and for thiiranes (16)
the most selective initiators were obtained when 1:1 ratio was
used. For ratio below 1, some cationic character appears with
partial ring-opening in α position.
Different soluble and insoluble species are produced during the
reaction between diethylzinc and water and their respective role
is not yet clearly understood. Very recently Colclough (17) was
able to isolate a soluble specie with $Et_8Zn_{13}O_9$ stoechiometric
composition which was monomeric in tetrahydrofuran and hexameric
in cyclohexane. The conditions of preparation of the initiator
influence its behaviour : for example, when it is prepared in
the presence of monomer it could lead to non stereoregular polymers as established in case of methylthiirane (18). The short
description of these two particular initiator systems illustra-
tes the numerous parameters involved. In the preparation of
stereospecific initiators, most of the latter being still heterogeneous. As seen, the main requirement for obtention of a good
initiator is the producing of M-X-M groupings (X being an heteroatom O,S or N) on which the monomer could be coordinated and
activated before the ring-opening step.

2-2) Mechanistic aspects of the stereospecific polymerization

2-2-1) Coordination step   As pointed earlier polymerizations involving such initiators and leading to stereoregular polymers are called "coordinated". The polymerization is considered to proceed on metallic atoms having vacant coordination sites
on which the monomer "coordinates". There is only a very small
amount of such sites as well in heterogeneous systems as in homogeneous.
In order to obtain a regular ring-opening in β-position and a
stereoselective chain propagation a strict control of the monomer complexation on the site is necessary.
Such a complexation is difficult to demonstrate directly in ca-
talytics conditions.
Several indirect proofs could be established. Tsuruta (19) had
studied the complexation of different donors including methyloxirane on $ZnEt_2$ and $CdEt_2$ by infrared spectroscopy. Complexation of methylthiirane by $ZnEt_2$ and by $CdMe_2$ was investigated
respectively by [1]H (20) and [13]C (21) spectroscopy. The complexa-

tion of different oxiranes and thiiranes on metal porphyrins was
established by visible spectroscopy (22). All these examples involve organometallic compounds which does not initiate in usual
conditions the polymerization of cyclic monomers. However, it is
also known that the addition of donnor compounds such as ethers,
or amines can inhibit or at least reduce the selectivity of stereospecific initiators (23). This can be satisfactorily explained by a competition for the coordination on the active site
between the monomer and the additive.
The coordination is depending on the nature of the monomer and
on the metal of the catalyst and follows generally the hard-soft
acid-base rule of the Pearson's classification.

2-2-2) <u>Ring-opening</u>   The ring-opening reaction of monosubstituted oxiranes and thiiranes, when using stereospecific
initiators, occurs almost exclusively in β position i.e. on
the less substituted carbon and therefore is similar to a typical anionic process. For this reason such polymerizations are
often called "anionic-coordinated" one . In few cases however,
namely with $ZnEt_2/H_2O$ or $AlEt_3/H_2O$ initiator with a ratio of
components higher than 1, α-sission occurs in some amount and
this can be established by NMR spectroscopy, selective degradation by ozonolysis or organometallic compounds. One can use also optically active monomers as it was done for methyloxirane(2)
and for methylthiirane (24). The optical activity of polymers is
lowered due to the inversion of configuration of the asymmetric
carbon as result of α-sission.
The latter property was clearly established by Vandenberg (4) in
his work on cis and transdimethyl-oxiranes polymerizations. He
had demonstrated the complete inversion of configuration during
the ring-opening particularly on the example of the optically
pure (R,R) trans dimethyl oxirane which produces an optically
inactive polymer with -RS- type configurational monomeric
units. We shall see later on the contrary that from cis type monomers it is possible to obtain optically active polymers as a
result of a regioselective attack with chiral initiators.
The results of Vandenberg were fully confirmed by Price (25) and
Tsuruta (26) who have studied deuterated polyoxiranes by IR and
NMR.

2-2-3) <u>Mechanism of the chain growth</u>   There is good proofs that the chain propagation proceeds through insertion of monomer units in the metal-heteroatom bond. The coordination mechanism necessitates the presence of two neighbouring metal
atoms, one of them binding the chain to the initiator, the other
one coordinating the monomer and orienting it in convenient direction. A mechanism called "flip-flap" was proposed by
Vandenberg (4) in the case of alkyl alumina initiators and seems
to be the most satisfactory one which can be applied to most of

the cases (scheme II). In this mechanism the chain is alternatively attached to two neighbouring metal atoms.

The termination or transfer reactions proceed generally on the
alcoxy or hydroxy groups of the polymer or of the initiator.
As most of these polymerizations are run in heterogeneous conditions it was not always easy to study the kinetics of the reaction. There is indications that it proceeds through a "living"
type process (23). Different type of kinetics were proposed involving 1st and 2nd order in monomer and in initiator (27). Most
frequently one finds $R_p = K |M| \ |C|$ as kinetics equation.

2-3) Stereoselectivity of the process

The preparation of stereoregular polymers depends on different
parameters, such as the nature and the structure of initiators
and monomers, which will be now briefly discussed.

2-3-1) Enantiomorphic sites concept   The stereospecific
polymerization of racemic methyloxirane produces generally a polymer with some cristallinity which can be fractionated by selective solubility (using acetone for example) in a crystalline
fraction and an amorphous one (an intermediate "semi-crystalline"
fraction can be also isolated).
The crystalline fraction is purely isotactic, that means that it
is composed of poly R and poly S chains as a mixture or eventually as a stereoblock copolymer. The amorphous fraction is heterotactic.

Tsuruta (2) had proposed that the presence of chains of different tacticities is due to active sites of various stereospecificity. There is probably a full spectra of sites in the initiator with a more or less pronounced R and S character. The
sites with pure R and S character produce isotactic fraction. This hypothesis which satisfactoraly explains most of the
results was substantiated by optical resolution of racemic polymers. Thus, fractions of one sign were isolated from polymethyloxirane by preferential complexation (28) and fractions of both
sign of very low optical activity were obtained using sucrose
(29).
Very recently using the chromatographic technique on optically
active supports elaborated by italian school, Marchetti and
Chiellini in collaboration with us (30) performed the separation
of different samples of stereoregular polymethylthiiranes and
obtained significant separations in two fractions of opposite
sign.

2-3-2) <u>Dependence of selectivity on the physical structure
of the initiator</u>   The structure of initiators we have discussed
in chapter 2-1 was rather complex for the reason that most of
them were heterogeneous. The physical properties in solution of
these initiators are of great interest for the understanding of
stereospecificity and require a systematic study of different
parameters.
Such a study was achieved by Teyssié and coworkers (31) on the
example of bimetallic μ-oxoalkoxides. These initiators were
synthesized purposely as well-defined compounds containing several metal atoms linked by oxygen bridges : a trivalent metal
bearing alkoxidesgroup on which the polymerization occurs and a
divalent metal necessary for the coordination step.
One of the methods of preparation of such compounds consists in
a controlled hydrolysis of a Meerwein's double alkoxide in an
alcoholic medium.

$$
\begin{array}{c}
\underset{\underset{R}{|}}{\overset{\overset{R}{|}}{\text{RO}}} \quad \underset{R}{\overset{R}{|}} \\
\end{array}
\qquad \xrightarrow[\text{ROH} \atop (\text{excess})]{2\ H_2O} \qquad
\tag{3}
$$

The central metal is usually Zn, Co, Fe or Mo. These compounds
of general formula $|(RO)_4 Al_2 O_2 M^{II}|_n$ are associated, the degree
of association depending on the nature of the central atom, on
the nature of R group and on the nature of the solvent.
Typically n vary between 2 and 8, which corresponds to clusters containing 6 to 24 metal atoms.
It was shown that one of the OR group plays a particular role

and that one catalytic unit initiates only one polymeric chain. The polymerization proceeds without dissociation of the aggregates. These initiators have remarkable activity for polymerization of lactones, thiiranes and oxiranes. For example, a half-reaction time of 20 mn is observed for methyloxirane at 30°C in heptane solution when using $Al_2ZnO_2(OBu^n)_4$ initiator. With the later monomer the formation of two main products is observed : a low molecular weight oligomer (DP between 4 and 12) and a high molecular weight polymer with a broad distribution, partially isotactic. The proportion of oligomers could be modified on a very broad range from 3 % to 70 % depending on the nature of the central metal and the degree of association as indicated in table I. It is also clear from this table that the selectivity of the process, i.e. the proportion of isotactic fraction, is depending on the same parameters.

Table 1. Different products obtained in the polymerization of methyloxirane using various bimetallic μ-oxoalkoxides as initiators.

Metals	- R	n (C_6H_6)	Oligomers (%)	h.m.w. fraction cryst.(%)	amorph.(%)	ref.
Zn-Al	nBu	8.0	25	75	25	(32)
	iPr		20	45	55	
Co-Al	nBu	4.1	7 (5)	67 (10)	33 (90)	(32)
	iPr	2.0	4 (3)	36 (22)	64 (78)	
Mn-Al	nBu		70	3	97	(32)

( ) in brackets results obtained when adding alcohols to the system.

However, contrarily to the living type relation $|M|/|C|$ observed in the ε-caprolacton polymerization the number of active centers in the case of methyloxirane represents only a small proportion of the total initiator introduced according to the very high DP observed in the h.m.w. fraction. The stereospecificity could be completely lost when using alcohols which dissociate the catalytic aggregate (table 1). Therefore this study is suggesting that the more or less stereospecific character of enantiomorphic sites depends quite closely of the degree of association but still the active sites are due to a particular coordination or transformation reaction in the aggregate. Another study which must be mentioned concerns the polymerization of methylthiirane using metallic thiolates as initiators (33). This homogeneous

polymerization of a living-type leads to stereoregular products
in certain conditions. Cadmium thiolates are stereospecific at
temperatures below 10°C while zinc derivatives are not. Allyl-
thiolates are the most efficient and the polymerization in bulk
is the most selective as compared to that one in toluene or in
tetrahydrofurane solution. The temperature plays an important
role : in toluene solution using cadmium allylthiolate a lowering
of the temperature from 20°C to 10°C is enough to change the nature of obtained polymer from amorphous to crystalline.
Although the structure of the initiator was not studied in this
case, the presence of associations of polymeric chains in reaction media was established.

2-3-3) Dependence of selectivity on the monomer As already
reported (12) the same initiator proceeds with different stereospecificity depending on the monomer. For example with $ZnEt_2$-
H_2O (1:1) system one finds the following characteristics with
three monomers :

	Me oxirane	Me thiirane	tBu thiirane
tacticity (i %):	61	76	90
relative crys-tallinity (a) :	27	70	95

(a) based on the TAD area of the most crystalline isotactic
sample obtained by stereospecific initiation.

The same difference in behaviour was observed for a serie of oxiranes when using Al-Zn oxoalkoxides (31). It was also shown that
the selectivity increases with increase of enantiomeric purity
of monomer.
A non linear increase of the optical activity of the crystalline
fraction was observed by Tsuruta (2) when polymerizing methyloxirane of increasing optical purity by $ZnEt_2$-MeOH initiator system. Optically pure product was obtained from a monomer which was
only 80 % optically pure. At the same time the optical activity
of residual monomer was not changed and no preferential consumption of one enantiomer observed. According to the explanation
proposed, the cross propagation on sites of one type takes over
the homopropagation, due to increasing concentration of enantiomer of opposite chirality. It seems also possible that the overall spectrum of sites is modified towards an enrichment in more
specific sites.
Never substantial preferential consumption of one enantiomer had
been observed with such achiral stereospecific initiators. This
means, that although it is possible to modify the character and
or the number of sites as a whole, it is not possible to change
the internal enantiomorphic distribution. Complexation with external chiral agents or working in chiral media has no effect in

contrast with what would be observed when chiral agents are used.
The active sites must have therefore a very strong enantiomorphic
character.

3 - STEREOSPECIFIC POLYMERIZATION USING CHIRAL INITIATORS

The optical activity is a very helpful tool in the study of mechanism of polymerization. The first optically active polypropylene oxide was prepared by Price (34) starting from the pure
enantiomer. The preparation of optically active polymers is also
possible when using the racemic monomer and proceeding in a stereocontrolled reaction. The pioneering work in this field was
done by Inoue, Tsuruta and Furukawa (35) who demonstrated the
possibility to obtain optically active polypropylene oxide from
the racemic monomer using diethylzinc-d-borneol initiator.
Most of the chiral initiators used up to now derived from the
reaction between an organometallic compound and an optically active reagent bearing an hydrogen active group. The latter could
be an alcohol, a glycol, an amino-acid etc, the former ZnR_2 or
CdR_2. Organomagnesium compounds were found to be of very poor
activity for such reactions (36). Organoalumina compounds, while
very active, are not leading to optically active compounds (37).
A process in which a single stereoisomer of a mixture is polymerized giving macromolecules containing one type of configurational base units is called "stereoelective polymerization" and
it can be illustrated by the following scheme :

$$
\begin{array}{c}
\text{racemic} \\
\text{monomer} \\
\text{(R=S)}
\end{array}
\xrightarrow[\text{choosing S}]{\text{optically active initiator}}
\begin{array}{c}
\text{polymer} \\
\text{poly S}
\end{array}
+
\begin{array}{c}
\text{unreacted} \\
\text{monomer} \\
\text{S/R} \rightarrow 0
\end{array}
\quad (4)
$$

As result of such a polymerization an optically active polymer
with single stereoisomer units is produced and the unreacted monomer is enriched in antipode of opposite configuration. In ideal
case the reaction should stop at 50 %yield and a pure enantiomer
isolated. In most of the cases the reaction is not as perfect
and only a preferential polymerization of one of the enantiomers
from a mixture is observed. As seen, such a process could be a
potentially interesting method for resolution of racemic monomers (38).
It is clear that in this process the predominant element is the
enantiomorphic choice of the initiator. It is mainly depending
on the nature of the latter, but also on the nature of the monomer.
The stereoelective choice is characterized by the sign i.e. the
configuration of the elected antipode and by the magnitude or
"stereoelectivity" i.e. the optical purity of the resolved monomer.
The main known features in such stereoelective processes were

recently described in few papers (9, 11, 12, 38) and therefore
we shall only briefly emphasize on the most important of them
reporting more extensively some very recent results.

3-1) Dependance of the sign of stereoelective choice on the composition of initiator

Three kinds of chiral components were mainly used in stereoelective type initiators : alcohols, glycols and aminoacids.
In most of the cases the following configurational correlations
were established : the choice of the initiator is called "homosteric" if the preferentially chosen enantiomer has the same absolute configuration as the chiral component of the initiator.
This is the most usual case and it was well established with
1,2 diols and aminoacids for both oxiranes and thiiranes. One
must be careful, in order to establish correlations with other
series of chiral compounds such as alcohols, in the choice of
groups of comparison for a given configuration (12,38).
It was found however, that in some other cases the choice corresponds to opposite configuration, that is to an "antisteric"
type process.
We were able to establish that such a choice was depending only
on the chemical composition of the initiator.
Thus, if one considers the resulting product of alcoolysis reaction described in chapter 2-1, which can be characterized as a
mixed species of the type $|EtMetOR|_x$ $|Met(OR)_2|_y$.
We have demonstrated in the case of few thiiranes, oxiranes and
recently for α,α disubstituted β propiolactones (36) that the
type of the choice (homosteric or antisteric) is depending only
on the x/y ratio i.e. the ratio of alkylalkoxide species over
dialkoxide species.
The following rule was found :

when x/y < 2 the choice is homosteric,
when x/y > 3 the choice is antisteric

This rule is enough general to be applied to initiators resulting from $ZnEt_2$, $CdMe_2$ and $CdEt_2$ as organometallic compounds and
alcohols and 1,2 diols as chiral components. It is interesting
to notice that in the copolymerization of racemic methylthiirane
with achiral monomers such as ethylene sulfide or isobutylene
sulfide, the choice is not modified this strongly substantiating
the coordination interaction between the initiator and the monomer and the lack of existence of chain effect (39).
In some cases both type of species could be isolated under a soluble form. For example, when reacting diethylzinc with (+)3,3
dimethyl-2-butanol, antisteric species with a composition close
to $Et_6Zn_7(OR)_8$ were found (40) which are similar in composition
to those reported for the methanol derivatives (13). The homosteric species corresponded to $Zn(OR)_2.EtZnOR$ composition.
In practice we have prepared the homosteric initiator by reacting at room temperature diethylzinc with (-) 3,3 dimethyl 1,2
butanediol and the antisteric initiator by reacting the latter
diol with dimethylcadmium in the same conditions.

Thus, with one chiral component, it is possible to obtain monomers enriched in enantiomers of opposite configuration depending
on the conditions of preparation of the initiator which are determining the composition of the latter.

3-2) Stereoelectivity of the process

The magnitude of stereoelective choice i.e. the stereoelectivity
could be determined from the optical purity of recovered monomer.
A comparison at a given conversion could be used as a criterion
of efficiency of resolution. As expected, for a given monomer the
stereoelectivity increases with the bulkiness of the substituent
in the chiral component of the initiator e.g. 1,2 diol with t-
butyl group is much more efficient than one with methyl group.
Reciprocally for a given initiator the monomer with the bulkier
group seems to be the best resolved. It appears, however, when
studying the resolution reaction on all the conversion scale,
that monomers differentiates very much in their behaviour according to their nature. The experimental data collected up to day
indicate three types of kinetical equations which define three
clases of monomers.
The ratio of relative rates of enantiomer consumption could take the following expressions :

$$\text{1st class} : \quad \frac{d\,|R|}{d\,|S|} = r_R \frac{|R|}{|S|} \tag{5}$$

$$\text{2nd class} : \quad \frac{d\,|R|}{d\,|S|} = \rho_R \frac{|R|^2}{|S|^2} \tag{6}$$

$$\text{3rd class} : \quad \frac{d\,|R|}{d\,|S|} = \gamma_R \,|R|^2 \tag{7}$$

r_R, ρ_R and γ_R are stereoelectivity ratios related to the preferential choice of the initiator to R enantiomer. Most of experimental data reported here were obtained with the best homosteric type initiator system choosing R enantiomer prepared by reacting diethylzinc with R(-) 3,3 dimethyl 1,2 butanediol (DMBD).
The stereoelectivity ratios were found to be constant throughout
the polymerization, therefore equations (5), (6) and (7) can be
integrated and after introducing the optical purity of residual
monomer α/α_o and the conversion x , they give respectively
the following general equations applied when the starting mixture is racemic :

$$(1 - x)^{r-1} = \frac{1 + \alpha/\alpha_o}{(1 - \alpha/\alpha_o)^r} \tag{8}$$

$$\frac{1}{(1-x)(1-\alpha/\alpha_0)} = \frac{\rho}{(1-x)(1-\alpha/\alpha_0)} + 1 - \rho \qquad (9)$$

$$\frac{1}{(1-x)(1+\alpha/\alpha_0)} = \frac{\gamma}{4}|M_0|^2(1-\alpha/\alpha_0)(1-x) + \frac{\gamma}{4}|M_0|^2 + 1 \qquad (10)$$

The experimental data for corresponding monomers are fitting well on these theoretical curves as seen in Fig. 1.

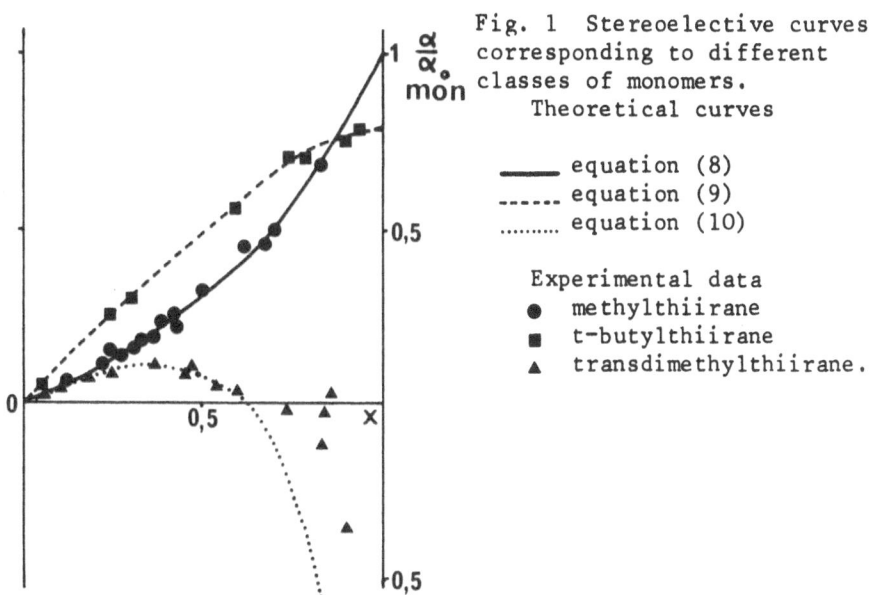

Fig. 1 Stereoelective curves corresponding to different classes of monomers.
Theoretical curves

——— equation (8)
------ equation (9)
.... equation (10)

Experimental data
● methylthiirane
■ t-butylthiirane
▲ transdimethylthiirane.

The monomers who belongs to these three kinetical equations are the following :
1st class monomers : they correspond to equation (8) called previously of first order. This family is the most numerous one and include oxiranes and thiiranes bearing not very bulky substituents. The values of stereoelectivity ratio r_R , obtained with the initiator cited above in polymerizations carried out in bulk at 25°C for thiiranes and 80°C for oxiranes are listed below in table 2.
The stereoelectivity ratio was not modified when running a copolymerization reaction in the presence of achiral monomer e.g. ethylene sulfide. This confirms again the strong catalyst control process involved in these polymerization.

Table 2. Stereoelectivity ratios r_R for various oxi-ranes and thiiranes.

Subst.[*]	Me	Et	C M	M M	H M	TMS	DEA
oxir.	1.8	-	1.1	2.0	1.02	1.5	-
thiir.	2.3	2.1	-	1.6	n.d.	-	2.0
ref.	(38)	(40)	(41)	(39)	(42)	(41)	(41)	(43)

[*]Me : CH_3- Et : C_2H_5- CM : $ClCH_2-$ MM : CH_3OCH_2-
HM : $HOCH_2-$ TMS : $(CH_3)_3SiOCH_2-$ DEA : $(C_2H_5)_2NCH_2$

<u>2nd class monomers</u> : correspond to equation (9), called previous-
ly of second order. Two monomers fitted with this equation : isopropyl and t-butylthiiranes (44). For both of them the stereoelectivity ratio approaches the value $\rho_R = 8$ at 25°C. As seen
from fig. 1 at complete conversion the optical activity of un-
reacted monomer reaches a limit value equal to $\alpha_o \cdot \dfrac{\rho_R - 1}{\rho_R + 1}$.
The main caracteristics of this process is that it is strongly
depending on the temperature. The stereoelectivity ratio ρ_R is
doubled in value when the temperature lowers from 25° to -7°C
and on the contrary a raising in temperature decreases ρ_R and
even inverted the enantiomeric choice ($\rho_R < 1$) as reported below:

Table 3. Dependence of stereoelectivity ratio ρ_R on
the temperature of polymerization.

T°C	-7	-3	20	63	135
ρ_R	16	14	8	2.5	0.9

<u>3rd class monomers</u> : corresponds to equation (10). One represen-
tative of this class, the transdimethyl-thiirane, was recently
studied by one of us (45). The originality in behaviour of this
system consists in enrichment at the beginning with one enantiomer (R), then change of the sign of unreacted monomer at given
conversion and enrichment in the opposite antipod. As seen in
Fig. 1, experimental data are fitting well with equation (10) up
to 60 % of conversion, but after scatter occurs. It seems possible that another type of kinetics is superimposed. The stereoelectivity is sensitive to temperature effects. The main features
of these processes are presently under the study.

3-3) <u>Stereoselectivity of the process</u>

The study of the stereoregularity of polymers gives informations
on the stereoselectivity of the process. The crystallinity of polymers is depending on the initiator, but for a given initiator

system, for example, the one cited above, it depends on the nature of the monomer. Monomers with bulky groups such as t-butyl
thiirane produce purely isotactic crystalline products. It must
be mentioned that poly t butylthiirane obtained by stereoelec
tive polymerization can be separated in two fractions one of them
identified as pure poly R polymer, the other one as the optically inactive racemate (poly R + poly S) (44). In this case the
stereoelective polymerization could be considered as an original
method for preparation optically pure polymers from racemic monomers.
Polymethyloxirane samples obtained with the same catalyst can be
fractionated by the usual method in crystalline and amorphous
parts. It was found (46) that the stereoelectivity ratio corresponding to crystalline part and amorphous part are quiet different, respectively equal to r = 2.6 and r = 1.6, this supporting again the concept of a large spectra of enantiomorphic sites of different stereospecificities present in the initiator.

3-4) Ring opening reaction

The polymers prepared by stereoelective method with usual initiators were studied by NMR, ORD and degradation methods and it
was demonstrated that the ring-opening occured at least at 95 %
in β position. The situation is completely changed if one use
chiral initiators derived from disubstituted 1,2 diols of the

type $H - \overset{R}{\underset{OH}{C^*}} - \overset{R}{\underset{OH}{C^*}} - H$ where R can be methyl or phenyl group.

These compounds contain two asymmetric carbons of the same configuration in neighbouring position and the results of polymerization with such initiators show an unusual situation in the ring
-opening. The sign of optical activity of obtained polymers is
opposite to that found in ordinary cases and according to ORD
and CD studies one must consider that ring-opening occured at
50 % in α position with inversion of configuration of the asymmetric atom.
The stereoelective choice of monomers is however in agreement
with configuration rules as seen from the sign of optical activity of unreacted monomer. Homosteric and antisteric processes are
observed with considerable amount of α-scission (39). According
to the chemical composition of these initiators a cationic character of the latter seems to be excluded and therefore this particular behaviour could be due to some steric reasons which are
not yet completely understood. New studies are now in progress.

4 - NEW WAYS IN STEREOSPECIFIC POLYMERIZATION

Three main directions seem presently to be of interest in this
field :

- Findings of new highly active initiators with well defined
  structure to be used in homogeneous processes.
- Improvement of the best initiators known by using them in more
  favorable conditions, for example in chiral media.
- Use of new routes for preparation of stereoregular polymers
  such as asymmetric polymer synthesis.

Tentative studies in the first direction were already discussed
in chapter 2-3-2), thus we shall now describe the two other ways.

4-1) Polymerization in chiral media

In the case of monomers which obey to the first order consumption law, it is possible to increase stereoelectivity by using
chirally modified stereoelective initiators. The optically active modiying agents can be classified into two groups : those
which directly participate in the polymerization reaction, like
the monomer itself, enriched in one of its enantiomers or
bearing another chiral center in the side chain, and those which
do not participate to the reaction but are able to interact more
or less with metal atoms of the initiator like alkylsulfides,
ethers, amines and terpenes. These external agents can be used
in small amounts as additives or in large amounts as chiral media. We have observed, using our standard homosteric initiator,
that whatever the prevailing configuration of the monomer or of
the additive chiral agent may be, the configurational type of
the choice of the initiator is not changed and remains the same
as that observed with racemic monomer or in the absence of chiral agents. Thus, $ZnEt_2$/R(-) 3,3 dimethyl 1,2-butanediol system
will choose preferentially R(+) isomer of methylthiirane even if
the starting monomer is strongly enriched in S enantiomer. The
magnitude of this choice, however, is, in all the cases, significantly enhanced.

With enantiomerically enriched monomers, the stereoelectivity
ratio r_R , calculated from an equation similar to that proposed
for stereoelective process of racemic monomer, is constant during all the course of the polymerization. This result is con-
sistant with the assumption that the active sites are formed in
the presence of the monomer in an irreversible way at the first
beginning of the reaction. For methyl- and ethyl-thiiranes, the
stereoelectivity ratio r_R is a linear function of the initial
composition $\frac{S_0}{R_0}$ or $\frac{R_0}{S_0}$ of the monomer (Fig. 2) and almost equal

to : $r_{rac} \cdot \frac{S_0}{R_0}$ for an S enriched monomer and to : $r_{rac} \cdot \frac{R_0}{S_0}$

for an R enriched one ; r_{rac} being the stereoelectivity ratio observed with racemic monomer. In the case of methyloxirane
the use of optically active monomer led to a decrease of the
stereoelection (47). However, very recently, a stereoelectivity
ratio as high as 4 has been observed for an oxirane having a second asymmetric carbon atom in the side chain, namely the

(2RS,3S) 3-methyl 1,2-epoxypentane (48).

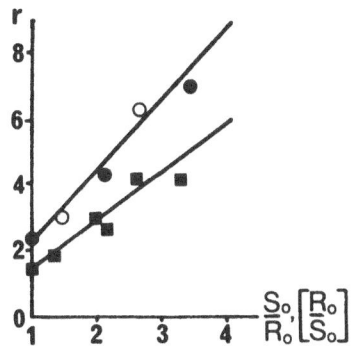

Fig. 2 Dependence of stereoelectivity ratio on enantiomeric composition of initial monomer.
Methylthiirane (polymerizations in
toluene solution at 25°C)
    ● S-enriched
    ○ R-enriched
Ethylthiirane (polymerizations in
bulk at -30°C)
    ■ S-enriched.

The effect of increase of stereoelection can be used to prepare
monomers of very high optical purity up to 98 % in a "super stereoelective" process (12,38) in which successive polymerizations
are carried out using unreacted monomer recovered from one step
as initial monomer in the next step. This particular process can
also be used as an auxiliary method for determination of the optical activity α_o of a pure enantiomer in the case of monomers
not easily or impossible to prepare by usual organic synthesis.
Drastic differences are observed for the resulting polymers according to the initial enrichment of the monomer. In the case of
R predominant configuration one can obtain highly optically active and crystalline polymers. For example starting from a (+)
methylthiirane (o.p. : 45 %) it is possible to isolate in a large domain of conversion polymers with optical purity close to
85 %. With monomer enriched in S enantiomer both the optical
activity and the crystallinity of resulting polymers are very
low. This important difference in crystallinity (up to a ratio
of ten) could be partly due to kinetically unfavored and favored
cross-propagation at R sites predominant in the catalyst.
With racemic methylthiirane, the stereoelectivity ratio r_R obtained with the $ZnEt_2$/R(-) DMBD system modified with chiral additives such as (S) 2-methylbbutyl disulfide, (R) 1,2-dimethoxy
3,3-dimethyl butane and sparteine of good optical purities, is
practically the same for all additives and equal to 4, i.e. almost the double of that obtained for unmodified initiator (49).
In limonene, as chiral medium, r depends on the relative enantiomeric composition of the limonene used. It was found that the
variation of r versus the composition $\frac{|R_L|}{|S_L|}$ of(+) limonene pas-
ses through a minimum value (r = 2.4 practically the same as ob-
served without limonene) for $\frac{|R_L|}{|S_L|}$ = 15. The maximum value obser-

ved at $\left|\dfrac{R_L}{S_L}\right|$ = 33 (r = 4.3) is only slightly higher than that ob-
tained in racemic limonene (r = 3.65). As concerning the poly-
mers no increase in crystallinity seems to be produced by the increase of stereoelectivity.

Thus the chiral agents seem to promote two effects not necessarily correlated, firstly a modification of the chirality around
the active sites leading to an increase of their stereoelectivity and secondly a more or less important modification of their
selectivity.

The attempts to obtain stereoelection with a chirally modified
achiral stereospecific initiator such as $ZnEt_2$-MeOH generally
failed except in the case of polymerization of diastereoisomeric
mixtures of 3-methyl 1,2-epoxy pentane or 3-methyl 1,2-epithio-
pentane (48). Yet in this last case, the observed election is
rather due to different consumption rates for the two diastereoisomers present in the mixture than to a real modification of
the spectrum of catalytic sites.

4-2) <u>Asymmetric polymer synthesis</u>

Optically active polymers with new structures could be obtained
when polymerizing symmetrical achiral monomers using chiral initiators. It was pointed in chapter 2 that the ring-opening of
cis and trans dimethyloxiranes proceeds with complete inversion
of corresponding asymmetric carbons as demonstrated by Vandenberg (4). Thus, from a cis monomer with -RS- configurational
structure, one obtains polymers with -RR- and -SS- configurational units. It was interesting to know if the direction of ring-
opening could be preferentially orientated on one of the asymmetric carbons by the chiral choice of an optically active initiator. In this case one should expect to produce optically active polymers due to the prevalence of one type of configurational units.

optically active if n ≠ m

The polymerization of cis dimethyl thiirane (Y = CH_3) and cyclohexene sulfide (Y $\subset$ Y = cyclohexyl) was studied for this purpose
using the standard $ZnEt_2$-R(-) DMBD (1:1) initiator (50). Polymers of high optical activity were obtained, as shown in table 4,
demonstrating the preferential attack of one of the asymmetric
carbons. This is obviously supported by the opposite sign found
for polymers obtained with $CdMe_2$-R(-) DMBD an "antisteric" initiator (table 4).

Table 4. Polymerization of cisdimethylthiirane using MR$_2$-(-)DMBD initiator (1:1). In bulk at 25°C - C/M = 4.5 (moles %)

MR$_2$	Conv. (%)	whole polym. $\|\alpha\|$ (a)	A (b) %	$\|\alpha\|$ (a)	B (c) %	$\|\alpha\|$ (a)
	30	+41	29	+20	71	+50
ZnEt$_2$	54	+42	33	+30	67	+55
	100	+45	35	+38	65	+45
CdMe$_2$	72	-42	100	-42 (d)		

(a) in chloroform
(b) fraction sol. in toluene ; m.p. ~ 90°C
(c) fraction insol. in toluene ; m.p. ~ 126°C with a weak peak at ~ 190°C
(d) m.p. ~ 126°C.

For the understanding of the mechanism it was necessary to know, first, what configurational unit was predominant in the polymer, second, the relative amount of the two configurational units present and their distribution in the chain.
The first point could be answered using following considerations. In stereoelective polymerizations of monosubstituted monomers with the same zinc initiator, polymers with predominant R unit were obtained. They resulted from a preferential choice of the initiator for the R-enantiomer with an attack on the primary carbon, i.e. not involving the asymmetric carbon. Thus by analogy, one can predict for the cis monomer, a polymer with predominant -RR- structure due to a ring-opening in β position on a side opposite to the chosen -R- type asymmetric carbon. This is also in agreement with the negative sign of the Cotton effect at 235 nm in CD spectrum (10).
Informations on the distribution of both configurational units in the chain could be obtained from ^{13}C NMR study. Both methine and methyl carbons are stereosensitive but show complex patterns with several peaks. In the methine pattern the upfield peak located at 45.8 ppm is well separated and increases with the optical activity. It was assigned to isotactic triads of configurational units of the type : -RR-RR-RR-. Using a Bernouillian statistics a tentative correlation between optical activity and tacticity was established and the distribution in different triads, i.e. enantiomeric composition, evaluated. It was concluded that polymers with -RR-/-SS- ratios up to 10 are obtained. The structure of the polymer chain however was found to be much more complex than the pure diisotactic and disyndiotactic one proposed earlier by Vandenberg in the case of inactive polymers

(51).
A correlation of this findings with results obtained by cationic
degradation of the polymers, an elegant method elaborated by
Goethals and his group (52-54), is presently under study and is
reported in a communication (55).
The preliminary results show that a linear correlation exists
between optical activity of initial polymers and their degradation products and thus, the latter reflect the structure of the
polymer chain. From mechanistic point of view this study of cis
monomers supports the important concept that the chiral initiator is not only able to distinguish between two enantiomeric molecules (stereoelective process) but is able to recognize an
asymmetric carbon of parent configuration and to orientate the
attack on the neighbouring one. This is the caracteristics of a
"regioselective" process and it opens a field of the study of
differently disubstituted monomers leading to various diastereoisomeric structures.

5 - <u>CONCLUSION</u>

The present examination of main features in stereospecific polymerization of three-membered ring heterocycles has shown that
still many questions remain unresolved.
Among the most important one are the localisation of primary catalytic sites, the reason of the poor efficiency of most of initiators, even homogeneous one, and the variation of the efficiency and the stereospecificity with the nature of monomer.
Apparently a primary transformation or coordination is responsible for the creation of active sites on some sterically pre-
viligiate positions.
In mechanistic studies, the use of optically active reagents
seems to be a rather elegant and useful tool. It allows to follow the stereocontrolled reactions simply by looking on optical
activity of the monomer or the polymer. Steric effects are thus
substantially amplified, but still many chiral recognitions or
interactions are not yet completely understood.
When compared, optically pure and racemic polymers reveal some
significant differences in their properties such as crystallization, solubility or crystalline structure. The way of racemization of a polymer could be realized either by intercrystallite
compensation, as in the case of polymethylthiirane (56) or by
formation of a racemic lattice, as observed for monomers with
bulky substituents such as t-butylthiirane (57). The properties
of the racemic polymer are then very different of those of the
optically pure one, for example, the melting points could differ
of more than 50°C. A similar behaviour was recently observed in
the case of substituted β propiolactones (58-60). Therefore the
preparation of pure optically active polymers remains of inte-

rest. As discussed above, methods are now available to obtain optically active monomers from racemic mixtures with reasonable yields and the field of studied monomers could be enlarged namely by those containing functional groups.
Finally one can expect to obtain products of new tailoring by copolymerization and transformation reactions which were relatively few studied up to now.

REFERENCES

1. J. Furukawa, T. Saegusa, Polymerization of Aldehydes and Oxides, J. Wiley et Sons, N.Y. (1963)
2. T. Tsuruta, Stereochemistry of Macromolecules, Ed. by A.D. Ketley, vol. 2, p. 177, M. Dekker, N.Y. (1967)
3. Y. Ishii, S. Sakai, Ring-Opening polymerization, Ed. by K.C. Frisch and S.L. Reegen, vol. 2, p. 13, M. Dekker, N.Y. (1969)
4. E.J. Vandenberg, J. Polym. Sci., A1, 7, 529 (1969)
5. H. Tani, Adv. Polym. Sci., 11, 57 (1973)
6. P. Sigwalt, Ring-Opening polymerization, Ed. by K.C. Frisch and S.L. Reegen, Vol. 2, p. 191, M. Dekker, N.Y. (1969)
7. P. Sigwalt, Int. J. Sulfur Chem., C7, 83 (1972)
8. L.A. Korotneva, G.P. Belonovskaya, Uspekhi Khim. XLI, 150 (1972)
9. T. Tsuruta, J. Polym. Sci., D, 179(1972)
10. N. Spassky, P. Dumas, M. Sépulchre, P. Sigwalt, J. Polym. Sci., Symposium n° 52, 327 (1975)
11. P. Sigwalt, Pure and Appl. Chem., 48, 257 (1976)
12. N. Spassky, ACS Symposium Series, n° 59, Ring-Opening Polymerization, T. Saegusa and E. Goethals Ed., p. 191, (1977)
13. M. Ishimori, T. Hagiwara, T. Tsuruta, Y. Kai, N. Yasuoka, N. Kasai, Bull. Chem. Soc. Jap., 49(4) 1165 (1976)
14. M. Ishimori, T. Tomoshige, T. Tsuruta, Makromol. Chem. 120, 161 (1968)
15. S. Inoue, I. Tsukuma, M. Kawaguchi, T. Tsuruta, Makromol. Chem. 103, 151 (1967)
16. J.P. Machon, P. Sigwalt, C.R. Acad. Sci. Paris, 260, 549 (1965)
17. R.O. Colclough, J.R. Quijada, International Symposium on macromolecules, Dublin, Preprints, p. 145 (1977)
18. Yu. P. Kuznetsov, G.P. Belonovskaya, B.A. Dolgoplosk, Vysokomol. Soed. A, XX (3), 551 (1978)
19. O. Nakasugi, M. Ishimori, T. Tsuruta, Bull. Chem. Soc. Jap., 47 4), 871 (1974)
20. V.M. Denisov, Yu. P. Kuznetsov, Izv. Ak. Nauk SSR, Ser. Khim. 2710 (1975)
21. Ph. Guérin, Doctorat d'Etat, Paris (1976)
22. Ph. Dumas, Ph. Guérin, Can. J. Chem., 56, 925 (1978)

23. M. Ishimori, G. Hsiue, T. Tsuruta, Makromol. Chem., 128, 52 (1969)
24. N. Spassky, P. Sigwalt, Tetrahedron Letters, 3541 (1968)
25. M. Yokoyama, H. Ochi, H. Tadokoro, C.C. Price, Macromolecules, 5, 690 (1972)
26. M. Ishimori, K. Tsukigawa, T. Tsuruta, Makromol. Chem., 177, 1221 (1976)
27. J. Furukawa, Y. Kumata, Makromol. Chem., 136, 147 (1970)
28. T. Tsuruta, S. Inoue, I. Tsukuma, Makromol. Chem., 84, 298 (1965)
29. J. Furukawa, S. Akutsu, T. Saegusa, Makromol. Chem., 94, 68 (1966)
30. M. Marchetti, E. Chiellini, M. Sépulchre, N. Spassky, Makromol. Chem. 180, 1305 (1979)
31. Ph. Teyssié, T. Ouhadi, J.P. Bioul, Int. Rev. Sci., Vol. 8, Butterworths, London, (1975), p. 191
32. J.P. Bioul, PhD Thesis ; T. Ouhadi, PhD Thesis. Liège (1973)
33. Ph. Guérin, S. Boileau, P. Sigwalt, European Pol. J., 10, 13 (1974)
34. C.C. Price, M. Osgan, J. Amer. Chem. Soc., 78, 4787 (1956)
35. S. Inoue, T. Tsuruta, J. Furukawa, Makromol. Chem., 53, 215 (1962)
36. A. Leborgne, N. Spassky, to be published
37. Ph. Dumas, N. Spassky, P. Sigwalt, Makromol. Chem., 156, 65 (1972)
38. M. Sépulchre, N. Spassky, P. Sigwalt, Israel J. of Chem., 15, 33 (1976/77)
39. N. Spassky, unpublished results.
40. A. Deffieux, M. Sépulchre, N. Spassky, P. Sigwalt, Makromol. Chem. 175/4, 339 (1974)
41. A. Khalil, N. Spassky, to be published
42. N. Spassky, A. Pourdjavadi, P. Sigwalt, Europ. Pol. J., 13, 467 (1977)
43. M. Sépulchre, N. Spassky, J. Huguet, M. Vert, P. Granger, Polymer, 20, 833 (1979)
44. P. Dumas, N. Spassky, P. Sigwalt, J. Polym. Sci., Polym. Chem. Ed., 17, 1583 (1979)
45. A. Momtaz, Doctorat 3e cycle, Paris (1978)
46. C. Coulon, N. Spassky, P. Sigwalt, Polymer, 17, 821 (1976)
47. M. Sépulchre, C. Coulon, N. Spassky, P. Sigwalt, 1st International Symposium on ring-opening polymerization, Jablona, (1975) preprints p. 80
48. M. Goguelin, M. Sépulchre, Makromol. Chem. 180, 1231 (1979)
49. M. Sépulchre, P. Sigwalt, N. Spassky, IUPAC Internationl Symposium on Macromolecules, Dublin (1977), Preprints p.141
50. A. Momtaz, M. Reix, N. Spassky, P. Sigwalt, New Developments in ionic polymerization, Strasbourg (1977) Preprints p. 84
51. E.J. Vandenberg, J. Polym. Sci., A1, 2, 329 (1972)
52. W.Van Crayenest, E.J. Goethals, Europ. Polymer J., 12, 859 (1976)
53. E.J. Goethals, Advanc. Polym. Sci., 23, 103 (1977)

54. R. Simonds, H. Nuytten, E.J. Goethals, New Developments in ionic polymerization, Strasbourg (1977) Preprints p. 106
55. E. Goethals, R. Simonds, N. Spassky, NATO Advanced Institute on Stereoregular Polymers, Tirrenia (1978), Communication.
56. H. Sakakihara, Y. Takahashi, H. Tadokoro, P. Sigwalt, N. Spassky, Macromolecules, 2, 515 (1969)
57. H. Matsubayashi, Y. Chatani, H. Tadokoro, Ph. Dumas, N. Spassky, P. Sigwalt, Macromolecules, 10, 996 (1977)
58. C.G. D'Hondt, R.W. Lenz, J. Polym. Sci., Polym. Chem. Ed., 16, 261 (1978)
59. N. Spassky, A. Leborgne, M. Reix, R.E. Prud'Homme, E. Bigdeli, R.W. Lenz, Macromolecules, 11, 717 (1978)
60. R.E. Prud'Homme, J. Noah, NATO Advanced Institute on Stereoregular polymers, Tirrenia (1978), Communication.

PRESENT UNDERSTANDING OF THE STEREOSPECIFIC POLYMERIZATION OF
CYCLIC MONOMERS: LARGE RINGS

Jørgen Kops

Instituttet for Kemiindustri, Technical University of
Denmark, DK-2800 Lyngby, Denmark

Various types of cyclic monomers may undergo polymerization by
ring-opening reaction. These include the following types of compounds: oxacyclic (ethers), 1,3-dioxacyclic (formals and acetals),
higher oxacyclics, lactams, lactones and compounds containing
other heteroatoms. The present survey will cover the stereochemical aspects of polymerization of only the first two of the above-
mentioned types although the stereochemistry in connection with
ring-opening polymerization of bicyclic lactones also will be
mentioned.

The stereochemistry is associated with the presence of substituents on the cyclic monomers. However, it is a well known fact[1]
that substituents will stabilize cyclic monomers and the free
energy of polymerization may become very small or even attain a
positive value. In fact, very few substituted monocyclic larger
rings have been found to undergo polymerization. In order to com-
pensate for the deactivating effect of substitution the ring to
be opened in the polymerization step may be made part of a bicyclic structure. This can induce considerable strain into the ring
and may promote the polymerization. The effect of ring substitution on polymerizability and the behaviour of bicyclic monomers
of the above-mentioned types have previously been described[2].

MONOCYCLIC ETHERS

Stereochemical results have so far not been reported regarding the
polymerization of cyclic ethers other than epoxides. Oxacyclobu-
tane (1) (oxetane) is known to easily undergo polymerization and
compounds substituted at the 3 positions have been extensively

225

R. W. Lenz and F. Ciardelli (eds.), Preparation and Properties of Stereoregular Polymers, 225-241.
Copyright © 1979 by D. Reidel Publishing Company.

studied, e.g. the kinetics of polymerization of the 3-methyl (2)
the 3,3-dimethyl (3) substituted compounds[3].

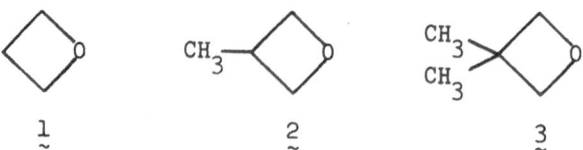

Obviously, with the symmetrical substitution the gross chemical
structure will be the same whether one or the other of the two
bonds to oxygen is being opened in connection with the polymerization. None of the monomers contain an asymmetric center, however,
upon ring opening of 2, pseudoasymmetric centers are formed and
in principle, tacticity should be possible, although not actually
reported.

Substitution at the 2 position (4) yields a chiral center and
stereoselection in the polymerization step could be visualized
leading to the formation of two enantiomeric isotactic forms:

However, such polymerization processes have so far not been studied. The 2-methyloxacyclobutane have been shown to undergo polymerization[4] and very recent results[5] have shown that the ring-opening which occurs by breaking one of the bonds O-C^2 or O-C^4 may take place randomly or more specificly depending on the conditions. This leads to polymer structures with varying amounts of head-to-tail, head-to-head, and tail-to-tail structures:

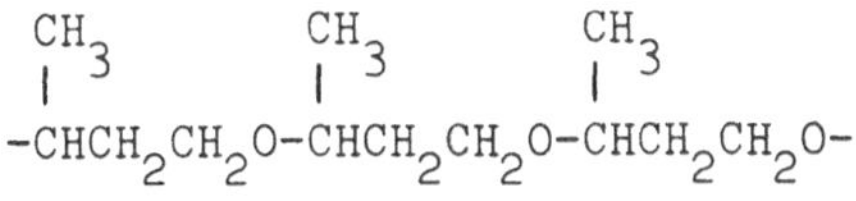
CH3 CH3 CH3
-CHCH2CH2O-CHCH2CH2O-CHCH2CH2O-

Head-to-tail

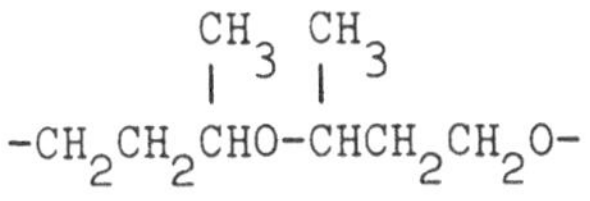
CH3 CH3
-CH2CH2CHO-CHCH2CH2O-

Head-to-head

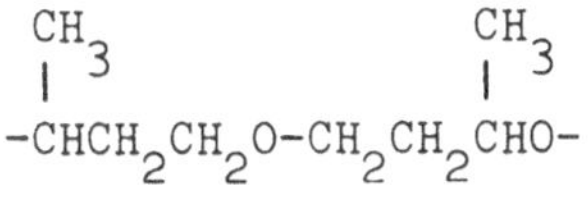
CH3 CH3
-CHCH2CH2O-CH2CH2CHO-

Tail-to-tail

Substitution of the higher cyclic ethers generally leads to deactivation. Thus oxacyclopentane (tetrahydrofuran) will form only oligomers in case of methyl substitution at the 2-position and low polymers from 3-methyltetrahydrofuran[6]. However, stereochemical results have not been rapported and studies concerning other substituted larger ring cyclic ethers have not been carried out.

MONOCYCLIC ACETALS AND FORMALS

The substituted dioxacyclic compounds show a somewhat greater tendency to form polymers. In the dioxacyclopentane series, derivatives which are substituted in the 2-position generally will not polymerize although 2-methyl-1,3-dioxolane yielded oligomers[7]. The 4-substituted 1,3-dioxolanes generally have a greater tendency to form polymers, however, not with high molecular weights. In case of 4-methyl-1,3-dioxolane the structure of the resulting polymers were investigated[8,9]. The ring opening may occur in two ways:

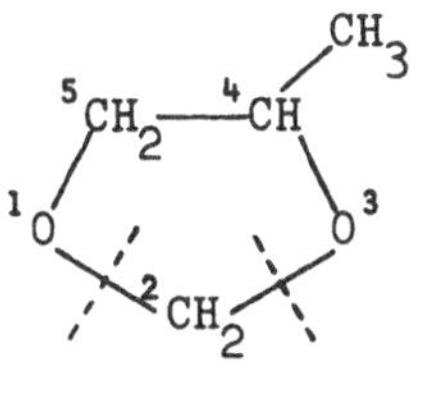
CH3
5 CH2
4 CH
1 O
O 3
2 CH2

The signal from the formal protons in ^{1}H-NMR is split into three peaks (τ = 5.34, 5.39 and 5.44 ppm) with relative intensities approximately 1:2:1. These were assigned to the three diad structures:

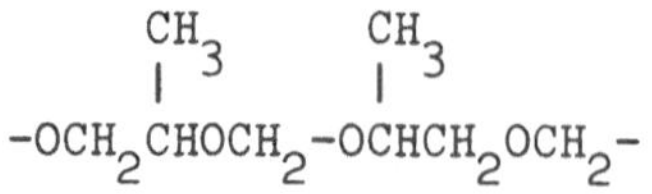

$$-OCH_2CHOCH_2-OCHCH_2OCH_2- \quad (CH_3, CH_3)$$

A

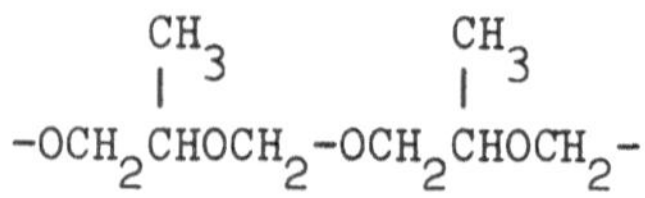

$$-OCH_2CHOCH_2-OCH_2CHOCH_2- \quad (CH_3, CH_3)$$

B

$$-OCHCH_2OCH_2-OCH_2CHOCH_2- \quad (CH_3, CH_3)$$

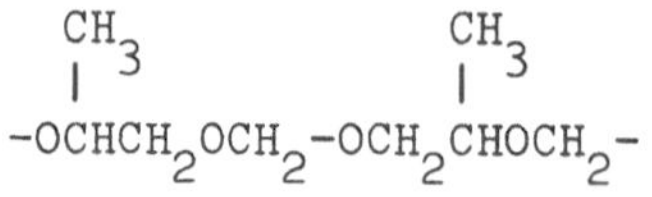

C

The diad structure B results from consecutive bond scissions of O^1-C^2 or O^3-C^2 while structures A and C result when the scissions alternate between the two types of bonds. The relative intensities of the signals indicate bond breakage in a random fashion.

Quite similar results were obtained in the study of the polymerization of 4-ethyl-1,3-dioxolane[10] and the 7-membered ring compound 4-methyl-1,3-oxepane[11]. In both cases the bond cleavage occurs nearly randomly at the two formal O-C bonds. The authors report in both cases multiplet signals in ^{1}H-NMR for the formal protons in the various diad configurations produced by the nearly random cleavage. The multiplicity is interpreted as being due to the effect of the asymmetric carbon atom located at the β-position from the formal protons, illustrated by the following which can be considered the head-to-head structure:

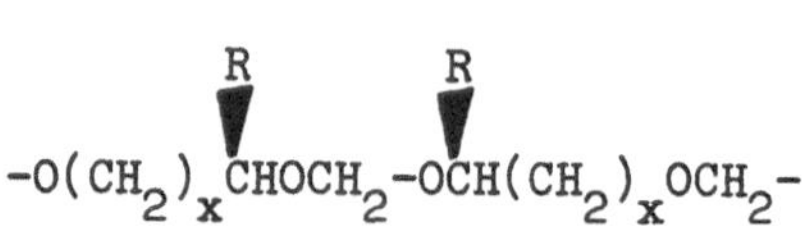

$$-O(CH_2)_xCHOCH_2-OCH(CH_2)_xOCH_2- \quad (R, R)$$

A'I

$$-O(CH_2)_xCHOCH_2-OCH(CH_2)_xOCH_2- \quad (R, R)$$

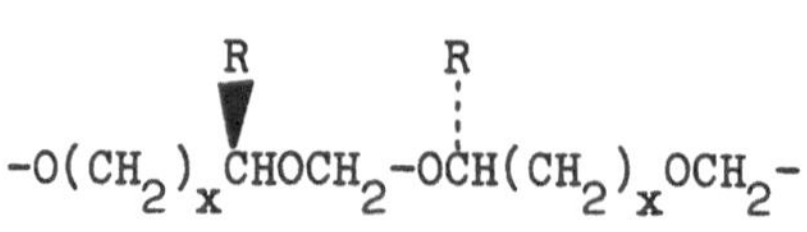

A'II

where $(x,R) = (1,C_2H_5)$ or $(3,CH_3)$.

In structure A'I the formal protons appear as a quartet since the
protons are non-equivalent when the alkyl groups are located on
the same side of the planar zigzag chain. In structure A'II these
protons become equivalent and appear as a singlet. In the structures A'I and A'II the asymmetric carbons are of the opposite and
of the same configuration respectively. Only if the asymmetric
carbon is positioned no further away than the β-position it will
be able to influence the spectrum of the formal protons. This is
noticeable for the head-to-tail structure B' and tail-to-tail
structure C' which appear as a quartet and a singlet respectively.

$$-O(CH_2)_x\overset{\overset{\displaystyle R}{|}}{C}HOCH_2-O(CH_2)_x\overset{\overset{\displaystyle R}{|}}{C}HOCH_2-$$

B'

$$-O\overset{\overset{\displaystyle R}{|}}{C}H(CH_2)_xOCH_2-O(CH_2)_x\overset{\overset{\displaystyle R}{|}}{C}HOCH_2-$$

C'

The nearly random cleavage of the formal bonds was substantiated
by ^{13}C-NMR measurements on the polymer of 4-methyl-1,3-oxepane[11].

Previously, the 2-methyl and the 2,2-dimethyl-1,3-dioxepane were
investigated[12] and only the monosubstituted derivative were found
to undergo polymerization. The polymer had a regular structure as
expected from the two equivalent acetal bonds in the monomer. Stereospecificity of this polymer is in principle possible, however,
results concerning this have not been reported.

Even larger ring compounds of this type may undergo polymerization, for example, the 11-membered 1,3,6,9-tetraoxacycloundecane[13,14].
However, substituted ring compounds have not been investigated
and therefore stereochemical aspects have so far not been considered for the very large rings.

BICYCLIC ETHERS

The stereochemistry of the ring-opening polymerization of 1,4-
bridged bicyclic ethers derived from cyclohexane have been studied.
The endo- and exo-2-methyl-7-oxabicyclo[2.2.1]heptane (5) were
found[15,16] to polymerize exclusively through nucleophilic attack
at the C^4 bridge head carbon in the S_N2 propagation step leading

to regular structures of the polymers:

$$\underset{\sim}{5}$$

Each of the polymers obtained from the two monomers was build up
of only one type of repeat unit. In case of endo $\underset{\sim}{5}$ all the substituents on the cyclohexane ring were equatorially positioned and
an essentially fixed conformation of the repeat unit was found:

endo $\underset{\sim}{5}$

When the exo $\underset{\sim}{5}$ was used as the starting monomer the CH_3 group was
positioned axially in the conformer corresponding to the one shown
above. However, in this case flipping between the two chair conformations of the repeat unit was possible.

Two dimethyl substituted derivatives have been studied that is
endo,exo- and exo,exo-2,6-dimethyl-7-oxabicyclo[2.2.1]heptane[16].
Only the former monomer yielded a polymer. The exo,exo-isomer ($\underset{\sim}{6}$)
will not polymerize since the conformational forms of the repeat
unit are unfavourable. In the two chair conformations two substituents will be in axial positions and in one form a 1,3-diaxial
interaction will exist between the methyl groups:

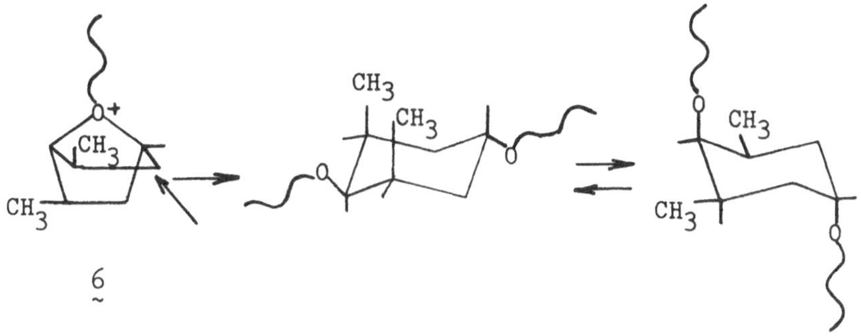

6

The polymerization of endo- and exo-2-tert.-butyl-7-oxabicyclo-
[2.2.1]heptane have been studied[17]. As expected the large substituent very effectively shields the neighbouring carbon from nucleophilic attack of the monomer and the ring opening occurs only
at the C^4 bridge head carbon.

When the bridge connecting the 1,4-positions of the cyclohexane
ring contains a methylene group in addition to oxygen, i.e. 2-
oxabicyclo[2.2.2]octane (7) the polymer structure will vary depending on how the ring-opening occurs. It turns out that the propagation occurs predominantly through nucleophilic attack at the
tertiary carbon in spite of the accessibility of the methylene
carbon for a similar attack. Thus a regular structure of the polymer is obtained[18]:

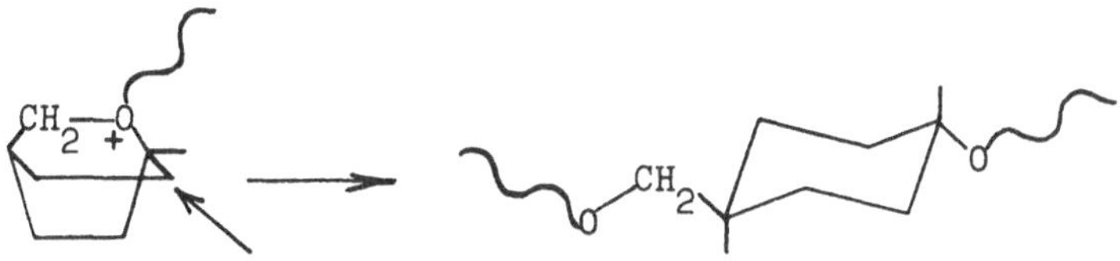

7

The formation of the polymer having a predominantly trans-1,4-di-
substituted cyclohexane ring structure was shown by comparing the
^{1}H-NMR spectrum with spectra of model compounds.

We have studied the polymerization behaviour of several bicyclic
ethers in which an oxacyclopentane ring is fused with alicyclic
rings of different sizes. The following compounds with the rings
fused into trans position were found to undergo polymerization
under the influence of cationic initiators:

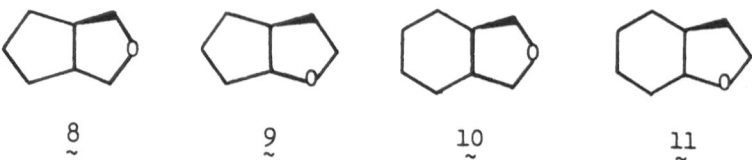

8   9   10   11

trans-3-Oxabicyclo[3.3.0]octane (8) is highly strained and it was
found to be extremely reactive[19]. Recent results[20] have shown that
trans-2-oxabicyclo[3.3.0]octane (9) also is very reactive although
not quite as reactive as (8). Previously it has been shown that
trans-8-oxabicyclo[4.3.0]nonane (10) quite readily form high polymers[21] and the reversible polymerization of trans-7-oxabicyclo-
[4.3.0]nonane (11) was demonstrated[22]. The structure of the polymers of these bicyclic ethers have been investigated by ^{13}C-NMR
and on this basis the mechanisms of the ring-opening reactions
were established[23].

The ring-opening polymerization of compounds 8 and 10 occur by
nucleophilic attack at one of the two equivalent methylene carbons
adjacent to the oxonium ion:

8 or 10

The symmetrical structure of the polymer corresponding to the aliphatic rings connected by $-CH_2OCH_2-$ linkages results for each of
the polymers in a simple ^{13}C-NMC spectrum with only four different
signals. By comparison with model ether compounds the structures
were firmly established.

The propagation step in case of compounds 9 and 11 may also be
considered to take place by nucleophilic attack of monomer on one
of the two carbons adjacent to the oxonium ion. However, the two
carbons are not equivalent in this case and the ^{13}C-NMR spectra
of the polymer were in agreement with an attack only at the methylene carbon:

9 or 11

With reference to Table I an excellent agreement is seen between
the spectral data for the polymers and for the model compounds
(the signals for a cis isomer have been found to be significantly
different). The mode of ring-opening is just the opposite of that
found in case of the polymerization of 7 where the tert. carbon
is subjected to the nucleophilic attack. The tertiary position in
case of 9 and 11 are severely shielded and for that reason the
attack occurs only on the methylene carbon.

The presence of fine splittings (0.1-0.5 ppm) of some of the signals from the polymers of 9 and 11 is seen in Table I. The splitting consisted in each case of two closely spaced peaks of very
nearly equal intensity. The splittings are observed for the carbons β and δ in both polymers and also of carbon γ in case of the
polymer of 11. The splittings reflect different stereochemical
relationships between neighbouring repeat units:

m   r

In principle the same effect should be observable in case of the
polymers of 8 and 10. However, the placement of the heteroatom in
these polymers between two methylene groups appears to weaken the
configurational effect.

The presence of m and r diads in the same proportion indicate a
random incorporation of R and S monomers. Thus no stereoselection
takes place in connection with the chain propagation.

TABLE I

Poly(trans-2-oxabicyclo[3.3.0]octane) and Poly(trans-7-oxabicyclo[4.3.0]nonane)
Comparisons with Model Compounds
^{13}C Chemical Shifts (ppm from TMS)

α	β	γ	δ	β'	γ'	δ'	γ''
86.8	42.72 / 42.56	34.7	68.40 / 68.26	31.2	30.3	22.5	
86.6	42.7	34.4	69.7	31.4	30.4	22.6	
82.2	40.50 / 40.29	33.10 / 33.00	67.52 / 67.04	31.1	30.7	25.3	24.5
82.3	40.7	32.8	69.4	31.5	31.2	25.5	24.8

BICYCLIC ACETALS AND FORMALS

The polymerization of compounds having the following ring structures will be discussed:

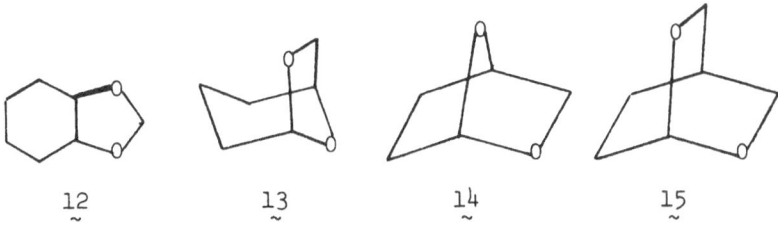

12   13   14   15

The 7,9-dioxabicyclo[4.3.0]nonane (12) have been shown to undergo
polymerization very readily[24]. The two formal bonds are equivalent
and therefore the chemical structure of the polymers is simple,
consisting of cyclohexane units connected with $-OCH_2O-$ linkages.
Actually the polymer is rather similar to that of the bicyclic
ether 10 which instead has $-CH_2OCH_2-$ linkages. 12 possesses asymmetric carbons and different stereochemical arrangements along
the chain are possible. Recent results have shown that the ^{13}C-NMR
spectrum of the polymers of 12 is rather complicated and reflect
the presence of different diad and higher structures[25].

The bicyclic acetal 6,8-dioxabicyclo[3.2.1]octane (13) have been
shown to yield polymers[26-28] although generally it is difficult
to obtain high molecular weights. The steric course of this polymerization has later been investigated[29]. In the ring-opening step
one of the two acetal bonds may be visualized to open while the
other two O-C bonds can immediately be ruled out. It was substantiated that the cleavage occurs exclusively of the C^5-C^6 bond
which leads to the formation of a polymer with oxacyclohexane
rings:

As indicated by the two structures shown for the repeat unit the
chain may be attached either trans or cis. By ^{1}H-NMR measurements
these two types of enchainments were found to vary with the polymerization conditions. The stereoregularity could be estimated
from the relative intensities of the signals from the equatorial
and axial acetal protons which appeared at τ = 5.15 and 5.60, respectively.

At low temperature (-78°C) a highly stereoregular structure was
formed as indicated by the practically complete absence of equatorial acetal proton. The structure corresponds in this case to
the one shown above with trans enchainments. The propagation occurs at the low reaction temperature exclusively by back side attack with inversion at the acetal carbon by an S_N2 type mechanism.
At higher temperatures for example at 0°C a considerable amount
of repeat units with a cis structure is formed. The authors explain this by suggesting that at the higher temperatures and also
in polar solvents the propagating species attain carbenium ion
character and the free ion may be attacked by monomer from either
of the two sides of the ring with a resulting loss of stereoregularity of the polymer. An alternative explanation for this is the
occurence of oxonium ion exchange whereby the penultimate unit
from the terminal unit is attacked by monomer at the neighbouring
position to the oxonium ion. This displacement reaction of the
terminal unit with inversion would also lead to formation of cis
units.

Stereospecific polymerization of triethers of 1,6-anhydroglucopy-
ranoses (16) was early reported by Schuerch[30,31] and a large number of different anhydrosugars of this generel structure has later
been studied with various coworkers[32]. The influence of structure
on the reactivity of 1,6-anhydrosugars has been discussed[33]. The
reactive intermediate propagating stereospecifically was suggested
to be a trialkyloxonium ion which as already discussed above later
has been used in explaining the polymerization behaviour of the
unsubstituted 6,8-dioxabicyclo[3.2.1]octane. The stereospecificity
of the polymerization of the 1,6-anhydrosugars may vary with conditions and the species propagating without steric control are
carbenium ions stabilized by the neighbouring oxygen[35]:

Similar species had previously been suggested to account for the non-stereospecificity in the polymerization of certain 1,4-anhy-drosugars[36] which will be discussed below.

2,7-Dioxabicyclo[2.2.1]heptane 14 was recently synthesized[37] and its polymerization behaviour studied[38]. At low temperature (-78°C) the polymer was exclusively composed of oxacyclopentane rings in-dicating that the propagation occurs by breaking the C^1-O^2 bond. At higher temperature a tendency to C^1-O^7 bond breaking was noted and a mixture of oxacyclohexane and oxacyclopentane rings was found in the polymer. The NMR spectra were complicated and no de-tailed analysis of the isomer composition could be carried out and the cis:trans ratio for the chain attachments to the rings could not be determined.

Polymerization of 1,3,3-trimethyl-2,7-dioxabicyclo[2.2.1]heptane was recently attempted[39], however, only low polymers were formed. The direction of the ring-opening was not actually established by analysis of the oligomeric products, however, in studying the pro-ducts formed by the ring-opening of the monomer by hydrogenation it was inferred that C^1-O^7 scission should predominate over C^1-O^2 scission. This mechanism leads to a polymer structure containing mainly six-membered rings.

1,4-Anhydrosugar derivatives which have a bicyclic skeleton corre-sponding to 14 have been shown to undergo polymerization[36]. 1,4-Anhydro-2,3-di-0-methyl-L-arabinose (16) and 1,4-anhydro-2,3,6-tri-0-methyl-D-galactose (17) yielded under the influence of Lewis acid initiators amorphous polymers containing both furanose and pyranose rings.

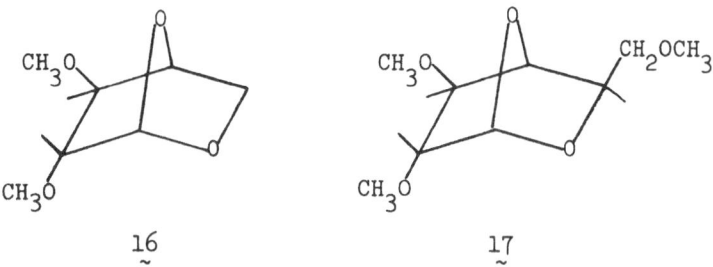

16   17

From studies of the changes in optical rotation and reaction rate during hydrolysis and methanolysis of the polymers it was estab-lished that the chain propagation proceeded in each case with ne-arly equal amounts of five- and six-membered ring-openings. The optical rotation for the polymers varied with the polymerization temperature and with the solvent used. This reflected the diffe-rences in configuration at the anomeric carbons. These observa-tions could not be explained by propagation via a trialkyl oxonium ion mechanism and a carbenium ion mechanism was suggested which, as already mentioned above, has been useful in explaining the stereochemistry of the chain propagation also for other bicyclic monomers.

In case of 16 it turns out that at low temperature the predominant linkage is of the β-L type at the anomeric center of the furanose units. This linkage although it is less stable than the α-L type is found because the counterion which is closely associated with the anomeric carbenium center forces attachment to take place in the β-L position. At higher temperature there is enough kinetic energy to partially overcome the electrostatic attraction between the carbenium ion and the counterion and α-L linkage predominates. The pyranosidic units have α-L linkages which are favoured both thermodynamically and kinetically. In case of 17 the question is how the large C^5 substituent will influence the stereochemistry of the chain propagation. This substituent will effectively shield one side of the furanosidic carbenium ion and for that reason all the linkages become of the β-D type which means attachment to the opposite side of the ring compared with the result observed in case of polymerization of 16. The six-membered galactopyranosidic units have β-D linkages, a result which may be rationalized on the basis of the structure of the carbenium ion intermediate and conformational analysis.

The bicyclic acetal 2,6-dioxabicyclo[2.2.2]octane 15 has been found to be highly reactive[40]. In this case the ring-opening reaction yields only one ring size since the two acetal bonds are equivalent. Stereoregular propagation by S_N2 displacement on the bicyclic oxonium ion occured under certain conditions at very low reaction temperature (-78°C). A stereorandom propagation occurs at +28°C and this behaviour is ascribed to propagation via an S_N1 reaction of an intermediate carbenium ion similarly to the mechanism for some of the other dioxabicyclic compounds already discussed above. The structure of the polymers could be established by measurements of the ratios of axial to equatorial acetals by NMR. The measured values were in an accord with those calculated from the conformational equilibria of the chain repeat units in the polymers prepared under the various conditions. A compound with a rather similar structure 2,6-dioxabicyclo[2.2.1]heptane has later been studied[38]. This compound which also has two equivalent acetal proton will yield only five-membered rings upon ring-opening. Both ^1H- and ^{13}C-NMR measurements indicated the presence of approximate a 60:40 mixture trans to cis linkages.

It should be added that polymers have been prepared from unsaturated dioxabicyclic monomers. These polymers may be hydroxylated to form polysaccharide-type polymers. 6,8-Dioxabicyclo[3.2.1]oct-3-ene (18) will undergo selective ring-opening polymerization[41]:

18

The unsaturated polymer has via epoxidation and hydrolysis been converted into a polysaccharide type polymer[42].

BICYCLIC LACTONES

Finally the preparation of polyesters by ring-opening polymerization of bicyclic lactones will briefly be mentioned. This type of polymerization has previously been investigated[43]. However, it is only recently that the importance of controlling the stereochemistry has been demonstrated[44]. In this study the polymers obtained from 2-oxabicyclo[2.2.2]octan-3-one (19) were characterized by NMR.

19

The anionic polymerization of 19 may involve acyl-oxygen or alkyl-oxygen cleavage. The propagating species in the two cases are alkoxide and carboxylate ions respectively. When n-butyl lithium is used as initiator the ring-opening occurs by acyl-oxygen cleavage and the ring-substituents will be cis-positioned. With Na-K alloy initiator the propagation also occurs with inversion at C^5 and cyclohexane units with some of the bonds in trans position will appear. The stereocopolymer was more tractable than the stereoregular cis-polyester and it had better properties than the stereoregular polymer.

The bicyclic lactone trans-7-oxabicyclo[4.3.0]nonan-8-one in which a five-membered ring is opened in the polymerization step has proved to polymerize very slowly and yields only two polymers[45]. This is in accord with the general inactivity of a monocyclic γ-lactones.

Bicyclic lactones containing the δ-lactone ring are more likely
candidates as suitable monomers and the stereochemical aspects of
these polymerizations warrent further investigation.

REFERENCES

1. F.S. Dainton and K.J. Ivin, Quart. Rev. (London) 12, 61 (1958).
2. J. Kops, Polym. Prepr. 17 (1), 182 (1976).
3. J.B. Rose, J. Chem. Soc. 542 (1956).
4. J.J. Stratta, F.P. Reding and J.A. Faucher, J. Polymer Sci.,
    Part A, 2, 5017 (1964).
5. J. Kops and H. Spanggaard, Unpublished results.
6. R. Chiang and J.H. Rhodes, J. Polymer Sci., Part B, 7, 643
    (1969).
7. Y. Firat and P.H. Plesch, J. Polymer Sci., Polymer Letters
    13, 135 (1975).
8. M. Okada, K. Mita and H. Sumitomo, Makromol. Chem. 176, 859
    (1975).
9. Y. Firat and P.H. Plesch, Makromol. Chem. 176, 1179 (1975).
10. M. Okada, K. Mita and H. Sumitomo, Makromol. Chem. 177, 895
    (1976).
11. M. Okada, T. Hisada and H. Sumitomo, Makromol. Chem. 179, 959
    (1978).
12. M. Okada, K. Yagi and H. Sumitomo, Makromol. Chem. 163, 225
    (1973).
13. C. Rentsch and R.C. Schulz, Makromol. Chem. 178, 2535 (1977).
14. Y. Yamashita, Y. Kawakami and K. Kitano, J. Polymer Sci.,
    Polymer Letters 15, 213 (1977).
15. T. Saegusa, M. Motoi, S. Matsumoto and H. Fujii, Macromolecu-
    les 5 (3), 233 (1972).
16. J. Kops and H. Spanggaard, Makromol. Chem. 151, 21 (1972).
17. T. Saegusa, M. Motoi and H. Suda, Macromolecules 9 (2), 231
    (1976).
18. T. Saegusa, T. Hodaka and H. Fujii, Polymer J. 2 (5), 670
    (1971).
19. J. Kops, S. Hvilsted and G. Sørensen, Proceedings of 1st Int.
    Symposium on Polymerization of Heterocycles (Ring Opening)
    Jablonna 1975, p. 132.
20. S. Hvilsted and J. Kops, Unpublished results.
21. J. Kops and H. Spanggaard, Makromol. Chem. 175, 3077 (1974).
22. J. Kops, E. Larsen and H. Spanggaard, J. Polymer Sci. Symp.
    56, 91 (1976).
23. J. Kops, S. Hvilsted and H. Spanggaard, Proceedings of the
    European Conference on NMR of Macromolecules, Sassari, Sardi-
    nia, 1978.
24. J. Kops and H. Spanggaard, Makromol. Chem. 176, 299 (1975).
25. J. Kops and H. Spanggaard, Unpublished results.
26. H. Sumitomo, M. Okada and Y. Hibino, J. Polymer Sci., Part B,
    11, 871 (1972).

27. J. Kops, J. Polymer Sci., Part A-1, <u>10</u>, 1275 (1972).
28. H.K. Hall, Jr., M.J. Steuck, J. Polymer Sci., Polym. Chem. Ed., <u>11</u>, 1035 (1973).
29. M. Okada, H. Sumitomo and Y. Hibino, Polymer J., <u>6</u> (3), 256 (1974).
30. E.R. Ruckel and C. Schuerch, J. Org. Chem. <u>31</u>, 2233 (1966).
31. E.R. Ruckel and C. Schuerch, J. Am. Chem. Soc. <u>88</u>, 2605 (1966).
32. C. Schuerch, Advan. Polym. Sci. <u>10</u>, 173 (1972).
33. C. Schuerch, Accounts Chem. Res. <u>6</u>, 184 (1973).
34. K. Kobayashi and C. Schuerch, J. Polymer Sci., Polym. Chem. Ed. <u>15</u>, 913 (1977).
35. J. Zachoval and C. Schuerch, J. Am. Chem. Soc. <u>91</u>, 1165 (1969).
36. J. Kops and C. Schuerch, J. Polymer Sci., C 11, 119 (1965).
37. H.K. Hall, Jr., and F. DeBlauwe, J. Am. Chem. Soc. <u>97</u>, 655 (1975).
38. H.K. Hall, Jr., F. DeBlauwe, L.J. Carr, V.S. Rao and G.S. Reddy, J. Polymer Sci., Symp. <u>56</u>, 101 (1976).
39. M. Okada, H. Sumitomo and S. Irii, Makromol. Chem. <u>177</u>, 2331 (1976).
40. H.K. Hall, Jr., L.J. Carr, R. Kellman and F. DeBlauwe, J. Am. Chem. Soc. <u>96</u>, 7266 (1974).
41. M. Okada, H. Sumitomo and H. Komoda, Makromol. Chem. <u>178</u>, 343 (1977).
42. M. Okada, H. Sumitomo and H. Komoda, Makromol. Chem. <u>179</u>, 949 (1978).
43. H.K. Hall, Jr., J. Am. Chem. Soc. <u>80</u>, 6412 (1958).
44. C. Ceccarelli, F. Andruzzi and M. Paci, Proceedings of the European Conference on NMR of Macromolecules, Sassari, Sardinia, 1978.
45. J. Kops and J.P. Tovborg Jensen, Unpublished results.

STEREOREGULAR POLYMERS FROM SUBSTITUTED β-LACTONES AND β-LACTAMS

Robert W. Lenz

Materials Research Laboratory
Chemical Engineering Department
University of Massachusetts,
Amherst, Mass. 01003   USA

R. W. Lenz and F. Ciardelli (eds.), Preparation and Properties of Stereoregular Polymers, 243–257.

The mechanism of stereoregulation in the polymerization of β-lactones and lactams is a subject in which very little definitive information is available and, consequently, many unanswered questions exist. Little is known about the molecular basis for the apparent stereoregulation in these ring-opening polymerization reactions, and unfortunately there is even very little quantitative information on tacticity in these families of polymers. Because of this lack of reliable tacticity information (as a result of the insensitivity of NMR spectra to differences in configurational sequences), stereoregularity is often inferred from crystallinity measurements. Unfortunately, however, there are many anomalies in the relationships between polymer crystallinity and apparent stereoregularity in the polyesters and polyamides obtained from β-lactones and lactams.

This chapter will be concerned with the anionic polymerization reactions of five specific types of monomers, including both β-lactones and β-lactams substituted in either the α- or β-positions or both, as shown in Figure 1.

In most cases these monomers have chiral centers, so both the monomers and the repeating units in the polymers will contain asymmetric carbon atoms and will be capable of existing as different optical isomers. Also in most cases investigated to date, racemic mixtures of the monomers were polymerized to give optically inactive polymers, so tacticity could not be determined by, or correlated with, optical activity. Nevertheless, stereoregularity in either the α- or β-substituted polyesters and polyamides prepared from these monomers can be discussed in terms of tacticity in the normal sense as illustrated in Figure 2 for the polymers with either isotactic or syndiotactic dyads in which the substituents are either on the α- or β-carbon position (or both).

α,α-DISUBSTITUTED-β-PROPIOLACTONES

It has been known for approximately 15 years that the homogeneous, anionic polymerization of a racemic mixture of a chiral α,α-disubstituted-α-propiolactones, illustrated below, yields crystalline polymers (1,2), as shown in Equation 1.

One unusual aspect of these polymers is that the repeating units in the crystalline regions can exist in either a helical or a planer, zig-zag conformation, with the two conformations being interchangeable mechanically (1,3). Secondly, there appears to be a most unusual and unexpected linear relationship between polymer melting points and the relative sizes of the two substituents in the α-position as shown by the plot in

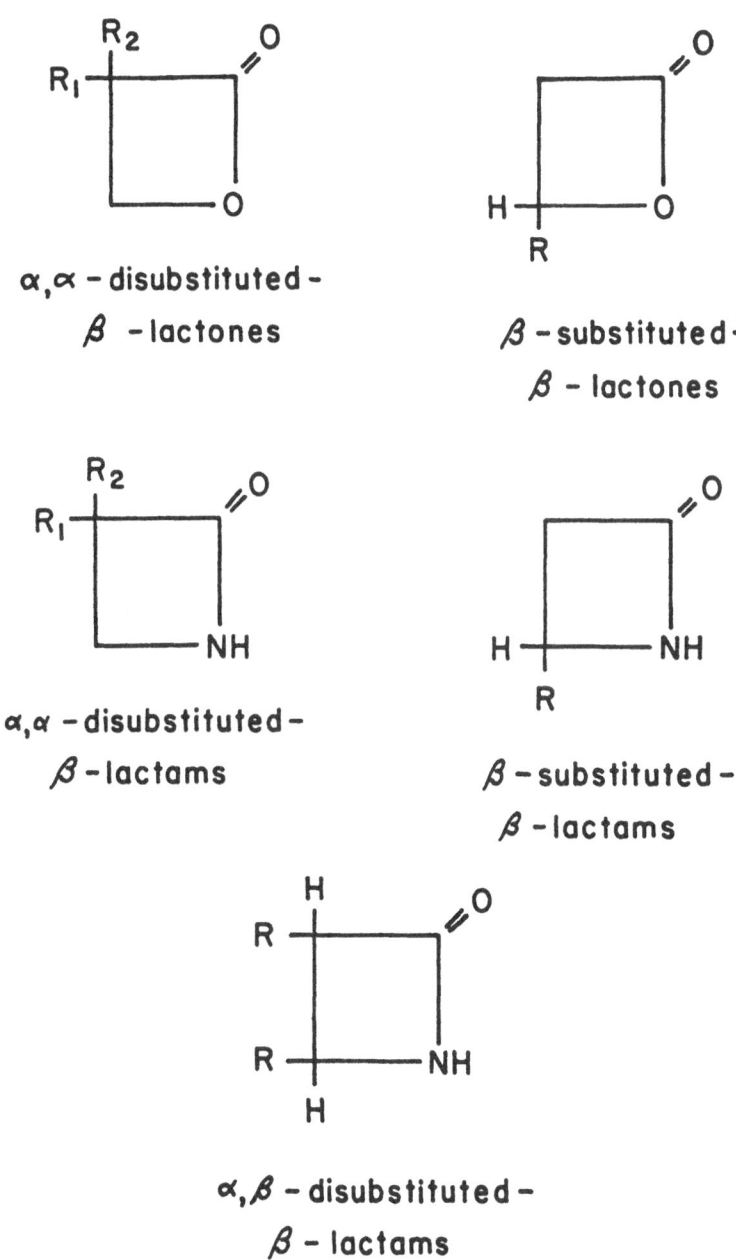

α,α - disubstituted -
β -lactones

β -substituted-
β - lactones

α,α -disubstituted -
β-lactams

β -substituted -
β -lactams

α,β - disubstituted -
β - lactams

Figure 1.   Lactone and lactam monomers discussed in this
             chapter.

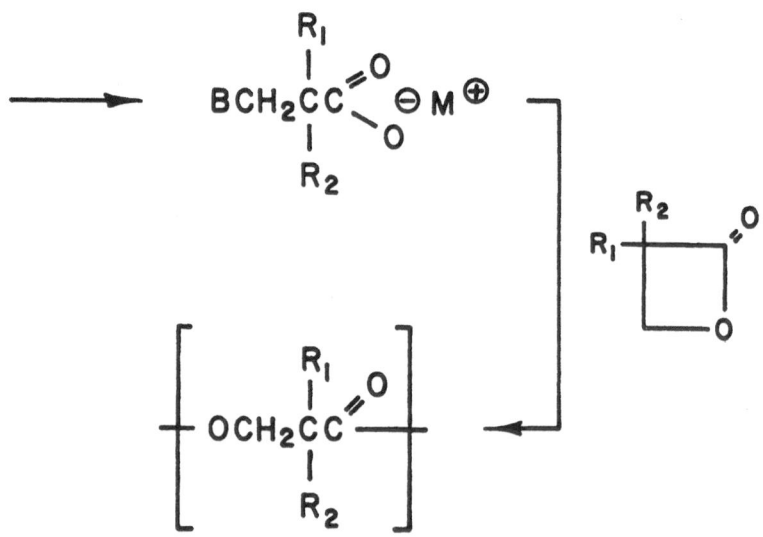

Equation 1

Figure 2 (1).   Thirdly, it was recently shown in the polymerization of α-ethyl-α-phenyl-β-propiolactone that the racemic
monomer mixture, on the one hand, and the optically-active
monomer, on the other, gave polymers with entirely different crystalline properties in these homogeneous anionic polymerization reactions (4).

   In the latter investigation, it was concluded that the
optically-active monomer gave the expected isotactic polymer,
which crystallized with a different unit cell and a different
melting point then the polymer from the racemic monomer.
Unfortunately, NMR characterization has not yet been capable of
specifying the tacticity of the racemic polymer, but recent
wide-angle x-ray investigations have been interpreted to
indicate that this polymer may well have a syndiotactic structure (5).   If so, the syndiotactic sequence may indicate
that an asymmetric induction or stereoelection can occur in the
polymerization reaction due to the steric interaction of
the two substituent groups at the α-position of the monomer
with those on the end-group of the active polymer chain.   This
suggested steric interaction is shown in Equation 2.

   While this proposal for a syndiotactic-type of stereoregularity in these polymers has not been unequivocally demonstrated, the possibility of a sterically-induced, or asymmetric,
selection in these reactions seems very favorable because of
the close proximity of the chiral centers in the active endgroup relative to those in the monomer.   That is, in the
transition state, the two chiral centers would be only five
atoms apart and could easily interact as indicated in
Equation 2.

β-ALKYL-β-PROPIOLACTONES

   Chiral β-substituted-β-propiolactones have also been
found to yield crystalline polymers, and recent NMR investigations have shown these products to be isotactic (6).   The
polymerization reaction systems used to date for these
monomers, however, have involved heterogeneous catalysts,
specifically tralkylaluminum-water systems activated by
third components.   In this type of polymerization reaction,
the ring-opening is believed to occur at the carbon-oxygen
position.

   Again, the crystalline properties of these polymers
were found to be very sensitive to the structure of the
substitutents at the β-position as indicated by the melting
points of the stereoregular polymers obtained from the
following β-alkyl-β-lactone monomers (7):   β-methyl, 175°;

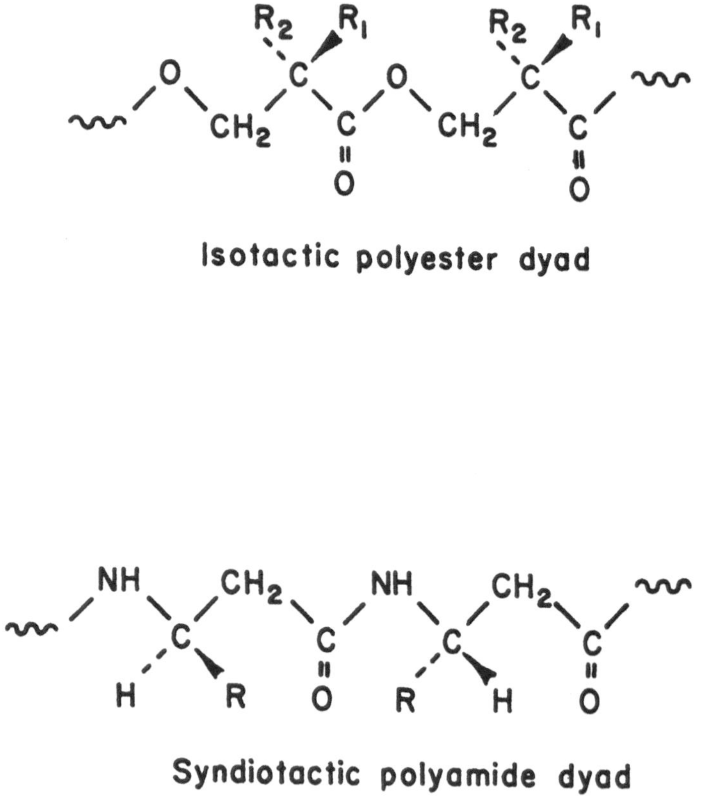

Isotactic polyester dyad

Syndiotactic polyamide dyad

Figure 2. Stereochemical structures of dyads.

Equation 2

β-ethyl, 100°; β-isopropyl, 90°; compared to a melting point
of 120° for poly-β-propiolactone itself. It was also concluded in a study of the crystalline properties of the polymers that the isotactic polymers contained a physical mixture
of the poly-R and the poly-S-β-alkylpropiolactones (8).

The trialkylaluminum-water catalyst used in these systems
is believed to initiate cationic polymerization reactions. Two
possible mechanisms have been suggested for these reactions,
either the S_N1 and S_N2 paths, as shown in Equation 3.

By either mechanisms, S_N1 or S_N2 , either retention
or inversion can occur depending upon whether the nucleophilic
attack is at the carbonyl or the β-alkyl position, respectively,
as shown above. The most likely mechanism appears to be the
S_N2 reaction with carbonyl-oxygen scission and retention of
configuration at the β-position for both the lactone and
lactam. If so, the stereoregularity of the polymer formed
would be primarily a function of the structure of the monomers
used, and isotactic polymers would be readily obtained from
optically-active monomers.

The effective polymerization of β-substituted-β-lactones
seems to be limited to cationic initiators in most reports to
date. However, it was recently observed that activation of the
β-position by a carboalkoxy substituent makes anionic polymerization possible, specifically in the case of the esters of
malolactone (9), as shown in Equation 4.

For this monomer, the crystalline properties of the
polyesters obtained were quite different for the products of
the polymerization reactions initiated with anionic compared to
the cationic initiators.

α,α-DISUBSTITUTED-β-PROPIOLACTAMS

The anionic polymerization of these monomers most likely
proceeds through a mechanism involving carbonly-oxygen scission
as generally proposed for the same type of reaction applied to
the polymerization of either pyrrolidone or caprolactam.
Indeed identical initiation systems may be used as with the
latter monomers consisting of a lactam anion salt as the
initiator and an acyllactam as coinitiator, as shown in
Equation 5.

There is little information about the application of this
type of polymerization reaction to α,α-disubstituted-β-lactams,
but what is known indicates the crystalline structure-property
relationships of these polymers are quite different from those

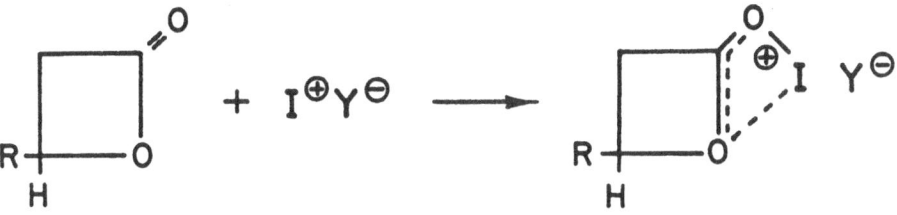

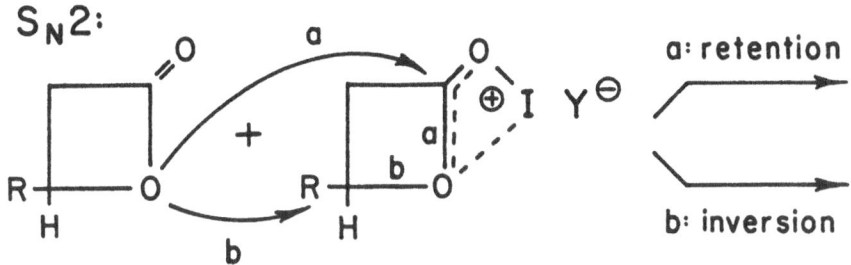

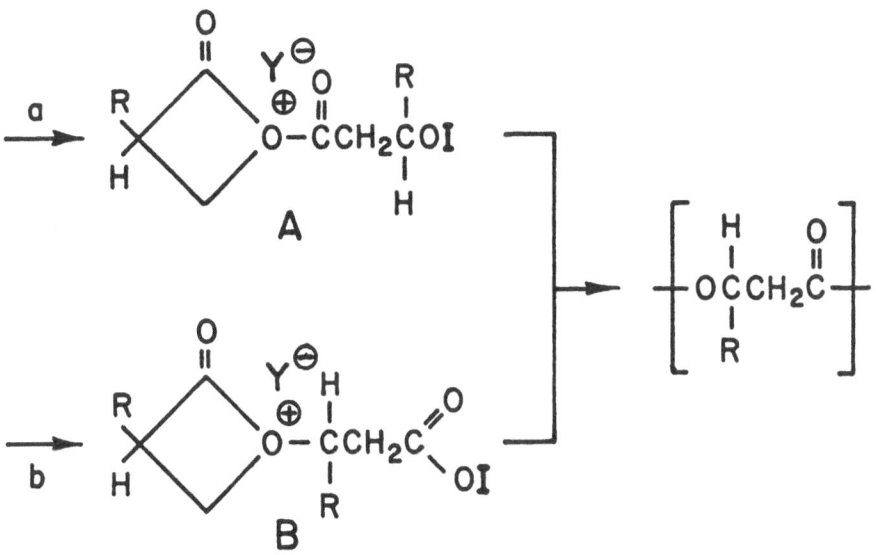

Equation 3

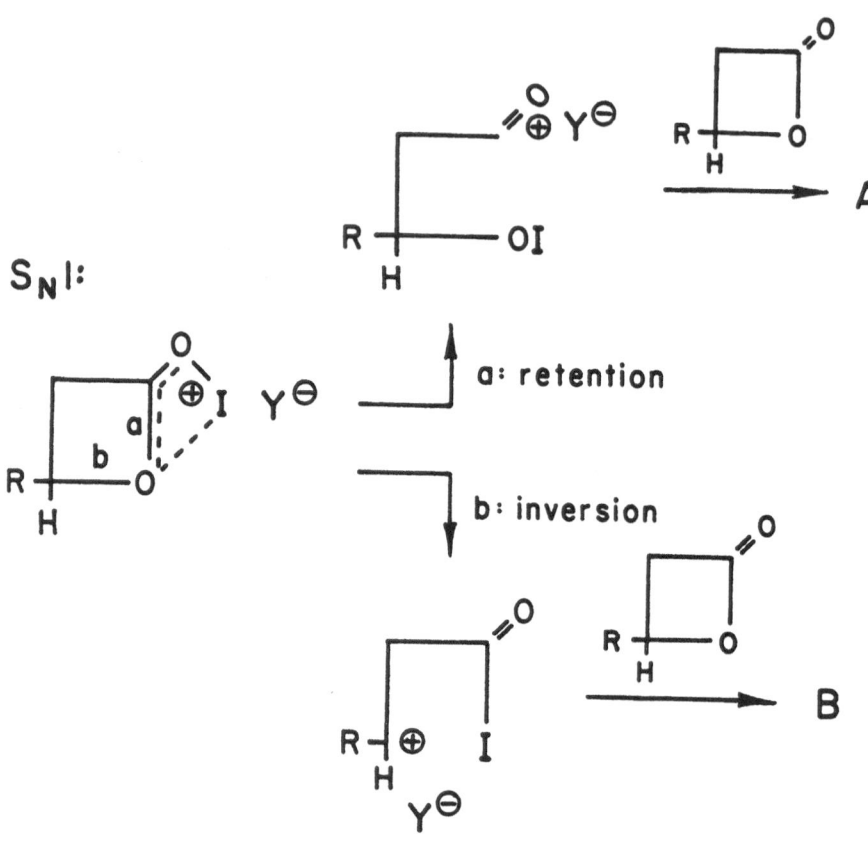

Equation 3

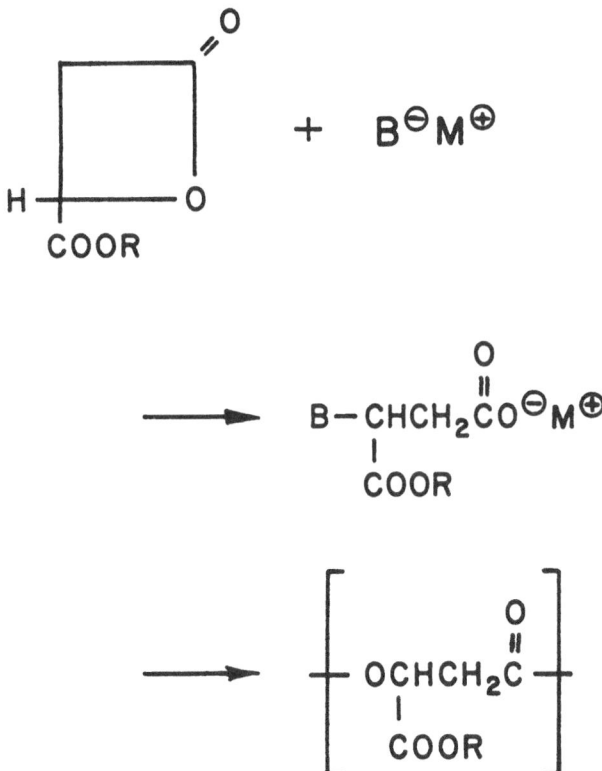

Equation 4

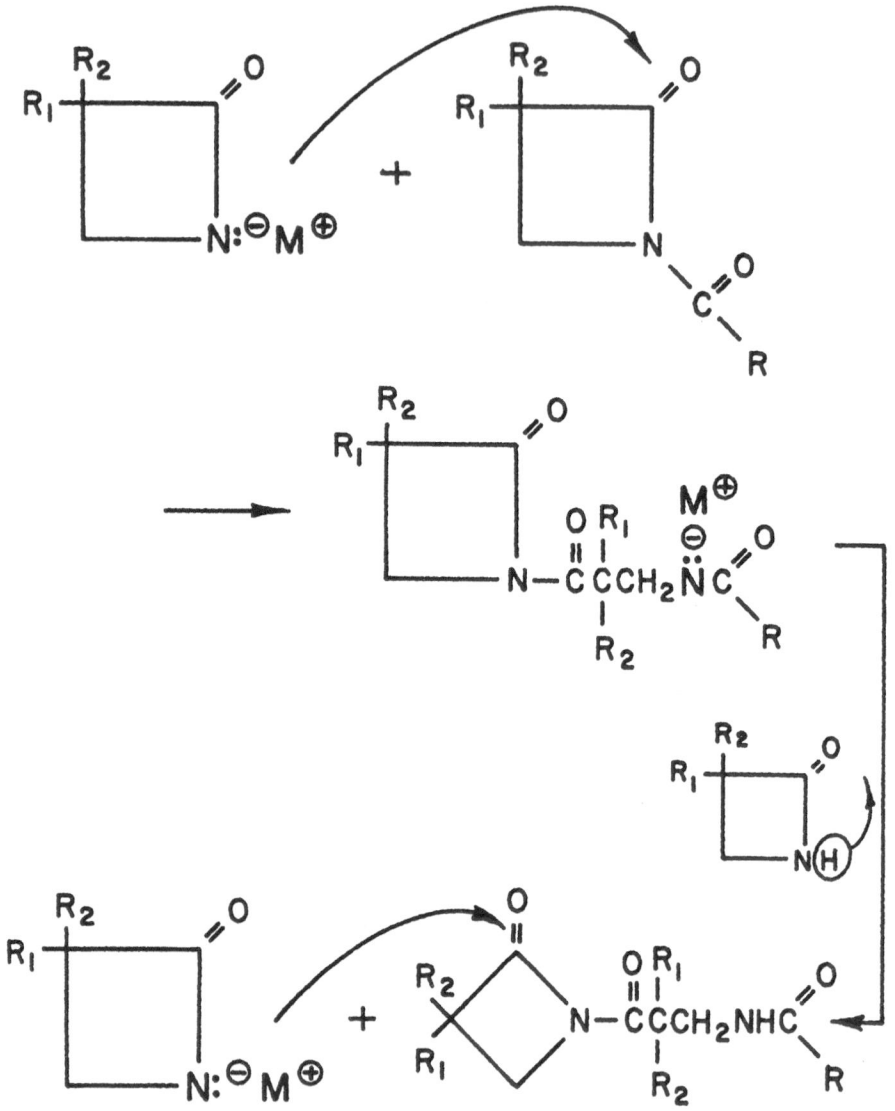

Equation 5

of the equivalent series of polyesters from α,α-substituted-α
lactones (10). Unlike the polyesters from the substituted
lactones, the melting points of the polyamides did not show
a linear correlation with substituent size, but instead, for
the polymers obtained from chiral monomers, the melting
points were essentially insensitive to the type of substitutent
over the following series: α-methyl-α-ethyl, 76°; α-methly-α-
propyl, 74°; α-methyl-α-butyl, 72°; compared to 268° for the
α,α-dimethly polymer (polypivalolactam). Unfortunately, as
with the polyesters from the β-lactones, NMR characterization
has not been capable to date of quantitatively determining
tacticity, and little is known about the relationship
between crystallinity or crystalline properties and stereoregularity.

β-SUBSTITUTED-β-PROPIOLACTAMS

The preparation and polymerization of β-substituted-β-
propiolactams, with formation of crystalline polymers under
apparently homogeneous conditions, has been known for many
years, but no definitive study has been reported on the polymerization reaction involved. The crystalline properties of
the polymers obtained, as a function of the type and location
of substituents in the monomer, are shown in Table 1 (11).
These polymers were prepared with anionic initiators, and it is
likely that the mechanism of polymerization is the same as that
shown above for the α-substituted-β-lactams; that is, only
carbonyl-nitrogen cleavage occurred with no effect on the
configuration of the chiral center so the repeating unit
configurations of the polymers in Table 1 were the same as
those of the monomers used for their preparation.

The differences in the crystalline properties of the
optically active polymers and the racemic polymers again
indicated, as with the α,α-disubstituted-β-propiolactones, that
the racemic monomers formed polymers with different stereo-
regularities than the optically active monomers in these
homogeneous anionic polymerization reactions (11). Since
the optically-active monomers could only form the isotactic
polymers, it can be concluded tentatively that the racemic
monomers formed syndiotactic polymers, possibly because of
steric interactions between the substituents on the β-carbons
of both the attacking monomer and the lactam end-group of the
growing polymer chain. Again, in the transition state, these
chiral centers would be only five atoms apart, and steric
interactions could well account for the stereoregularity
observed.

TABLE 1 - Crystalline Properites of Poly-β-Substituted-
β-Propiolactams (11)

$$\left[-NHCH-CH \overset{\overset{O}{\|}}{\underset{R_2\ \ R_1}{C}} \right]$$

R_1	R_2	Monomer Configuration[a]	Degree of Crystallinity[b]	Polymer Melt Temp, °C
H	CH_3	(-)S	++	345
		(±)	++	328
H	$CH=CH_2$	(-)R	++	340
		(±)	+	320
H	C_2H_5	(-)S	+	380
		(±)	+	330
CH_3	CH_3			
	threo	(-)αS, βS	+	395
		(±)	+	340
	erythro	αR, βS	++	410
		(±)	++	376

[a] Either optically-active, (-)S or (-)R , or racemic
(±) monomers were polymerized.

[b] ++ is for more highly crystalline and + less highly
crystalline polymers.

CONCLUSIONS

All of these results still leave unanswered two major
questions for most of these polymer systems: (1) are the
polymers obtained from racemic monomers by homogeneous polymerization reactions stereoregular, and if so, what is the
extent of the tacticity and how is the stereoregularity obtained; and (2) does the observed crystallinity in these
polymers require stereoregularity, and if so, what degree of
tacticity is required and what is the relationship between
tacticity and crystalline structure and properties? These
unanswered questions indicate that there is still much to be
learned in the study of the mechanisms of polymerizations of
substituted β-lactones and lactams.

LITERATURE CITED

1. R. W. Lenz, Bull. Soc. Chim. Beograd., 39, 395 (1974).
2. R. T. Thiebaut, N. Fisher, Y. Etienne and J. Coste, Ind.
    Plastique Mod., 2, 1 (1962).
3. J. Cornibert, R. H. Marchessault, A. E. Allegrezza, Jr.,
    and R. W. Lenz, Macromolecules, 6, 676 (1937).
4. C. G. D'Hondt and R. W. Lenz, J. Polymer Sci., Polymer
    Chem. Ed., 16, 261 (1978).
5. R. H. Marchessault, J. St. Pierre, M. Duval, and S. Perez,
    Macromolecules, 11, 1281 (1978).
6. M. Iida, S. Hayase, and T. Araki, Macromolecules, 11, 490
    (1978).
7. M. Iida, T. Araki, K. Teranishi, and H. Tani, Macro-
    molecules, 10, 275 (1977); K. Teranishi, M. Iida, T.
    Araki, S. Yamashita, and H. Tani, Macromolecules, 7, 421
    (1974).
8. M. Yokouchi, Y. Chatani, H. Tadokoro, and H. Tani, Polymer
    J., 6, 248 (1974).
9. M. Vert and R. W. Lenz, Polymer Preprints, 19, 608,
    (1979).
10. C. D. Eisenbach, R. W. Lenz, M. Duval and R. H. Marchessault,
    Makromol. Chem., in print.
11. E. Schmidt, Angew. Makromol. Chem., 14, 185 (1970).

PHYSICAL AND MECHANICAL PROPERTIES OF OPTICALLY ACTIVE AND RACEMIC POLY(α-METHYL-α-n-PROPYL-β-PROPIOLACTONE)

Robert E. Prud'homme and Joseph Noah

Chemistry Department, Laval University, Québec 10, P.Q., Canada, G1K 7P4

INTRODUCTION

In a previous paper presented by Dr. Leborgne and Spassky, from the Université Pierre et Marie Curie, in Paris, it has been shown that it is now possible to prepare optically active polylactones by a "stereoelective" polymerization process. It is the purpose of the present paper to indicate important differences in properties between the optically active and the racemic poly(α-methyl-α-n-propyl-β-propiolactone) (PMPPL). Results will be presented concerning the crystallization and melting properties, as well as the dynamic and transient mechanical properties of these polymers.

EXPERIMENTAL

The optically active and racemic polymers used in the present study, hereafter called PMPPL-OA and PMPPL-R, respectively, are described in Table I. Both samples have high molecular weights. The PMPPL-OA sample shows a significant value of optical rotation while the PMPPL-R sample does not. Optical rotations were measured in chloroform for concentrations of 0.4 g/dl.

The differences in physical properties reported in the next sections of this paper for the PMPPL-OA and the PMPPL-R samples are truly related to a difference in optical activity and not to a difference in initiator. This point is shown in Table II where PMPPL were prepared using five different initiators. All samples presenting no optical rotation have a melting point of 357-359 K, while that being optically active, has a melting point of 369 K.

259

R. W. Lenz and F. Ciardelli (eds.), Preparation and Properties of Stereoregular Polymers, 259-262.
Copyright © 1979 by D. Reidel Publishing Company.

Sample	Initiator	M_{osmo}	$[\alpha]$
PMPPL–OA	ZnEt$_2$–(–)–3,3–dimethyl–1,2–butanediol	26,000	–0.6
MPPL–R	Tetrahexylammonium benzoate	88,000	0.0

Table I: PMPPL characterization data

The melting points reported in Table II were recorded at the end of the melting process, made at a heating rate of 40 K/min, on a DSC-IB Perkin-Elmer calorimeter for samples prepared by isothermal crystallization at 273 K, during 4 h.

Sample	Initiator	M_{osmo}	$T_m(K)$	$[\alpha]$
1	Tetrahexylammonium benzoate	88,000	359	0.0
2	Sodium mirror	600,000	359	0.0
3	ZnEt$_2$–H$_2$O	–	358	0.0
4	Spartéine	340,000	357	0.0
5	ZnEt$_2$–(–)DMDB	–	359	0.0
6	ZnEt$_2$–(–)DMBD	26,000	369	–0.6

Table II: Influence of initiator on MPPL polymerization

CRYSTALLIZATION AND MELTING

Isothermal crystallization experiments were conducted on the DSC-IB apparatus, with both samples and the crystallization was followed by a melting experiment at a heating rate of 40 K/min. Enthalpies of fusion were then measured. Some results obtained from these experiments are presented in Table III.

It is seen that when the crystallization is conducted at the same temperature for both samples, 273 K, the PMPPL-OA sample crystallizes much faster than the racemic one and that it also reaches higher values of enthalpy of fusion and thence higher degrees of crystallinity. But it has been shown earlier that the equilibrium melting point of the optically active sample is 13 degrees larger than that of the racemic polymer (383 and 370 K, respectively). It is then necessary to compare samples prepared at the same degree of supercooling. In Table III,

Crystallization temperature (K)	Time of crystal-lization (min)	ΔH_m (J/g)	
		PMPPL-R	PMPPL-OA
273	5	0.3	8.1
273	15	0.8	10.0
273	25	0.8	12.0
273	45	1.7	12.7
273	60	2.5	12.7
273	90	6.5	-
273	120	9.3	-
273	240	10.3	-
285	5	-	6.8
285	10	-	7.6
285	15	-	9.0
285	25	-	9.8
285	45	-	9.6
285	60	-	10.5
313	5	-	7.7
313	35	-	12.9
313	60	-	14.1

Table III: PMPPL crystallization behavior

one can compare the crystallization data obtained for PMPPL-OA
at 285 K to those obtained for PMPPL-R at 273 K. Still the
optically active sample crystallizes faster and reaches higher
degrees of crystallinity.

MECHANICAL PROPERTIES

Relaxation experiments have been made for both samples in the
temperature range 230-320 K. The relaxation results can be reduced to a master curve by using an horizontal shift uniquely.
The shift factor obeys the WLF equation in the Tg temperature
range but important deviations are observed at low temperatures
due to the higher viscosity of the samples in the glassy region
and at high temperatures, in the rubbery region, where chain
entanglements play a significant role. These deviations are different for the racemic and the optically active polymers. Both
polymers have an apparent activation energy, ΔH, which varies
with temperature. As is shown in Table IV, the maximum value of
ΔH, ΔH_{max}, is much larger for PMPPL-R than for PMPPL-OA. Similarly, the moduli values (E_r, E', E''), the relaxation function
H(t), the loss tangent Tg, tan δ_{max}, are larger for PMPPL-R than
for the optically active sample. This behavior is certainly

Property	PMPPL-R	PMPPL-OA
Tg, obtained from a dynamic experiment, at 3.5 Hz (K)	285	283
Tg, obtained from a relaxation experiment (K)	270	266
Activation energy at Tg (kJ/mol)	132	127
Elongation at rupture (%)	≈750	≈750
log E_r(100 s) at 298 K (in Pa) at 251 K	8.0 9.21	7.94 9.04
ΔH_{max} (kJ/mol)	610	480
H_{max}(t) (Pa)	8.00	7.60
log E''_{max} (Pa)	8.30	7.92
tan δ_{max}	0.26	0.18

Table IV: Summary of the main mechanical properties
measurements

related to the higher degree of crystallinity of the latter sample.

In this study, Tg was measured by various means. Each method
used indicates a larger Tg value for PMPPL-R than for PMPPL-OA.
This behavior is different from that found for the melting points
(Table II). It seems to indicate that the optical activity in-
fluences mainly the crystalline part of the polymer and not very
much its amorphous fraction. The small difference in Tg between
the two polymers can be related to their difference in molecular
weight.

CONCLUSIONS

Important differences are observed between the properties of
optically active and racemic PMPPL. These results, and particularly the crystallization and the melting data, can be interpreted in terms of a block structure for the two samples. But
our work does not permit one to exclude completely other kinds
of structures.

NMR ANALYSIS OF STEREOREGULAR HOMOPOLYMERS

H. James Harwood

Institute of Polymer Science
University of Akron, Akron, Ohio 44325

The nuclear magnetic resonance spectra of stereoregular polymers are useful in many ways. The spectra
of almost perfectly stereoregular polymers can be
interpreted in terms of chemical shifts and coupling
constants associated with the various nuclei present
and these provide information about the conformations
of the polymers in solution. Studies on partially
deuterated polymers are often necessary to obtain complete interpretations of the spectra and these can
provide information about the mode of bond opening in
stereoregular polymerizations. When resonances due
to structural imperfections can be detected, they provide a means for characterizing the steric purity of
the polymers and careful analysis of them often provides information about the stereoregulating process.
End group studies on living chains often provide information about the nature of the propagating species
in stereoregular polymerizations.

The first portion of this article will concern
vinyl polymers which contain predominantly meso (isotactic) or racemic (syndiotactic) placements. Attention will then be given to the study of homopolymers
containing appreciable amounts of both meso and racemic
placements, since the study of such materials is necessary if trends toward stereospecificity or stereoselection

263

R. W. Lenz and F. Ciardelli (eds.), Preparation and Properties of Stereoregular Polymers, 263-293.
Copyright © 1979 by D. Reidel Publishing Company.

in polymerization systems are to be ascertained.

$$\begin{array}{cc} X & X \\ | & | \\ -C-CH_2-C- \\ | & | \\ Y & Y \end{array} \qquad \begin{array}{cc} X & Y \\ | & | \\ -C-CH_2-C- \\ | & | \\ Y & X \end{array}$$

meso (m)   racemic (r)

The pmr spectra of stereoregular vinyl polymers are difficult to interpret because the intrinsic resonances of the protons often have very similar chemical shifts and because spin-spin coupling causes many lines to be observed for each type of proton. These multiplets are are often overlapped and this makes analysis of the spectra difficult. The analysis of spectra recorded with higher field spectrometers is simplified because the multiplets are less overlapped, however, and the analysis is also simplified if partially deuterated polymers can be studied.

As a starting point for discussing the spectra of stereoregular polymers, idealized spectra that are uncomplicated by spin-spin interactions will first be considered. (The individual lines considered are separately observable in the spectra of polymers in which all possible H-H spin-spin couplings have been eliminated by appropriate deuterium substitution.) Spin-spin coupling will be gradually introduced into the spectra until idealized completely coupled spectra are presented. Resonances due to protons of groups pendant from the backbone will not be considered at this point.

UNCOUPLED SPECTRA OF ISOTACTIC AND SYNDIOTACTIC POLYMERS

In the absence of spin-spin interactions, the spectrum of a syndiotactic polymer should consist of two lines, one for the methylene protons, which are magnetically equivalent and one for the methine protons. Since the methylene protons of isotactic polymers are not magnetically equivalent, three lines would be observed in the absence of spin-spin couplings.

It is convenient at this point to discuss a nomenclature for the methylene protons of isotactic polymers. Consider the following segment of an isotactic
polymer chain, the chain carbon atoms being written as
a planar zig-zag.

H_A is opposite (anti) to both R groups in this conformation and it bears a <u>threo</u> configurational relationship
to the methine protons. Thus, it is customary to refer
to H_A as the <u>anti</u>- or <u>threo</u>-meso proton. Since H_B is
on the same side of the planar zig-zag chain as the R-
groups, and since it bears an erythro configurational
relationship to the methine protons, it is referred to
as the <u>syn</u>- or <u>erythro</u>-meso proton. It is not easy to
assign methylene proton resonances to <u>anti</u>- or <u>syn</u>-
protons, but some assignments have been made based on
model compound studies [1], on considerations of bond
anisotropies [2], or on theoretical analyses of the
infrared spectra of partially deuterated polymers
[3-5]. X-ray diffraction analyses are also helpful for
similar assignments in polymers derived from α,β-di-
substituted monomers [6,7]. When assignments can be
made, they provide a basis for investigating the mode
of bond opening or the directions in which bonds are
established during stereoregular polymerizations.

COUPLED SPECTRA OF ISOTACTIC POLYMERS

Coupling between the methylene protons of isotactic
polymers results in a typical AB pattern which is usually complicated by additional fine structure due to
coupling with methine protons. The coupling constant
between methylene protons (J_{AB}) is ~13.5 Hz, typical
for geminal protons, whereas the couplings between
methine and either syn (J_{BC}) or anti (J_{AC}) methylene
protons in isotactic polymers are ~6-7 Hz. Since each
methylene proton is coupled to two methine protons,

each of the lines in the methylene quartet is further
split into a triplet. Since $J_{AB} \sim 2J_{BC} \sim 2J_{AC}$, the inner
and outer triplets are partially overlapped and a ten
line pattern is to be expected for the methylene proton resonances. Such a pattern is evident in the
spectra of polymers such as isotactic poly(isopropyl
acrylate) [8,9], isotactic 1,2-polybutadiene [10] and
isotactic poly(3,3,3-tri-deuteriopropylene) [11] where
there is an appreciable difference between the chemical
shifts $(\Delta\delta=\delta_A-\delta_B)$ of the two methylene protons. When
$\Delta\delta$ is less than $\sim$30 Hz (0.1 ppm at 300 MHz or 0.5 ppm
at 60 MHz), the ten line pattern overlaps itself.
Thus, only eight methylene proton resonances are evident in the 220 MHz pmr spectrum of isotactic polystyrene $(\Delta\delta_{AB}=0.06$ ppm) [12] and in the 300 MHz pmr
spectrum of poly(butene-1 $(\Delta\delta=0.07$ ppm) [13].

    The methine protons in isotactic polymers are
coupled to four methylene backbone protons and may be
coupled to protons on the pendant R group as well. If
we avoid the latter possibility, and remember that
$J_{BC} \sim J_{AC}$ for isotactic polymers, the methine resonance
pattern may be expected to be a quintet, a result of
three overlapping triplets. The scheme shown in
Figure 1 shows how the typical backbone resonance
pattern of isotactic polymers arises from spin-spin
interactions. The methine proton resonances of isotactic polymers can thus be analyzed as the A-part of
an AB_2C_2-spin system and the methylene proton resonances can be analyzed as the BC part of an A_2BC-spin
system. Some investigators prefer to interpret such
spectra in terms of the 6-spin system shown below, but
we have not found this to be of advantage in our own
work [10].

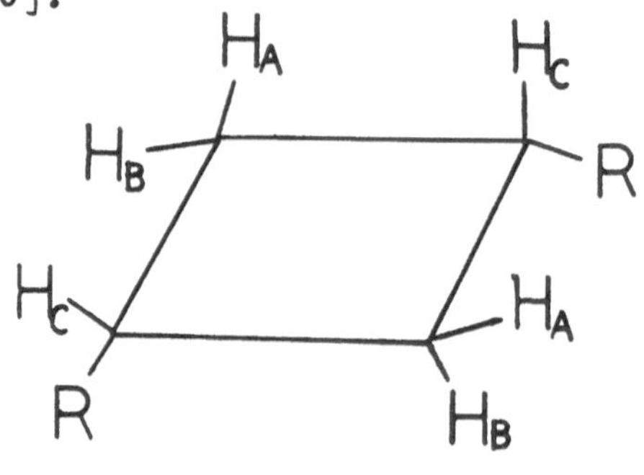

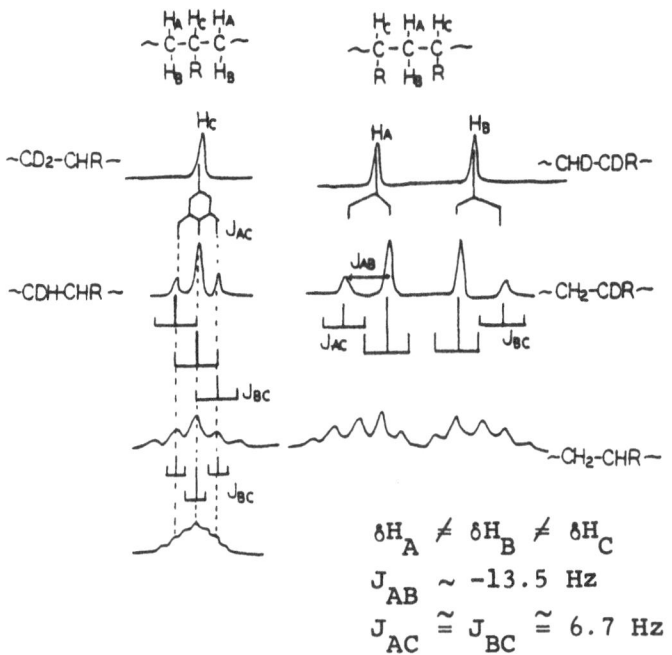

$$\delta H_A \neq \delta H_B \neq \delta H_C$$

$$J_{AB} \sim -13.5 \text{ Hz}$$

$$J_{AC} \cong J_{BC} \cong 6.7 \text{ Hz}$$

Figure 1.   Spin Systems in Isotactic Vinyl Polymers

The effects of spin-spin coupling on the spectra of polymers can be eliminated by decoupling experiments or by investigating partially deuterated polymers. This latter approach is particularly effective where the spectra are badly overlapped, as in the spectra of polyhydrocarbons. Zambelli, Segre and colleagues [11,14,15] have recorded the spectra of a large variety of deuterated isotactic polypropylenes and a number of these are compared in Figure 2. The spectrum of iso-tactic poly(1,2,3,3,3-pentadeuteriopropylene), Fig-ure 2C, contains only the resonances of anti-methylene protons. These are coupled with syn-methylene protons in isotactic poly(1,3,3,3-tetradeuteriopropylene), Figure 2D, and the expected quartet is seen. In the spectrum of isotactic poly(3,3,3-trideuteriopropylene), Figure 2F, coupling of methine and methylene protons is complete; the expected ten peak methylene resonance pattern and the methine quintuplet are clearly evident, although there is minor overlapping between methine and methylene proton resonances. Coupling between methine

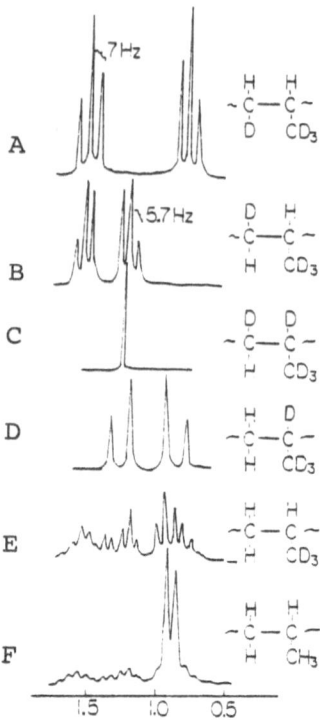

Figure 2. PMR Spectra of Partially Deuterated Isotactic Polypropylenes [12,14,15].

protons and either syn- or anti-methylene protons is
responsible for the patterns observed in Figures 2A
and 2B. Since each proton in isotactic poly(1,2-
dideuteriopropylene) is coupled to two other protons,
the individual proton resonances are observed as triplets from which the methine-methylene couplings can
be measured directly. Coupling between methine and
syn- or anti-methylene protons is thus seen to be 7
and 5.7 Hz respectively. Finally, the spectrum of isotactic polypropylene is shown for comparison in Figure 2F. The methyl proton resonances overlap with
the syn methylene proton resonances and coupling of
methyl protons with methine protons complicates the
methine proton resonance pattern even further. The

value of investigating partially deuterated polymers
in interpreting the nmr spectra of stereoregular polymers is clearly indicated by Figure 2.

Figures 3 and 4 compare the 300 MHz spectra of
isotactic poly(butene-1) and isotactic polystyrene
with the spectra of partially deuterated analogs. The
resonances of methine, methylene and methyl protons
are adequately resolved in these spectra. An AB pattern is observed for the methylene proton resonances
in Figure 3A, as expected, but is barely evident in
Figure 4C, due to the fact that the methylene protons
in polystyrene have very similar shifts.

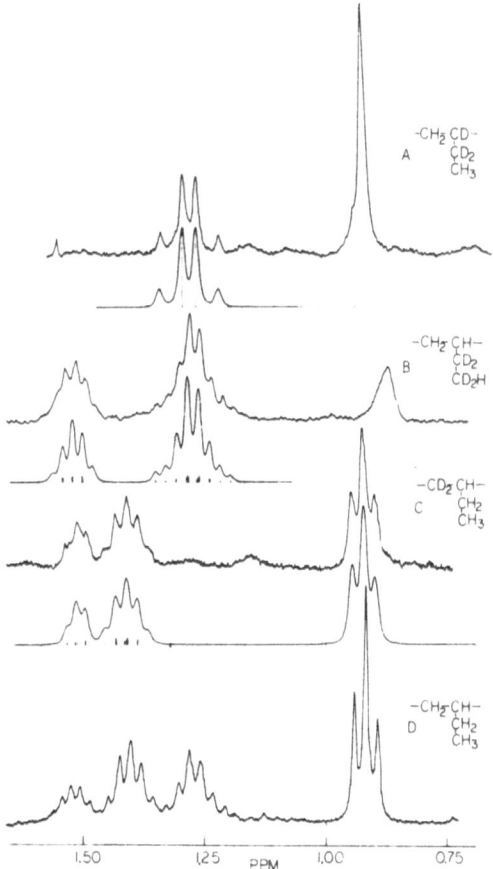

Figure 3. 300 MHz PMR Spectra of Partially Deuterated
Isotactic Polybutenes [13].

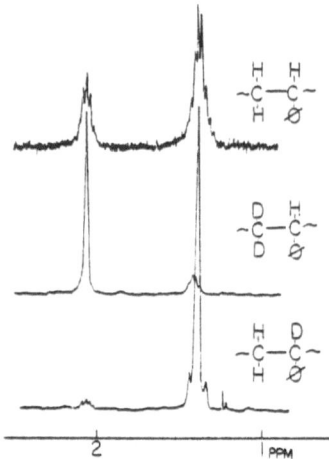

Figure 4.   300 MHz PMR Spectra of Partially Deuterated
Isotactic Polystyrenes [16].

COUPLED SPECTRA OF SYNDIOTACTIC POLYMERS

The methylene protons in vinyl polymers containing
only racemic (r) placements (syndiotactic), have the
same chemical shift (homosteric) and are observed as
singlets when they are not coupled to methine protons
[e.g., syndiotactic poly(methyl methacrylate), syndiotactic poly(α-deuteriopropylene)]. However, the
methylene protons in syndiotactic polymers are not
coupled equally to methine protons, and this leads to
their being observed as four-line patterns in polymers
such as syndiotactic poly(1-deuteriopropylene), where
they are not simultaneously coupled to the same methine
proton. The origin of the four-line patterns is explained in Figure 5. However, when the methylene
protons are simultaneously coupled to methine protons,
they behave as though they are equally coupled to the
methine protons and the spectra becomes <u>deceptively</u>
<u>simple</u> [17,18]. Thus, in the spectrum of syndiotactic
poly(3,3,3-trideuteriopropylene, Figure 6B, the methylene proton resonance is observed as a triplet and the
methine proton resonance occurs as a quintuplet. Figure 6 shows the spectra of a series of partially deuterated syndiotactic polypropylenes that were reported
by Zambelli and coworkers [19,20].

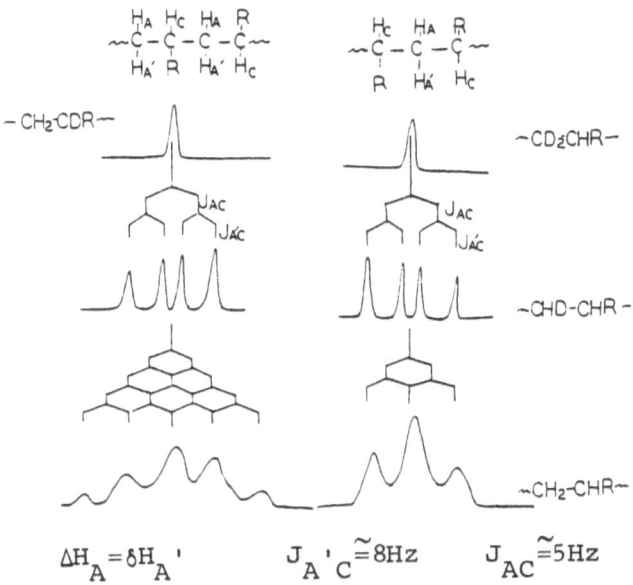

$$\Delta H_A = \delta H_A{}'   \qquad J_A{}'_C \cong 8Hz   \qquad J_{AC} \cong 5Hz$$

Figure 5.   Spin Systems in Syndiotactic Vinyl Polymers.

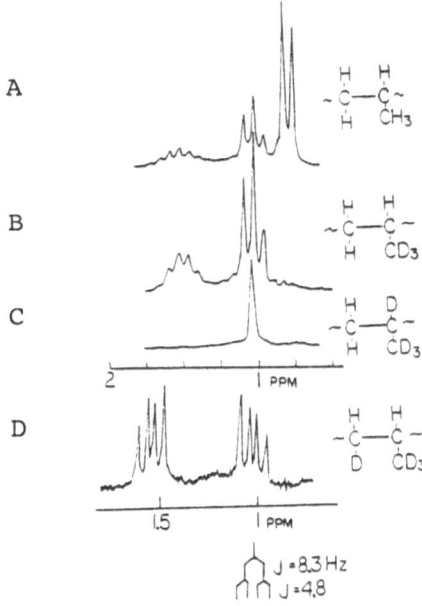

Figure 6.   PMR Spectra of Partially Deuterated Syndiotactic Polypropylenes [19,20].

The [1]H-spectra of isotactic and syndiotactic
polymers derived from cyclic ethers, thioethers and
esters should consist only of the resonances expected
for the spin systems present in the individual monomers.
AB patterns are typically observed for the backbone
protons in poly(α-deuterio-propylene oxide) [21],
poly(α-deuterio-propylene sulfide) [22], polylactide
[23], and related materials.

The ^{13}C-spectra of stereoregular polymers (decoupled from protons) consist simply of signals for
the individual atoms present in the repeat groups and
are not particularly interesting. However, relaxation
times can be determined for individual carbon nuclei
[24-26] and these provide information about the conformations of stereoregular polymers in solution. Similar
information is provided by ^{13}C chemical shifts, since
progress is being made in calculating them from conformational energies and bond anisotropies.

MODE OF BOND OPENING IN STEREOSPECIFIC
POLYMERIZATIONS [27]

Isotactic Polymers [28]

Provided that syn- and anti-meso protons can be
identified in the pmr spectra of isotactic polymers,
the mode of bond opening during their preparation can
be determined. Thus, as is shown below, the isotactic
polymer derived from cis-1,3,3,3-tetradeuteriopropylene
(cis in the sense that the two hydrogen atoms are cis
to each other) will contain only syn-meso methylene
protons if enchainment occurs by cis-addition. The
spectrum of such a polymer is depicted in Figure 2A.
On the other hand, the polymer would contain only anti-
meso methylene protons if enchainment occurs by trans-
addition. The spectrum of such a polymer is depicted
in Figure 2B. This same spectrum would be obtained for
poly(trans-1,3,3,3-tetradeuterio-propylene) if enchainment occurred by cis-addition.

By using this approach, Zambelli and coworkers [28],
have shown that cis-enchainment occurs during Ziegler-
Natta catalyzed polymerizations of propylene. Similar
studies on isotactic poly(acrylic esters) [1,29-33]
isotactic poly(ethyl cis-d_1-methacrylate) [34], predominantly isotactic poly(vinyl ethers) [2,7,21,35,36]
and poly(propylene sulfide) [22] have been reported.
It should be mentioned that nmr shows the final result
of monomer enchainment and does not necessarily provide
information about the stereochemistry of the transition
state for enchainment [34,37]. For example, the net
result of 1,2-cis addition, followed by inversion of
the configuration of the 2-carbon atom is apparent
trans-addition.

Syndiotactic Polymerizations

Since the methylene protons in syndiotactic polymers have the same chemical shift, it is not possible
to decide between cis- and trans-bond openings by
studying homopolymers. Zambelli and coworkers [38]
have shown that this difficulty can be solved by studying syndiotactic copolymers of completely and partially
deuterated monomers. For example, copolymerization of
perdeuteriopropylene with small amounts of either cis-
or trans-1,3,3,3-tetradeuteriopropylene yields a
polymer in which the syn and anti methylene protons
can be distinguished by the magnitude of their coupling
to the methine protons. Thus, an AB pattern will be
observed for the protons in the repeating group shown
below, which would be the result of a trans-opening of
the double bond in the cis-monomer.

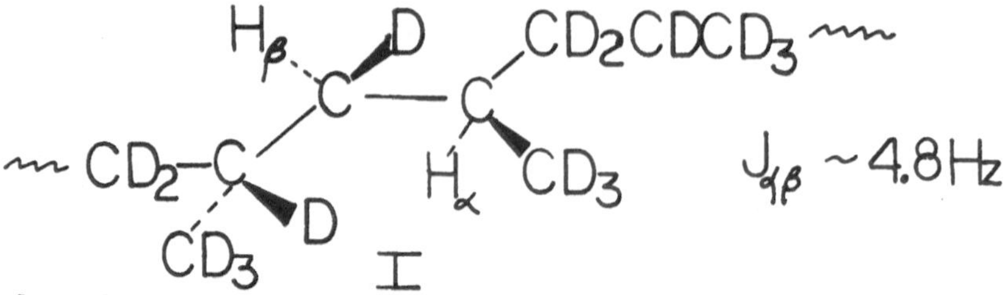

$$J_{\alpha\beta} \sim 4.8\,Hz$$

The structure shown below would result from cis-opening of the double bond in the cis-monomer or of trans-opening of the double bond in the trans-monomer.

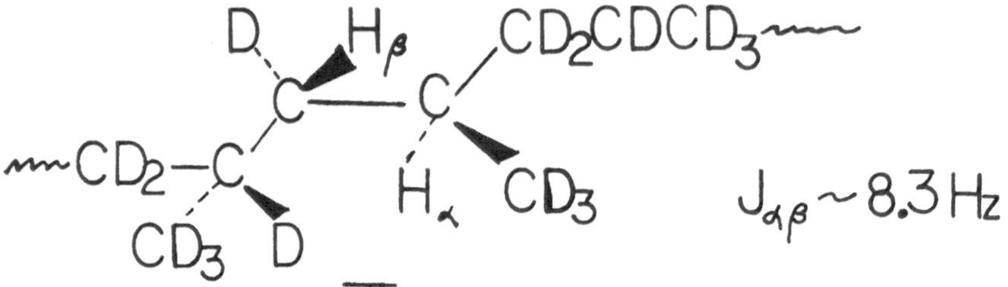

$$J_{\alpha\beta} \sim 8.3\,Hz$$

Since (TT) and (TG) conformations are the most favorable ones for syndiotactic polymers, the α and β protons will predominantly bear a gauche relationship to each other in Structure I whereas the relationship will be predominantly trans in Structure II. On the basis of Karplus relationship [39,40] it is to be expected that the coupling between the α- and β-protons ($J_{\alpha\beta}$) in I will be smaller than in II. Thus, if $J_{\alpha\beta}$ for the copolymer derived from the cis-monomer is less than that derived from the trans-monomer, trans opening of the double bond may be assumed. In fact, the opposite result was obtained by Zambelli and coworkers [38], indicating that cis-double bond opening occurs during the syndiotactic polymerization of propylene.

     The general approaches outlined here can also be applied to studies of bond opening in nonstereoregular polymerizations and copolymerizations. It is useful to note at this point that, in general, cis- and trans-openings are equally likely in free radical polymerizations and copolymerizations of vinyl monomers. This undoubtedly is a result of racemization of propagating radicals prior to subsequent propagation.

NMR STUDIES ON CONVENTIONAL POLYMERS

Since it is not always possible to prepare such highly isotactic or syndiotactic polymers as are obtained from hydrocarbon monomers, it is important to consider the spectral features of polymers containing appreciable amounts of both meso (m) and racemic (r) placements. Analyses of such polymers can be very valuable in evaluating tendencies for stereoregulation in polymerization systems. In addition, the assignments made in such studies are helpful for identifying the nature of imperfections in highly stereoregular polymers.

^{1}H-NMR and ^{13}C-NMR are very useful for characterizing sequences of m or r placements in polymers. Thus, the resonances of methylene protons or carbons can be used to characterize dyad (m,r), and tetrad (mmm, rmm, mrm, etc) stereosequence distributions, whereas the resonances associated with the other atoms can be used to characterize triad (mm, mr + rm, rr); pentad (mmmm, rmmm + mmmr, rmrm + mrmr, rmmr, etc) or higher distributions. Shown below is a representative segment of a vinyl polymer chain (Fischer projection), along with listings of the various dyads, triads, tetrads and pentads present.

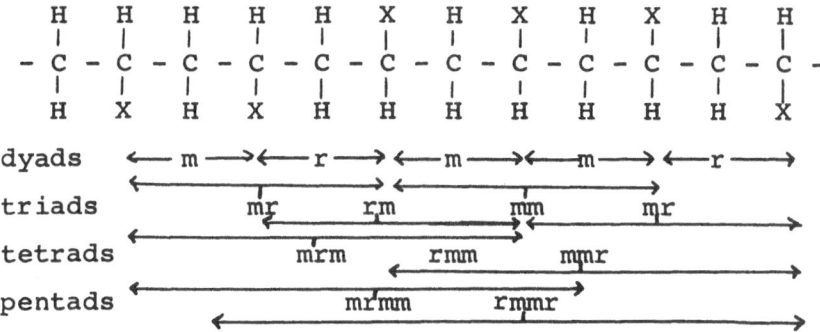

Spin couplings of protons in these various environments are similar to those encountered in isotactic and syndiotactic polymers. For example, AB quartets are observed for meso methylene protons and for some

r-methylene protons (e.g., mrr) in the spectra of
polymers derived from α-substituted vinyl monomers.
Multiplets are observed for the methylene and methine
protons in polymers derived from vinyl monomers.
These are usually badly overlapped and deuterated polymers must be studied if stereosequences are to be
evaluated. Although the chemical shifts of protons in
various stereosequences in polymers are very similar,
dyad and triad distributions are generally measurable
and studies on partially deuterated polymers often en-
able tetrad and pentad sequence distributions to be
measured [41-47]. In a few instances, shift reagents
have been helpful in separating resonances of various
stereosequences in polymers [48-50], but they are much
more valuable for studies on copolymers than for homopolymers.

     Hatada, et al, have also shown that interferring
resonances can be eliminated by a proper selection of
the pulse sequence when recording Fourier-transform
nmr spectra [51].

Statistical Relationships

     When the relative amounts of various stereose-
quences can be measured by nmr, it is possible to
discern what statistical relationships exist among the
various structural features and thus learn something
about the stereo-regulating processes occurring during
polymer formation. In such considerations, stationary
probabilities (e.g., P(m), P(mr), etc.) may be used to
represent the relative amounts of dyads, triads and
n-add placements and conditional probabilities (e.g.,
P(m/m), P(r/mm), etc.) may be used to represent the
probabilities that m or r placements follow particular
sequences in the polymer. The expression P(m/r), for
example, refers to the probability that a meso placement follows a racemic placement in a polymer chain.
It can be calculated from triad concentrations as
follows:

$$P(m/r) = \frac{P(rm)}{P(rm)+P(rr)}$$

Other conditional probabilities are defined and evaluated similarly.

When $P(m)=P(r)=P(m/m)=P(m/r)=P(m/....)=0.5$, then no stereoregulation occurs and the polymers are truly atactic. When $P(m)=P(m/m)=P(m/r)=P(m/mr)=P(m/....)$ and $P(r)=P(r/m)=P(r/r)=P(r/mr)=P(r/....)$ but $P(m)\neq P(r)$, then the polymerization process and the distribution of placements in the polymer is considered to obey Bernoullian or zero order Markoffian statistics. In such a case, a single parameter, $\sigma=P(m)$, is adequate to characterize the structure of the polymer. This is commonly the case for polymers prepared by free radical initiated processes.

When $P(m/m)=P(m/mm)=P(m/rm)=(1-\beta)\neq P(m/r)=P(m/mr)=P(m/rr)=\alpha$ and $P(r/m)=P(r/mm)=\beta\neq P(r/r)=P(r/mr)=P(r/rr)=(1-\alpha$, the placement distribution obeys 1st order Markoffian statistics. Then, two parameters (α,β) are necessary to characterize the polymer. First order and higher Markoffian statistics are often required for polymers prepared by ionic polymerization techniques. In cases where several propagating sites are involved in the polymerization process, more complicated statistical representations, such as that proposed by Coleman and Fox [52-53] may be required. The reader is referred to Bovey's text for more information on this point [41].

Assignment of Resonances

A major difficulty in interpreting the nmr spectra of polymers containing both meso and racemic placements is the assignment of observed resonances to structural features present in the polymers. When only resonances due to dyads or triads are observed, their assignment is rather straightforward, especially if spectra of stereoregular polymers are available for reference. ^{13}C-spectra or ^{1}H-spectra recorded with 220- or 300-MHz spectrometers often resolve resonances assignable to tetrad, pentad or even higher stereosequences, however.

Since there are six observable tetrads and ten
observable pentads, it is a formidable task to make
resonance assignments for these structural features.
However, there are a number of general approaches for
doing this. These include: (a) the use of stationary
or "necessary n-add relationships" to correlate the
relative intensities of resonances; (b) the development of empirical relationships to correlate differences in chemical shift due to structural differences;
(c) correlation of observed resonance areas with n-add
distributions calculated on the basis of a particular
statistical model (Bernoullian, Markhoffian, etc.);
(d) the synthesis and study of model structures; (e)
the study of epimerized polymers; and (f) theoretical
calculation of chemical shifts.

Since all of the methods discussed above have been
applied to analysis of the methyl carbon resonances in
polypropylene, the assignment of these resonances will
be covered in detail to illustrate the methods. Nine
methyl carbon resonances are observed (Figure 7), the
lowest field one of which has the same chemical shift
as the methyl carbons in isotactic polypropylene and

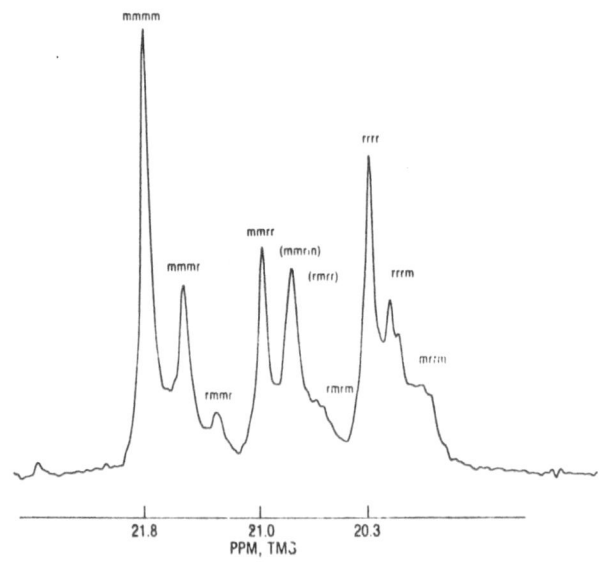

Figure 7. Methyl Carbon Resonances of Polypropylene
[55].

which may be assigned to mmmm pentads. On the same
basis the highest field methyl carbon resonance may be
assigned to rrrr pentads. Epimerization of isotactic
polypropylene to low extents yields a polymer whose
spectrum contains two additional signals of relative
intensity 1:2:2. These must necessarily be assigned
to mrrm (intensity=1) and mmrr (intensity=2) and mmmr
(intensity=2). Similarly, the spectrum of partially
epimerized syndiotactic polypropylene contains three
minor peaks of relative intensity 1:2:2 that may be
assigned to rmmr (intensity=1), rrmm (intensity=2) and
rmmm (intensity=2). Since rrmm is common to both epimerized isotactic and syndiotactic polymers, epimerization can be used for unequivocal assignments for six
of the ten possible pentads [55]. Tetrad distributions
for polypropylene can be measured from the methylene
proton resonance region. These distributions must be
consistent with the pentad distributions evaluated from
the methyl carbon resonances. "Necessary n-add relationships" such as those listed in Table I must therefore be obeyed. The requirement that these relationships hold can become the basis for making additional
structural assignments. These relationships are
discussed in a number of articles and texts [41,47,54,
56,57].

Table I

Typical "Necessary N-Add Relationships"

mmm = mmmm + mmmr = rmmm + rmmm

mmr = mmrm + mmrr = rrmm + mrmm

    = rmmr + mmmr = rmmm + rmmr

rmr = rmrm + rmrr = rrmr + mrmr

mm = rmm + mmm = mmr + mmm

r   = rm + rr = rr + mr

In another approach to assign the methyl carbon resonances of polypropylene, Zambelli and coworkers [58]

synthesized the following hydrocarbons in which the
central methyl carbon was enriched with ^{13}C.

$$
\begin{array}{c}
\text{H} \quad \text{CH}_3 \\
| \quad\quad | \\
\text{CH}_3\text{CH}_2\,\text{CCH}_2\,\text{CCH}_2\,\text{CHCH}_3\,\text{CH}_2\,\text{CHC}\overset{*}{\text{H}}_3\,\text{CH}_2\,\text{CHCH}_3\,\text{CH}_2\,\text{CCH}_2\,\text{CCH}_2\,\text{CH}_3 \\
| \quad\quad | \\
\text{CH}_3 \quad \text{H}
\end{array}
$$

I

$$
\begin{array}{c}
\text{H} \quad \text{CH}_3 \\
| \quad\quad | \\
\text{CH}_3\text{CH}_2\,\text{CCH}_2\,\text{CCH}_2\,\text{CHCH}_3\,\text{CH}_2\,\text{CHC}\overset{*}{\text{H}}_3\,\text{CH}_2\,\text{CHCH}_3\,\text{CH}_2\,\text{CCH}_2\,\text{CCH}_2\,\text{CH}_3 \\
| \quad\quad | \\
\text{CH}_3 \quad \text{H}
\end{array}
$$

II

The configurations of the 3,15,13 and 15 carbons were
fixed in these compounds but all possible configurations of the 7,9, and 11 carbon atoms were possible.
Thus, the mixture of diasterioisomers designated I
served as a model for rmmr, mmmm, (mmrr+rrmm), (mrmr+
rmrm), mrrm and rrrr pentads, whereas the diasterio-
isomer mixture II served as a model for (mmmr+rmmm),
(rrmr+rmrr), (mrrr+rrrm), and (mrmm+mmrm) pentads.
By examining the ^{13}C-spectra of these materials and
by assuming that the methyl carbon pentad shifts would
be grouped according to their central triad sequences,
assignments were made for the various pentad resonances.
Another approach was taken by Randall [59,60], who
developed empirical relationships to correct the
chemical shifts of the carbon resonances for steric
interactions caused by neighboring propylene units in
particular stereosequences. Taking -511 Hz as the
chemical shift expected for a methyl carbon in polypropylene in the absence of steric interactions, as
calculated from Grant-Paul parameters, he added corrections for steric interactions due to the mm, mr, rm
or rr dyads that flanked the central methyl carbon.
Values for these corrections were obtained empirically
from the chemical shifts of isotactic and syndiotactic
polypropylene as well as from the central resonances
of the mm- and rr-centered methyl pentad resonance
clusters. These assignments were checked by comparing

the relative intensities of the various methyl carbon resonance areas with calculated intensities based on the assumption of Bernoullian or Markoffian statistics. The assignments made were eventually confirmed by theoretical calculations [61,62] based on considerations of the conformations of polypropylene sequences and the shielding that results from γ-interactions.

The tetrad and pentad assignments made for polypropylene in this way have been very useful in defining the structure of imperfections in isotactic polypropylene [55,63,64] since this information has an important bearing on the mechanisms of stereospecific polymerization of polyolefins [65,66].

By use of the methods discussed here, ^{1}H and ^{13}C-assignments have been made for a large number of vinyl polymers. The reader is referred to a number of excellent texts and reviews for detailed information [41-47,54,67,68]. It seems that the stereosequence distributions of most vinyl polymers can now be characterized by nmr without difficulty. Polystyrene, one of the first polymers to be studied by nmr, and its analogs, represent a significant exception to this generalization. For this reason, our present understanding of the nmr spectra of polystyrene will be discussed next.

NMR Spectra of Aromatic Polymers

Although there have been many ^{1}H- and ^{13}C-studies on polystyrene and related polymers, present understanding of the spectra obtained is rather unsatisfactory. We will confine our discussion here to polystyrene and related polymers prepared by free radical initiated polymerization. The ^{1}H-spectrum of polystyrene is not rich in detail. The resonance of aromatic protons is observed in two areas, the ortho protons being observed upfield from the meta and para protons. In the 300 MHz spectrum of polystyrene or in the 100 MHz spectrum of poly(3,4,5-trideuterio-styrene) [69], the ortho proton resonance is observed in three areas (relative intensities: 9/6/1). Although

it is tempting to assign these to rr, (mr+rm) and mm
triads, and thereby deduce that polystyrene can be
characterized by a σ value of ~0.35, an alternative
interpretation will be presented later in this discussion. The methine and methylene proton resonances
of polystyrene are poorly defined even in 300 MHz
spectra, but pentads are partially resolved in the
spectra of poly(β,β-dideuteriostyrene) [70] and pentads
are partially resolved in the deuterium decoupled spectra of poly(β,β, 2,3,4,5,6-heptadeuteriostyrene) [15].

Figure 8 shows the deuterium decoupled spectrum of
this latter polymer. The three resonances observed
downfield from the principal methine resonance area
have been assigned in different ways.

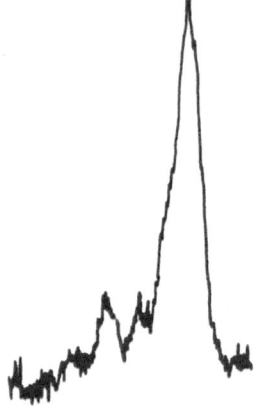

Figure 8. Methine Proton Resonances Observed for
Poly(β,β,2,3,4,5,6-heptadeuteriostyrene) in
O-Dichlorobenzene Solution at 160° with Deuterium
Irradiation [70].

Matsuzaki and coworkers [71] assign both mm and (mr+rm)
centered pentads to them and conclude that polystyrene
is highly syndiotactic. On the other hand, theoretical
calculations by Fujiwara and Flory [72] suggest that the
mm centered methine proton pentad resonances should
occur considerably downfield from the (mr+rm) and rr
centered pentads. On this basis the three lowest field

methine proton resonances. can be assigned to mm centered
pentads. Since they are responsible for 25 percent of
the total methine resonance area, polystyrene can be
considered to be atactic ($\sigma=0.5$).

A number of authors have reported ^{13}C-spectra of
polystyrene. The methylene and quaternary aromatic
(C-1) carbon atom resonances are rich in stereosequence
information, but they are difficult to interpret.
Several different tetrad assignments have been proposed
for the methylene carbon resonances. Except for the
mmm assignment, which could be made from the spectrum
of isotactic polystyrene, the assignments are based on
efforts to fit the observed resonance intensities to
calculated stereosequence distributions, assuming a
particular statistical model, coupled with the requirement that "necessary n-add relationships" were held in
all cases. One group [73] was led to conclude that
polystyrene contains about 30 percent meso dyads and
that 1st order Markoffian statistics are necessary to
characterize the stereosequence distributions. Randall
[74], however, concluded that polystyrene contains about
55 percent meso dyads and that Bernoullian statistics
are adequate to characterize the tetrad distributions.
These groups make totally different assignments for the
methylene carbon resonances.

The C-1 aromatic carbon resonances of polystyrene
are even more complex than the methylene carbon resonances. Randall reports the resolution of as many as
20 separate resonances and declined to attempt structure
assignments. Matsuzaki and coworkers [75] made pentad
assignments, assuming a Bernoullian distribution of
stereosequences ($\sigma=\sim0.3$). These differed from assignments made by Inoue, Nishioka, and Chujo [73] who
assumed 1st order Markoffian statistics (m=0.3). Model
compound studies [77] tend to support the Matsuzaki
assignments, but a convincing interpretation of the
^{13}C-spectrum of polystyrene is yet to appear.

In our own laboratories, Mr. L. Shepherd has
developed a procedure for epimerizing isotactic polystyrene. Isotactic polymers of $\beta,\beta-d_2$-styrene,
$3,4,5-d_3$-styrene and $2,4,6$-trideuteriostyrene, when

epimerized 3-10%, yielded materials whose ^{1}H-spectra
contained resonances assignable to nonads. Yoon and
Flory [77] have calculated chemical shifts for the
methine proton resonances observed in partially epimerized isotactic poly(β,β-d${}_2$-styrene). The observation of significantly different chemical shifts for
mmm centered nonads in these materials points out the
difficulty that will be encountered in interpreting
the nmr spectra of polystyrene. Figure 9 shows preliminary ^{13}C-resonances of methylene carbons in
partially epimerized isotactic polystyrene. As was
discussed earlier, the resonances of mrr and mmr tetrads
should be the first to be observed in epimerized polymers
and they should be equal in intensity at low degrees
of epimerization. Only one of these two resonances is
clearly evident in Figure 9. The other one is undoubt-
edly overlapped by the mmm resonance area. The highest

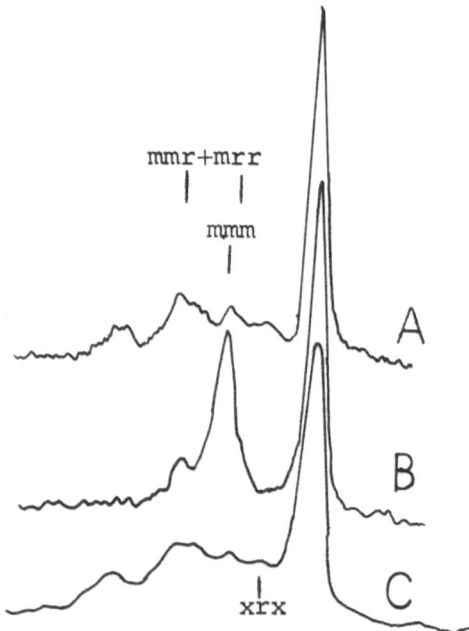

Figure 9. Aromatic C-1 Carbon Spectra of:
A-Atactic Polystyrene; B-Isotactic Polystyrene
Epimerized a Small Amount; C-Isotactic Polystyrene
Epimerized to a Large Extent [78].

field resonance observed in polystyrene appears only
in the spectra of polymers epimerized to high extents.
This suggests that this resonance is due to rmr or mrm
tetrads. Our preliminary results are incompatible with
the assignments of Randall but they are reasonably
consistent with the assignments of Inoue, Nishioka and
Chujo, although the rrr and mmr assignments should
probably be reversed. It is clear that the ^{1}H- and
^{13}C- magnetic resonance spectra of polystyrene are not
adequately understood at the present time and that the
stereochemical structure of this polymer has not been
defined.

This writer believes that there are many reasons
to expect radical initiated polystyrene to be atactic:
1) Related polymers, such as poly(vinylpyridine)
[79,80], poly(α,β,β-trifluorostyrene) [81], poly(p-
methyl-α,β,β-trifluorostyrene) [82], poly(vinyl
thiophene) [83] and poly(vinyl furan) [83] seem to be
atactic, based on their nmr spectra. 2) Alternating
copolymers of styrene with methyl acrylate [84],
methyl methacrylate [84], acrylonitrile [84] and
a large number of other monomers have equal amounts
of meso and racemic styrene-comonomer placements. If
styrene copropagation reactions occur without steric
preference, then it is reasonable to expect the styrene
homopropagation reactions to proceed likewise. Some
of the unusual aspects of the nmr spectra of polystyrene are observed for other monomer units in the
spectra of alternating styrene copolymers [84]; it
seems that the unusual shielding characteristics of
phenyl groups are responsible for the difficulty of
analyzing polystyrene spectra. 3) Practically all
other polymers derived from vinyl monomers by free
radical initiated techniques are atactic and there are
no compelling steric reasons to expect anything different for the polymerization of styrene. Chemical studies indicate that methyl acrylate units and styrene
units exert the same effect on the chemical reactivity
of methyl methacrylate units in copolymers [85]. This
implies that phenyl groups and carbomethoxy groups
have similar steric requirements in polymer chains.
If this is true, then it is reasonable to expect the
tacticity of polystyrene to be the same as that of

poly(methyl acrylate), and this latter polymer is
atactic. 4) Finally, the nmr spectra of polystyrene
prepared by free radical initiation at temperatures
ranging from -78 to +80⁰ are essentially independent of
polymerization temperature [75]. This is not the
behavior to be expected for a system that shows a
rather strong tendency at +80⁰ to yield syndiotactic
placements; it is more likely the behavior to be
expected for a system in which meso and racemic placements have similar enthalpies and entropies of activation. In this connection, it is important to note
that syndiotactic polystyrene has yet to be prepared.

At this point, preliminary studies on poly(vinyl
thiophene) will be reported. Figure 10 shows the
300 MHz pmr spectrum of poly(vinyl thiophene) prepared
by AIBN initiated polymerization. The aromatic proton
resonance is observed in three areas, the highest field

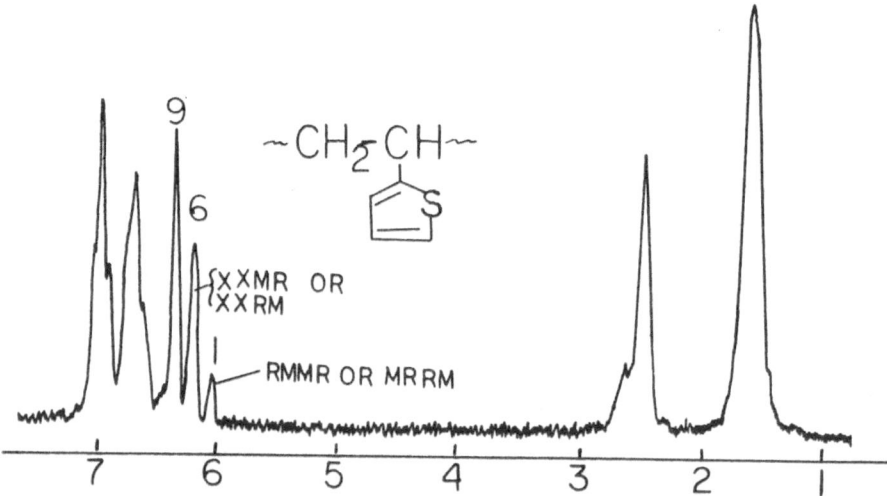

Figure 10. 300 MHz PMR Spectrum of Poly(2-vinyl
thiophene) in CDCl₃ at 28⁰.

area consisting of three well-resolved signals. These are similar to the o-proton resonances of polystyrene and poly(3,4,5-trideuteriostyrene) that were discussed earlier. The relative intensities of these resonances are almost exactly 9:6:1 in order of increasing field. Since there are 16 possible pentads, a possible interpretation of this pattern is that one of the pentad stereosequences is responsible for the highest field signal, that six of them are responsible for the central signal and that the remaining nine pentads are responsible for the lowest field area. The relative intensities would be exactly 9:6:1 if all pentads were equally likely, i.e., if poly(vinyl thiophene) were atactic. If this interpretation is correct, the highest field area is probably due to either mrrm, rmmr or rrrr and, of these, rmmr is preferred at the present time. This preference is based on results obtained in recent epimerization studies on isotactic polystyrene [77,78,86].

The methine proton resonance pattern of poly(vinyl thiophene) is the same as that observed for polystyrene. The lowest field resonance component amounts to 25 percent of the total methine proton resonance and is consistent with poly(vinyl thiophene) being atactic if this lower field component can be assigned to mm triads. Finally, the C-1 carbon resonance spectrum of poly-(vinyl thiophene) (Figure 11) seems to consist of triad resonances, with the lower field area being further subdivided into pentad resonances. This spectrum also suggests that poly(vinyl thiophene) is atactic. If we can conclude on the basis of the results presented here that poly(vinyl thiophene) is indeed atactic, then it seems reasonable to expect the same thing to be true for polystyrene, since the benzene and thiophene rings have very similar steric requirements and similar chemical behavior.

It is interesting that the pmr and cmr spectra of poly(α-methylstyrene) can be interpreted without apparent complication [87-88]. Much additional work needs to be done on polystyrene and related polymers before a clear understanding of their spectra and structure emerges, however.

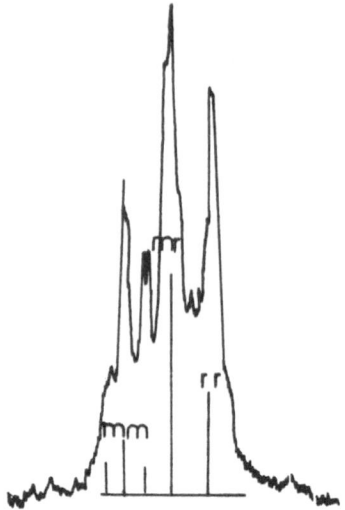

Figure 11. Quarternary Carbon Resonance Pattern of
Poly(2-vinyl thiophene).

Polymers Formed for Non-vinyl Monomers

Nuclear magnetic resonance spectroscopy is also
useful for characterizing the distributions of stereosequences in polyethers, polyesters, and polyamides,
although the method becomes difficult to apply when
the asymmetric atoms are separated by a large number
of atoms. Recent developments in this area include
[1]H- and [13]C- methods for characterizing polylactides
[23], a ^{13}C-method for characterizing substituted
poly(β-propiolactones) [50] as well as [13]C- and
[15]N-methods for characterizing stereosequence distributions in synthetic polypeptides [89-90].

REFERENCES

1. T. Yoshino, M. Shinomiya and J. Komiyama, J. Am.
    Chem. Soc., 87, 387 (1965).
2. J. R. Dombroski and C. Schuerch, Macromolecules,
    4, 447 (1971).

3. T. Miyazawa and Y. Ideguchi, J. Poly. Sci.,
    Part B, 1, 389 (1963).
4. M. Peraldo and M. Farina, Chim e Industria (Milan),
    42, 1349 (1960).
5. H. Tadakoro, M. Ukita, M. Kobayashi and
    S. Murahashi, J. Polym. Sci., Part B, 1, 405 (1963).
6. G. Natta, M. Peraldo, M. Farina and G. Bressan,
    Makromol. Chem., 55, 139 (1962).
7. Y. Oshumi, T. Higashimura, S. Okamura, R. Chujo
    and T. Kuroda, J. Polym. Sci., Part A1, 5, 3009
    (1967).
8. F. Heatley and F. A. Bovey, Macromolecules, 2,
    303 (1968).
9. C. Schuerch, F. Fowells, A. Yamada, F. A. Bovey
    and F. P. Hood, J. Am. Chem. Soc., 86, 4481 (1964).
10. J. Zymonas, E. R. Santee, Jr., and H. J. Harwood,
    Macromolecules, 6, 129 (1973).
11. G. Natta, E. Lombardi, A. L. Segre, A. Zambelli
    and A. Marinangeli, Chim e Industria (Milan), 47
    379 (1965).
12. F. Heatley and F. A. Bovey, Macromolecules, 1,
    301 (1968).
13. J. G. Murray, J. Zymonas, E. R. Santee, Jr., and
    H. J. Harwood, A.C.S. Polymer Preprints, 14, 1163
    (1973).
14. A. Zambelli, M. G. Giongo and G. Natta, Makromol.
    Chem., 112, 183 (1968).
15. A. L. Segre, Macromolecules, 1, 93 (1968).
16. Spectra recorded at The University of Akron by
    Dr. R. C. Chang and Dr. F. Shepherd.
17. R. J. Abraham and H. J. Bernstein, Can. J. Chem.,
    39, 216 (1961).
18. J. D. Roberts, "An Introduction to the Analysis of
    Spin-Spin Splitting in High Resolution NMR Spectra,"
    W. A. Benjamin, Inc., New York, N.Y., 1962,
    pp. 71-77.
19. E. Lombardi, A. Segre, A. Zambelli and
    A. Marinangeli, J. Polym. Sci., Part C, 16, 2539
    (1967).
20. A. Zambelli, M. G. Giongo and G. Natta, Makromol.
    Chem., 112, 183 (1968).
21. H. Tani, N. Oguni and S. Watanabe, J. Polym. Sci.,
    Part B, 6, 577 (1968).

22. K. J. Ivin and M. Navratil, J. Polym. Sci.,
    Part A1, 9, 1 (1971).

23. E. Lillie and R. C. Schulz, Makromol. Chem.,
    176, 1901 (1975).

24. K. Hatada, T. Kitayama, Y. Okamoto, K. Ohta,
    Y. Umenura and H. Yuki, Makromol. Chem., 179,
    485 (1978) and references cited therein.

25. K. Hatada, H. Ishikawa, T. Kitayama and H. Yuki,
    Makromol. Chem., 178, 2753 (1977).

26. W. Gronski and N. Murayama, Makromol. Chem.,
    179, 1509, 1521 (1978).

27. M. Goodman in Topics in Stereochemistry, edited
    by N. L. Allinger and E. L. Eliel, Interscience
    Publishers, New York, 1967, pp. 73-156.

28. A. Zambelli, A. L. Segre, M. Farina and G. Natta,
    Makromol. Chem., 110, 1 (1967).

29. T. Yoshino, J. Komiyama and M. Shinomiya, J. Am.
    Chem. Sci., 86, 4482 (1964).

30. T. Yoshino and K. Kuno, J. Am. Chem. Soc., 87,
    4404 (1965).

31. T. Yoshino and J. Komiyama, J. Polym. Sci.,
    Part B, 4, 991 (1966).

32. C. Schuerch, W. Fowells, A. Yamada, F. A. Bovey
    and F. P. Hood, J. Am. Chem. Soc., 86, 4481 (1964).

33. F. A. Bovey, Pure Appl. Chem., 15, 349 (1967).

34. W. Fowells, C. Schuerch, F. Bovey and F. P. Hood,
    J. Am. Chem. Soc., 89, 1396 (1967).

35. H. Tani and N. Oguni, J. Polym. Sci., Part B,
    7, 803 (1969).

36. H. Tani, N. Oguni, and S. Watanabe, J. Polym.
    Sci., Part B, 6, 577 (1968).

37. F. A. Bovey, Accounts Chem. Res., 1, 175 (1968).

38. A. Zambelli, M. G. Giongo and G. Natta, Makromol.
    Chem., 112, 183 (1968).

39. M. Karplus, J. Chem. Phys., 33, 1842 (1960).

40. M. Karplus, J. Am. Chem. Soc., 85, 2870 (1963).

41. F. A. Bovey, High Resolution NMR of Macromolecules,
    Academic Press, New York, 1972.

42. F. A. Bovey, Prog. Polymer Sci., 3, 1 (1971).

43. F. A. Bovey, Polymer Conformation and Configuration,
    Academic Press, New York, 1969.

44. W. Cooper, in Stereochemistry of Macromolecules, edited by A. D. Ketley, M. Dekker, Inc., New York 1967, Volume 2, Chapter 5.

45. J. C. Woodbrey, in Stereochemistry of Macromolecules, edited by A. D. Ketley, M. Dekker, Inc., New York 1967, Volume 3, pp. 61-145.

46. K. C. Ramey and W. S. Brey, J. Macromol. Sci. Rev. Macromol. Sci., C1, 263 (1967).

47. E. Klesper in Polymer Spectroscopy, edited by D. O. Hummel, Verlag Chemie, Weinheim, 1974, Chapter 3.

48. S. Amiya, I. Ando, and R. Chujo, Polymer J., 4, 385 (1973).

49. S. Amiya, I. Ando, S. Watanabe and R. Chujo, Polymer J., 6, 194 (1974).

50. M. Iida, T. Araki, K. Teranishi and H. Tani, Macromolecules, 10, 275 (1977); M. Iida, S. Hayase and T. Araki, Macromolecules, 11, 490 (1978).

51. K. Hatada, K. Ohta, Y. Okamoto, T. Kitayama, Y. Umemura and H. Yuki, J. Polym. Sci., Part B, 14, 531 (1976).

52. B. D. Coleman and T. G. Fox, J. Chem. Phys., 38, 1065 (1963).

53. B. D. Coleman and T. G. Fox, J. Am. Chem. Soc., 85, 1241 (1963).

54. J. C. Randall, Polymer Sequence Determination, Carbon-13 NMR Method, Academic Press, New York, 1977, pg. 21.

55. F. C. Stehling and J. R. Knox, Macromolecules, 8, 595 (1975).

56. B. D. Coleman and T. G. Fox, J. Polym. Sci., Part A, 1, 3183 (1963).

57. K. Ito and Y. Yamashita, J. Polym. Sci., Part A, 3, 2165 (1965).

58. A. Zambelli, P. Locatelli, G. Bajo and F. A. Bovey, Macromolecules, 8, 876 (1975).

59. J. Randall, J. Polym. Sci., Phys., 12, 703 (1974).

60. J. Randall, J. Polym. Sci., Phys., 14, 2083 (1976).

61. A. Provesoli and F. D. Ferro, Macromolecules, 10, 874 (1977).

62. A. Tonelli, Macromolecules, 11, 565 (1978).

63. A. Pavan, A. Provasoli, G. Moraglio and
    A. Zambelli, Makromol. Chem., 178, 1009 (1977).

64. T. Asakura, I. Ando, A. Nishioka, Y. Doi and
    T. Keii, Makromol. Chem., 178, 791 (1977).

65. A. Zambelli and C. Tosi, Adv. Polymer Sci.,
    15, 32 (1974).

66. A. Zambelli in NMR, Basic Principles and Progress,
    edited by P. Diehl, E. Fluck and R. Kosfeld,
    Springer Verlag, New York, 1971, Volume 4, 101-
    108.

67. V. D. Mochel, J. Macromol. Sci., Rev. Macromol.
    Sci., C8, 289 (1972).

68. J. Schaefer, in Topics in Carbon-13 NMR Spectro-
    scopy, edited by G. C. Levy, J. Wiley and Sons,
    New York, 1974, Volume 1, Chapter 4.

69. J. F. Kinstle and H. J. Harwood, A.C.S. Polymer
    Preprints, 10, 1389 (1969).

70. A. L. Segre, P. Ferruti, E. Toja and F. Danusso,
    Macromolecules, 2, 35 (1969).

71. K. Matsuzaki, T. Uryu, K. Osada and T. Kawamura,
    J. Polym. Sci., Part C, 12, 2873 (1974).

72. Y. Fujiwara and P. J. Flory, Macromolecules, 3,
    43 (1970).

73. Y. Inoue, A. Nishioka and R. Chujo, Makromol.
    Chem., 156, 207 (1972).

74. J. C. Randall, J. Polym. Sci., Polym. Phys. Ed.,
    13, 889 (1975).

75. K. Matsuzaki, T. Uryu, T. Seki, K. Osada and T.
    Kawamura, Makromol. Chem., 176, 3051 (1975).

76. B. Jasse, F. Laupretre and L. Monnerie, Makromol.
    Chem., 178, 1987 (1977).

77. D. Y. Yoon and P. J. Flory, Macromolecules, 10,
    562 (1977).

78. T. K. Chen and H. J. Harwood, to be published.

79. K. Matsuzaki, T. Kanai, T. Matsubara and S.Matsumoto,
    J. Polymer Sci., Polymer Chem. Ed., 14, 1475 (1976).

80. A. C. Watterson, results presented at the conference
    covered by this book.

81. F. M. Lin, F. T. Lin, J. Zymonas, C. W. Wilson III,
    and H. J. Harwood, to be published.

82. V. L. Maksimov, M. P. Votinov, A. F. Dokukina,
    Polymer Sci. (USSR), 8, 1230 (1966).

83. D. T. Trumbo, T. Suzuki and H. J. Harwood, to
    be published.

84. F. Blouin, Ph.D. Dissertation, University of Akron, 1975.

85. E. Thall, Ph.D. Dissertation, University of Akron, 1972.

86. L. Shepherd, Ph.D. Dissertation, University of Akron, 1978.

87. S. Brownstein, S. Bywater and D. J. Worsfold, Makromol. Chem., 48, 127 (1961).

88. K. Fujii, D. J. Worsfold and S. Bywater, Makromol. Chem., 117, 2751 (1968).

89. W. E. Hull and H. R. Kricheldorf, J. Polym. Sci., Polym. Lett. Ed., 16, 215 (1978).

90. H. R. Kricheldorf and W. E. Hull, Makromol. Chem., 178, 253 (1977).

NMR ANALYSIS OF STEREOREGULAR COPOLYMERS

H. James Harwood

Institute of Polymer Science
The University of Akron, Akron, Ohio 44325

Stereoregular copolymers are formed by the chemical modification of stereoregular polymers, by copolymerizations of α-olefins in the presence of stereoregulating catalysts and by ring opening copolymerizations of chiral monomers. Also appropriate for consideration are polydienes containing cis-1,4- and trans-1,4- units as well as those containing 1,2-(or 3,4-) and 1,4- units. Analysis of such copolymers by NMR spectroscopy provides information about their compositions and the arrangement of monomer units along their chains. Such information is valuable for understanding the chemical behavior of stereoregular polymers and the mechanisms of stereoregular polymerization processes.

STRUCTURAL FEATURES PRESENT IN
STEREOREGULAR COPOLYMERS

If a stereoregular copolymer containing A and B monomer units is considered, then some of its structural features that can be measured by NMR spectroscopy can be listed as follows.

295

R. W. Lenz and F. Ciardelli (eds.), Preparation and Properties of Stereoregular Polymers, 295-316.

COMPOSITION	A		B
DYADS	A A	AB+BA	BB
TRIADS	AAA		BBB
	BAA+AAB		BBA+ABB
	BAB		ABA
TETRADS	AAAA		ABBA
	$\left\{\begin{array}{c}\text{AAAB}\\ +\\ \text{BAAA}\end{array}\right\}$	AABA+ABAA BABA+ABAB AABA+ABAA	$\left\{\begin{array}{c}\text{ABBB}\\ +\\ \text{BBBA}\end{array}\right\}$
	BAAB	BABB+BBAB	BBBB
PENTADS	AAAAA		ABBBA
	$\left\{\begin{array}{c}\text{BAAAA}\\ +\\ \text{AAAAB}\end{array}\right\}$		$\left\{\begin{array}{c}\text{BBBBA}\\ +\\ \text{ABBBB}\end{array}\right\}$
	BAAAB		BBBBB
	etc.		etc.

ASSIGNMENT OF OBSERVED RESONANCES
TO STRUCTURAL FEATURES

The various structural features outlined above can give rise to a large number of resonance lines in the nmr spectrum of a stereoregular copolymer. It is not a simple matter to make assignments for these resonances since there are ten possible tetrads and twenty possible pentads, and since the various tetrad or pentad resonances are usually overlapped to a considerable extent. The case can be even more complicated if all of the nuclei in a given environment do not have the same chemical shift, as for example, the methylene protons in isotactic copolymers.

Resonance assignments are often made by empirical relationships or rules that relate structural differences to chemical shift differences. These can be based on extensive studies on model compounds as in the case of the Grant-Paul or Lindeman-Adams relationships [1-3] for predicting the chemical shifts of carbon atoms in hydrocarbons and polyhydrocarbons. These have been very valuable for assigning the resonances observed in the spectra of ethylene-propylene copolymers [4,5], propylene-butene-1 copolymers [6,7], hydrogenated polydienes [8-18], and hydrogenated alkene-diene copolymers [8]. Parameters are now also available for predicting the chemical shifts of carbon atoms in unsaturated polyhydrocarbons [19-26].

When sufficient model compounds are not available for parameterization, assignments can be made based on empirical rules which assume that constant chemical shifts can be expected between pairs of structures having common differences. For example, consider the chemical shifts of the nuclei associated with A units centered in a series of pentadic environments in an A-B copolymer. If the chemical shift difference between the nuclei in AAAAA and AAÅAB pentads is α Hz, it is reasonable to expect a similar difference between the chemical shifts of the nuclei in AAÅAB and BAÅAB pentads. Thus whatever shielding is exerted on the central A unit by a B unit two units away can reasonably be expected to be increased two-fold when there are two B-units so situated. When the A and B units have similar structures, as is often the case in polymer modification studies, then it is also a reasonable approximation to assume that there will also be a difference of α Hz between the chemical shifts of Å units in XXÅXA and XXÅXB pentads, where X = A or B. In other words, it may be assumed "that the chemical shift change, which occurs when a certain part of an <u>n-add</u> is altered to form another <u>n-add</u>, is independent in its magnitude from the nature of the unaltered part of the <u>n-add</u>. (<u>triad</u> replaced by <u>n-add</u>-HJH)" [27]. Following similar reasoning, a difference of β Hz may be considered to prevail between the chemical shifts of Å units in XAAXX and XBÅXX pentads. If the values of α and β are known, the chemical shifts of all the A-centered pentads may be related accordingly to the scheme shown below.

$$
\begin{array}{ccccccc}
\text{AAAAA} & \xleftrightarrow{\beta} & \text{ABAAA} = \text{AAABA} & \xleftrightarrow{\beta} & \text{ABABA} \\
\updownarrow{\alpha} & & \updownarrow{\alpha} & \updownarrow{\alpha} & \updownarrow{\alpha} \\
\text{BAAAA} & \xleftrightarrow{\beta} & \left\{\begin{array}{l}\text{BBAAA} = \text{AÅABB}\\ \text{ABAAB} = \text{BAABA}\end{array}\right\} & \xleftrightarrow{\beta} & \text{BBABA} \\
\updownarrow{\alpha} & & \updownarrow{\alpha} & \updownarrow{\alpha} & \updownarrow{\alpha} \\
\text{BAAAB} & \xleftrightarrow{\beta} & \text{BBAAB} = \text{BAABA} & \xleftrightarrow{\beta} & \text{BBABB}
\end{array}
$$

The above assignment scheme is based on two rules:

1. Changing the end unit of a pentad from an A unit to a B unit, or vice versa, causes a shift of α Hz in the resonance of the central A-unit of that pentad.
2. Changing the end unit of a triad from an A unit to a B unit, or vice versa, causes a shift of β Hz in the chemical shift of the central A-unit of that triad.

The evaluation of α and β for application of the above rules is accomplished by examining the spectra of a series of copolymers. The chemical shift difference between AAAAA and BBABB pentads affords a measure of $2(\alpha + \beta)$ and the spectra of polymers with either low or high A contents, where only a few pentads are present in appreciable concentration, can be used for the estimation of α and β. The validity of assignments made this way can be checked by determining whether or not relative pentad concentrations evaluated from the spectra obey necessary n-add relationships [28-30]. In addition, it is a good idea to use several solvent systems for the nmr measurements. Values of α and β may be different for different solvent systems, but pentad distributions evaluated from the spectra must be independent of the solvent employed.

Sometimes more than two empirical rules must be developed to make proper assignments, but it is seldom necessary to use more than three or four rules. This general approach can also be used for assigning resonances due to stereosequences in homopolymers, as has been mentioned in the discussion on stereoregular homopolymers. Klesper employed four rules to completely assign the α-methyl proton resonances of atactic methacrylic acid-methyl methacrylate copolymers [27]. Since he was the first to develop and apply rules of the sort considered here, it seems appropriate to refer to them in a general sense as "Klesper's rules."

Once assignments have been made for the resonances of stereoregular copolymers, the spectra of copolymers prepared under varying conditions can be analyzed to obtain information about stereoregular copolymerization systems or about the chemical behavior of stereoregular polymers.

UTILIZATION OF n-ADD DISTRIBUTIONS
MEASURED BY NMR

The microstructural features of stereoregular copolymers prepared directly from monomers can be related to conditional monomer placement probabilities, which are, in turn, related to monomer reactivity ratios and monomer feed ratios. Thus, the conditional probability that an A unit follows a B unit in a copolymer chain, $P(A/B)$, can be calculated from BA and BB dyad distributions or from the monomer reactivity ratio for B, r_B, and the ratio of monomers A and B in the feed (A_f/B_f), according to the following equations.

$$P(A/B) = [BA]/([BA] + [BB])$$
$$P(A/B) = 1/(1 + r_B B_f/A_f)$$

Conditional probabilities can then be used to calculate structural features of the copolymers or to evaluate statistical features of the copolymerization mechanism. If terms such as $P(A)$, $P(AA)$, and $P(ABA)$ are used to represent unconditional probabilities of finding monomer units, dyads, triads, etc. in the polymer chain, and if terms such as $P(A/A)$, $P(A/AB)$ and $P(B/BBB)$ are used to represent conditional probabilities that A or B units follow other monomer units or monomer sequences in the copolymer chain, a copolymer can be considered to obey Bernoullian statistics if $P(A) = P(A/A) = P(A/B) = P(A/AA) = P(A/BA) = P(A/anything)$. This is the situation commonly encountered in stereoregular copolymers. In some cases, however, first order Markoffian statistics are needed to characterize the copolymers. In such cases $P(A) \neq P(A/A) \neq P(A/B)$.

Dyad, triad and higher n-add distributions can be calculated using the above-mentioned quantities and used for interpreting copolymer spectra. The reader is referred to other articles for detailed information about the statistical calculation of copolymer

features [8,28-33], but a few representative relationships are provided below.

P(A) = P(AA) + P(AB)
P(AA) = P(A) P(A/A)
P(AB) + P(BA) = P(A) P(B/A) + P(B) P(A/B)
P(BAB) = P(B) P(A/B) P(B/A)

STUDIES ON STEREOREGULAR COPOLYMERS
OBTAINED BY CHEMICAL MODIFICATION

Copolymers are necessarily intermediates when
polymers are modified chemically. When the reactions
employed occur cleanly with few or no side reactions
and without altering the stereochemical configuration
of the polymer, it is possible to prepare stereoregular
copolymers from stereoregular homopolymers. The relative amounts of various structural features present
at any given stage in these copolymers are dependent
on the relative reactivities of functional groups in
different structural environments in the polymer chains.
Thus, in considering the conversion of a polymer containing A- units to one containing B- units, the rate
of the process and the relative amounts of the various
structural features present at a given conversion may
depend on the relative reactivities of A- units centered
in AÅA (k$_0$), (BÅA+AÅB)(k$_1$) and BÅB (k$_2$) triads. Theoretical [34-41], numerical [42] and Monte Carlo methods
[43-47] have been developed for calculating these
various quantities from values of k$_0$, k$_1$ and k$_2$. In
addition, convenient methods for deducing k$_0$, k$_1$ and k$_2$
from nmr spectral information are available [48,49].

~A - A - A - A - A - A - A - A - A - A - A~
    $\downarrow$k$_0$ $\downarrow$k$_0$ $\downarrow$k$_0$

~A - A - B - A - A - A - B - A - B - A - A~
    $\downarrow$k$_1$ $\downarrow$k$_2$

~A - A - B - B - A - A - B - B - B - A - A~

Klesper has pioneered the use of ^{1}H- and ^{13}C-
magnetic resonance to monitor chemical reactions of

polymers and has done extensive work on the hydrolysis of isotactic and syndiotactic polyMMA [50-54], on the reactions of diazoalkanes with isotactic and syndiotactic poly(methacrylic acid) [55,56], and on the esterification of poly(methacrylic acids) with alcohols in the presence of carbodiimides [57,58]. As an example of the results obtained in these studies, Figure 1 shows the 220 MHz PMR spectra of several syndiotactic methacrylic acid(A)-methyl methacrylate(M) copolymers prepared by the hydrolysis of syndiotactic poly(methyl methacrylate). Separate resonances are evident for the α-methyl protons centered in AAA, (MAA+AAM), MAM, AMA, (AMM+MMA) and MMM triads. Depending on reaction conditions, the relative intensities

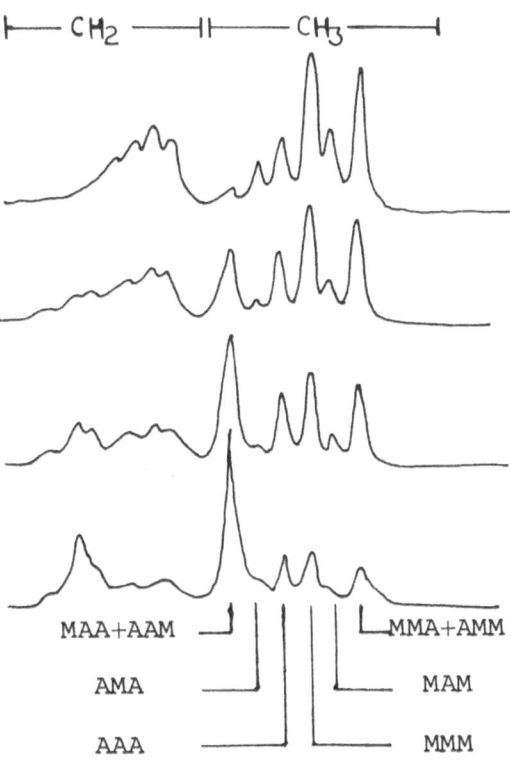

Figure 1. 220 MHz PMR Spectra of Syndiotactic Methyl Methacrylate(M)-Methacrylic Acid Copolymers in Pyridine at 100° [27].

of these resonances indicate random hydrolysis
($k_0 = k_1 = k_2$), a tendency for alternation
($k_0 > k_1 > k_2$) or a tendency for the methacrylic acid
units to occur in blocks ($k_0 < k_1 < k_2$). Plate and
coworkers have also investigated the hydrolysis of
stereoregular polymethacrylates and have developed
computer programs for analyzing the spectra of the
intermediate copolymers [59,60]. Studies of this type
contribute not only to our understanding of polymer
reactions, but they provide a basis for interpreting
the nmr spectra of complex polymers. Other stereoregular polymer modification reactions that have been
studied by nmr include transesterification of poly(aryl
methacrylates)[61,62], addition reactions involving polydienes [63,64] and isomerization of polydienes [65-71].

STEREOREGULAR COPOLYMERS PREPARED FROM ACHIRAL MONOMERS AND CHIRAL SPECIFIC CATALYSTS

Ethylene-Propylene Copolymers

Catalysts which are isospecific for the polymerization of propylene, butene-1, etc., can also be used
for the preparation of copolymers of these olefins with
ethylene and with each other. There is much evidence
to indicate that the chiral specificity of the catalyst
is retained during these reactions [72-76]. Thus, the
spectra of ethylene(E)-propylene(P) copolymers prepared
with an isospecific catalyst system are much simpler
than those prepared with a nonspecific catalyst [75].
Studies on copolymers of propylene with ^{13}C-enriched
ethylene prepared with an isospecific catalyst indicate
that the propylene-ethylene-propylene triads have a
single type of configuration, which is probably meso.
This is indicated by the fact that only two resonances
are observed for ethylene units in P-E-P triads.
Furthermore, there is no evidence for the presence of
ethylene units flanked by two tertiary carbon atoms
(due to inversions of propylene units) in the copolymers.
In contrast, copolymers prepared from nonspecific or
syndiospecific catalysts have been shown to contain
ethylene units centered in both meso- and racemic- PEP

triads. These polymers contain appreciable concentrations of ethylene units flanked by two tertiary carbon
atoms. Much work has been done on the spectra of
propylene- ^{13}C-enriched ethylene copolymers [72-74,76]
because the stereochemical information provided by
these spectra is valuable for understanding the mechanisms of Ziegler-Natta catalyzed polymerizations.
This work is covered in greater detail elsewhere in this
volume. Theoretical calculation of the chemical shifts
expected for P-E-P triads in these copolymers have led
Tonelli [77] to conclude that the nmr spectra of these
copolymers cannot be expected to provide the stereochemical information desired, however.

Ray, Johnston and Knox [75] investigated the [13]C-
spectra of ethylene-propylene copolymers prepared with
an isospecific catalyst system. The absence of propylene head-to-head or tail-to-tail inversions in these
copolymers and their essentially complete isotactic
enchainment enabled their spectra to be analyzed precisely. Assignments made previously for atactic
ethylene-propylene copolymers, together with new assignments made possible by the simplicity of the isotactic
copolymer spectra were used to obtain all n-add distributions through triads, P-P and E-E-centered tetrad
distributions, and a number of pentad distributions.
The analysis of these results is promised in a forth-
coming publication. A casual analysis of the results
suggests that the distributions are close to being
Bernoullian, although measured EEEE distributions are
larger than would be expected for Bernoullian behavior.

Procedures are available for preparing isotactic-
cis-, isotactic trans- and syndiotactic-cis- varieties
of poly-1,3-pentadiene. Hydrogenation of these
polymers yields polymers that may be considered to be
stereoregular alternating copolymers of ethylene and
propylene. The cmr spectra of such copolymers have been
investigated [14,15,18]. Differences observed in the
methyl and methylene carbon resonances were observed
for the isotactic and syndiotactic polymers. The parent
polymers may be considered to be alternating stereoregular copolymers of acetylene with propylene. Their
spectra also revealed stereochemical differences.

Propylene-Butene Copolymers

Brown and Cudby [6] and Randall [7] have analyzed
the ^{13}C-spectra of a series of propylene-butene-1 copolymers prepared using an isospecific catalyst system.
The enchainment of the monomer units was essentially
isotactic and head-tail. Resonances observed for the
homopolymers, assignments made previously by Fish and
Dannenberg [78], the Grant-Paul relationships and
variations in resonance intensity with copolymer composition were used to make assignments for the resonances
of methine, methylene and methyl carbon atoms. Triad
and some tetrad sequence distribution measured from the
spectra were consistent with Bernoullian distributions
over the entire range of copolymer compositions examined.

Ethylene-Butene Copolymers

Ziegler-Natta catalysts can be used for the
preparation of highly alternating copolymers of ethylene
with 2-butene. Copolymerization of ethylene with cis-
2-butene yields a crystalline erythro diisotactic copolymer, whereas the copolymer obtained from ethylene
and trans-2-butene is threodiisotactic. Cmr spectra
of these copolymers have been reported by Zambelli,
et al. [73]. Prud'homme, et al. [79] investigated the
pmr (220 MHz) and cmr spectra of hydrogenated 1,4-
poly-(2,3-dimethylbutadiene). Although the hydrogenated
products were not stereoregular, they can be considered
to be alternating ethylene-butene copolymers. A comparison of the spectra of the hydrogenated copolymer
with the spectra reported by Zambelli, et al. enabled
resonances assignable to sequence distribution effects
to be identified.

Isoprene-Butadiene Copolymers

Lobach and coworkers [80] prepared copolymers of
isoprene and butadiene in which monomer units had
either predominantly 1,4-cis- or 1,4-trans- configurations and investigated their cmr spectra. Assignments
were made for the various aliphatic carbon resonances

observed.   Only weak signals assignable to 1,4-4,1-
or 4,1-1,4- linkages involving isoprene units were
detected and it was concluded that the distributions
of monomer units in the copolymers obeyed Bernoullian
statistics.

Butadiene-Alkene Copolymers

Furukawa and coworkers [81-83] have developed
procedures for preparing alternating copolymers of
butadiene with propylene and other 1-olefins.   The
structures of these polymers have been studied by
220 MHz pmr [84] and 25 MHz cmr spectroscopy [85].
The butadiene units in the copolymer prepared with a
vanadium catalyst system have 1,4-trans configurations
exclusively, but no information is available concerning
the tacticity of the propylene units.

Copolymers of ethylene with butadiene in which the
butadiene units have the 1,4-trans configuration have
been studied by several groups, using cmr spectroscopy
[83,86].   In one case the copolymers had predominantly
alternating structures, but in another case the copolymers had block structures.

POLYMERS FROM CYCLIC OLEFINS

The ring-opening polymerization of cyclic olefins
can yield polymers containing cis and trans units.
Such polymers can be considered to be stereoregular
copolymers, and their nmr spectra can be analyzed to
obtain information about the relative amounts of cis
and trans units present and their arrangement along the
polymer chain.   The structures of polymers derived
from cyclopentene [87-89], cyclooctadiene [71,90] and
bicyclo-2,2-1-octene [89,91] have been studied by nmr
spectroscopy, for example.   The relative amounts of
cis and trans units present vary with reaction conditions, but the cis and trans units are generally
distributed randomly along the polymer chains.   Ivin
and coworkers [91] report that a tendency toward blocki-
ness develops at long reaction times, however.

Chen [92] has measured the cmr spectra of polybutenamer, polypentenamer, polyheptenamer and polyoctenamer but has not reported sequence distribution information. It should be noted that there are two ways of incorporating bicyclo-2,2,1-octene units into polymers. It is necessary to consider the distributions of the asymmetric centers present in polymers derived from this monomer, in addition to considering the cis and trans structures present, to completely interpret the spectra of polymers derived from bicyclo-2,2,1-octane. Polymers derived from unsymmetrical cyclic olefins, in general, should not be expected to be stereoregular. However, cmr and pmr studies of polymers derived from 1-methylcyclobutene and 1-methyl-trans-cyclooctene have predominantly head-tail structures [93,94]. Since they contain both cis- and trans- units, they can be considered to be stereoregular copolymers.

POLYMERS DERIVED FROM DIENES

Since dienes can be incorporated into polymers in several ways (e.g., 1,4-cis-, 1,4-trans-, 1,2- or 3,4-enchainments) it is quite possible to obtain stereoregular copolymers from dienes if the polymerization process induces the incorporation of only (or predominantly) two structural features into the polymers. For example, polydienes containing only 1,4-cis and 1,4-trans units may be considered to be stereoregular copolymers. Such polymers can be prepared by polymerizing dienes in hydrocarbon solvents using organolithium compounds or π-allyl nickel compounds as initiators, by isomerizing stereoregular 1,4-polydienes, or by ring-opening polymerization of cyclic olefins, such as cyclooctadiene. In addition, procedures are available for converting dienes into polymers that contain approximately equal amounts of 1,4-cis and 1,2- or 3,4-units, with very small amounts of 1,4-trans units. Such "equibinary copolymers' have been characterized extensively by nmr spectroscopy. Dienes can also be copolymerized under conditions that yield stereoregular homopolymers when polymerized individually. The products of such reactions are indeed stereoregular copolymers.

1,4-Cis/1,4-Trans Polydienes

The pmr and cmr spectra of isomerized 1,4-poly-
butadienes [65,66,69-71,90,95] and of isomerized
1,4-polyisoprenes [67,78], including the naturally
occurring polymers chicle [67] and gutta percha [67],
have been investigated by several research groups. In
the case of polybutadiene, dyad distributions are
obtainable from the methylene proton and olefinic
carbon resonances and triad distributions are obtainable from the olefinic proton resonances, provided
that the methine protons are decoupled from methylene
protons. The distributions are those expected for
polymers having a random arrangement of cis and trans
units. In the case of polyisoprene, dyad distributions
were measured from the resonances of the methylene and
quartenary olefin carbon atoms. These showed that
isomerized 1,4-polyisoprenes have random distributions
of cis and trans units. An interesting result obtained
by Tanaka and Sato [67] is that chicle is either a
blend of 1,4-cis- and 1,4-trans-polyisoprene or a
block copolymer.

The "equibinary" 1,4-cis/1,4-trans polybutadiene
prepared using a π-allyl nickel trifluoroacetate
catalyst [70,71,96,97] and the polybutadiene obtained
by polymerization of cyclooctadiene using an olefin
metathesis catalyst system where shown by nmr to have
random distributions of cis- and trans-units, although
there is some indication that "equibinary" copolymers
with non-Bernoullian structures are obtained in some
cases [96]. Polybutadienes prepared using alkyl
lithium initiators in hydrocarbon solvents have also
been shown to have random distributions of 1,4-cis
and 1,4-trans units [20,23,71,90].

Polyisoprenes prepared by alkyl lithium initiated
polymerizations conducted in hydrocarbon solvents
contain predominantly cis and trans units, but these
may be enchained in a 1,4- or 4,1- manner. Cmr studies
indicate that the cis and trans units are distributed
randomly and that negligible amounts of 4,1-1,4 and
1,4-4,1 linkages are present [16,92-98]. Polyisoprenes
prepared with radical or alfin catalysts contain

appreciable amounts of such linkages, however.

Similarly, poly(2,3-dimethyl-1,3-butadiene) prepared with a butyl lithium/hexane system was found to
contain predominantly 1,4-cis and 1,4-trans units,
arranged randomly [100,101]. The methylene carbon and
proton resonances were used to obtain dyad distributions in these studies, and the methyl proton resonances provided a measure of triad distributions.
Although polymers prepared from 2,3-dimethyl-1,3-
butadiene had about 7% 1,2-structure, these polymers
were also found to have Bernoullian structures.

Suzuki and coworkers [102] investigated the
structures of poly(1-phenylbutadienes) prepared by
anionic or coordinated polymerization. The polymers
contained predominantly 1,4-enchainments of the
1-phenylbutadiene units and dyad distributions were
measurable from the olefinic carbon resonances. A
predominantly trans-1,4-polymer was isomerized to
yield a 1,4-cis/1,4-trans copolymer in which the cis
and trans units were distributed randomly along the
polymer chain. The polymers obtained by anionic polymerization also had random structures.

Finally, polychloroprene [103] prepared by radical
initiated polymerization at 25⁰ has been shown by cmr
spectroscopy to contain 94% trans and 6% cis units,
but about 20 percent of the placements are of the
4,1-1,4 or 1,4-4,1 type.

1,2-/1,4-Cis- or 1,2-/1,4-Trans-
Polybutadienes

Several procedures are available for preparing
polybutadienes that contain equal amounts of 1,4-cis
and 1,2- units [104,105]. The structures of these
"equibinary" polymers have been investigated extensively by cmr spectroscopy [21,22,106,107]. The
olefinic carbons of the 1,4-cis units are particularly
useful for determining cis unit centered triad distributions, and some of the aliphatic carbon resonances
have provided measures of dyad distributions. The

1,2- and 1,4-cis units appear to be distributed randomly, but all resonances have not been assigned unequivocally and little is known about the configurations
of the 1,2-units. Conti and coworkers [20] investigated the cmr spectrum of a 1,2-/1,4-cis polybutadiene
that contained 85% 1,2-units as well as the spectrum
of a 1,2-/1,4-trans polybutadiene that contained 94%
1,2-units. Chemical shifts were assigned as described
previously but the relative intensities of the observed
resonances were not analyzed.

3,4-/1,4-Cis-Polyisoprene

Polyisoprene prepared with a $Ti(OR)_4$ $-Et_3Al$
catalyst system contains 3,4- and 1,4-cis-isoprene
units almost exclusively and these are enchained mostly
in a head-tail fashion [13]. The 3,4-3,4 dyads have a
predominantly meso configuration. The cmr spectrum of
this polymer has been interpreted in terms of a structure which obeys first order Markoffian statistics
[108]. This is an interesting result because the
structures of all other polymers prepared from dienes
seem to have Bernoullian structures, even when 1,4-cis,
1,4-trans, 1,2- or 3,4- units are present simultaneously
[11,13,23].

STEREOREGULAR COPOLYMERS PREPARED
FROM CHIRAL MONOMERS

When copolymers are prepared by ring opening polymerization of chiral monomers such as cyclic ethers
and thioethers, lactones, lactams, N-carboxy anhydrides,
etc., the chirality of the monomers is often retained
in the repeating groups incorporated into the copolymers. A few stereoregular copolymers prepared this way
have been analyzed by nmr spectroscopy, but much
remains to be done in this area.

Kricheldorf and Schilling [109-110] for example,
have shown that the carbonyl carbon resonances of
L-alanine-L-phenylalanine copolymers can provide a
measure of dyad distributions. Four carbonyl carbon

resonances are observed for such copolymers, corresponding to Ala·Ala, Ala·Phe, Phe·Ala and Phe·Phe dyads.
Similar results were also noted for the carbonyl carbon
resonances of L-alanine-glycine copolymers. Kricheldorf
and Hull [111,112] have studied the ^{15}N-resonances of
sequential copolypeptides derived from glycine and
other amino acids, such as alanine, valine, leucine,
phenylalanine or proline. Separate resonances are
observed for nitrogen atoms associated with various
peptide linkages. These preliminary results indicate
that ^{15}N-spectroscopy will be very useful for characterizing the structures of both natural and synthetic
polypeptides. The technique has already been applied
to polypeptides prepared by the N-carboxyl anhydride
technique [113]. Hiraoki and coworkers [114] have
investigated the ^{1}H- spectra of γ-benzyl-L-glutamate-
(γ-methyl-L-glutamate) copolymers that were prepared
by the N-carboxyl anhydride technique and found them
to have random structures.

Carbon-13 spectroscopy has been used very effectively by Corno and coworkers [115-117] to characterize
the distributions of monomer sequences in copolymers
derived from episulfides using anionic catalysts.
Although chiral monomers were not employed in these
studies, it is worth noting that tacticity effects
had a relatively small effect on the resonance patterns
observed, but that the chemical shifts of in-chain
carbon atoms in different sequences were substantially
different. On the basis of assignments and empirical
shift parameters developed by Corno, et al., the spectra of stereoregular ethylene sulfide-propylene sulfide
copolymers and propylene sulfide-isobutylene sulfide
copolymers should be readily analyzed. Studies on
copolymers derived from racemic monomers indicate
them to have random structures; a similar result can
be expected for copolymers derived from optically
active monomers.

REFERENCES

1. D. M. Grant and E. G. Paul, J. Am. Chem. Soc., 86, 2984 (1964).

2. L. P. Lindeman and J. Q. Adams, Anal. Chem., 43, 1245 (1971).

3. C. J. Carman, A. R. Tarpley, Jr., and S. H. Goldstein, Macromolecules, 6, 719 (1973).

4. C. J. Carman, R. A. Harrington and C. E. Wilkes, Macromolecules, 10, 536 (1977).

5. J. C. Randall, Macromolecules, 11, 33 (1978).

6. A. Bunn and M.E.A. Cudby, Polymer, 17, 548 (1976).

7. J. C. Randall, Macromolecules, 11, 594 (1978).

8. J. C. Randall, Polymer Sequence Determination - Carbon-13 NMR Method, Academic Press, New York, 1977.

9. C. J. Carman, Macromolecules, 7, 789 (1974).

10. J. C. Randall, J. Polym. Sci., Polym. Phys. Ed., 13, 1975 (1975).

11. Y. Tanaka, H. Sato and A. Ogura, J. Polym. Sci., Polym. Chem. Ed., 14, 73 (1976).

12. D. Khlok, Y. Deslander and J. Prud'homme, Macromolecules, 9, 809 (1976).

13. Y. Tanaka and H. Sato, Polymer, 17, 413 (1976).

14. K. F. Elgert and W. Ritter, Makromol. Chem., 177, 2781 (1976).

15. K. F. Elgert and W. Ritter, Makromol. Chem., 178, 2857 (1977).

16. H. Sato, A. Ono and Y. Tanaka, Polymer, 18, 580 (1977).

17. A. S. Khatchaturov, E. R. Dolinskaya, L. K. Prozenko, E. L. Abramenko and V. A. Kormer, Polymer, 18, 871 (1977).

18. L. Zetta, G. Gatti and G. Andiso, Macromolecules, 11, 763 (1978).

19. Y. Alaki, T. Yoshimoto, M. Imanari and M. Takeuchi, Kobunshi Kagaku, 29, 397 (1972); Eng. Ed., 1, 580 (1972).

20. F. Conti, M. Delfini, A. L. Segre, D. Pini and L. Porri, Polymer, 15, 816 (1974).

21. J. Furukawa, E. Kobayashi, N. Katsuki and T. Kawagoe, Makromol. Chem., 175, 237 (1974).

22. K. F. Elgert, G. Quack and B. Stutzel, Makromol. Chem., 176, 759 (1975).

23. K. F. Elgert, G. Quack and B. Stutzel, Polymer, 16, 154 (1975).

24. P. T. Suman and D. D. Werstler, J. Polym. Sci., Polym. Chem. Ed., 13, 1963 (1975).

25. H. Sato, A. Ono and Y. Tanaka, Polymer, 18, 580 (1977).

26. F. Conti, M. Delfini and A. L. Segre, Polymer, 18, 310 (1977).

27. E. Klesper, W. Gronski and A. Johnsen, in NMR-Basic Principles and Progress, Volume 4, ed. by P. Diehl, E. Fluck and R. Kosfeld, Springer-Verlag, Berlin-Heidelberg-New York, 1971, pp. 47-70, esp. pg. 55.

28. B. D. Coleman and T. G. Fox, J. Polym. Sci., Part A1, 1, 3183 (1963).

29. K. Ito and Y. Yamashita, J. Polym. Sci., Part A, 3, 2165 (1965).

30. E. Klesper in Polymer Spectroscopy, edited by D. O. Hummel, Verlag Chemie, Weinheim, 1974, Chapter 3.

31. G. S. Georgiev, J. Macromol. Sci., Chem., A10, 1081 (1976).

32. H. J. Harwood and W. M. Ritchey, J. Polym. Sci., Part B, 2, 601 (1964).

33. H. J. Harwood, Y. Kodaira and D. L. Newman, in Computers in Polymer Sciences, ed. by J. S. Mattson, H. B. Mark, Jr., and H. C. McDonald, Jr., Marcel Dekker, Inc., New York, 1977, Chapter 2.

34. N. A. Plate, A. D. Litmanovich, O. V. Noah, A. L. Toom and N. B. Vasilyev, J. Polym. Sci., Polym. Chem. Ed., 12, 2165 (1974).

35. E. Klesper, A. Johnsen and W. Gronski, Makromol. Chem., 160, 167 (1972).

36. J. J. Gonzalez and K. W. Kehr, Macromolecules, 11, 996 (1978).

37. J. J. Gonzalez, Macromol., 11, 1074 (1978).

38. N. A. Plate, A. D. Litmanovich and O. V. Noah, Makromolekulyarnye Reaktsii, Khimya, Moscow, USSR, 1977.

39. E. A. Boucher, Prog. Polym. Sci., 6, 63 (1978) and references cited therein.

40. H. K. Frensdorff and O. Ekiner, J. Polym. Sci., Part A-2, 5, 1157 (1967).

41. V. Stoy, J. Polym. Sci., Polym. Chem. Ed.,
    15, 1029 (1977).

42. B. J. Bauer, Macromolecules, to be published
    in 1979.

43. E. Klesper, W. Gronski and V. Barth, Makromol.
    Chem., 150, 223 (1971).

44. E. Klesper and A. O. Johnsen, in Computers in
    Polymer Science, ed. by J. S. Mattson, H. B.
    Mark, Jr., and H. C. MacDonald, M. Dekker, Inc.,
    New York, 1977, Chapter I.

45. O. V. Noah, A. D. Litmanovich and N. A. Plate,
    J. Polym. Sci., Part A2, 12, 1711 (1967).

46. A. D. Litmanovich, N. A. Plate, O. V. Noah
    and V. I. Golyakov, Europ. Polym. J. Suppl.,
    1969, 517.

47. T. Saito and Y. Matsumura, Polym. J., 4, 124
    (1973).

48. H. J. Harwood, K. G. Kempf and L. M. Landoll,
    J. Polym. Sci., Polym. Lett. Ed., 16, 91 (1978).

49. E. Klesper and V. Barth, Polymer, 17, 787 (1976).

50. E. Klesper, V. Barth and A. Johnsen, Pure Appl.
    Chem., 8, 151 (1971).

51. E. Klesper, W. Gronski and V. Barth, Makromol.
    Chem., 139, 1 (1970).

52. E. Klesper, D. Strasilla and V. Barth, in Reactions
    on Polymers, ed. by J. A. Moore, D. Reidel Pub.
    Co., Dordrecht, Holland, 1973, pp. 137-166.

53. V. Barth and E. Klesper, Polymer, 17, 893 (1976).

54. V. Barth and E. Klesper, Polymer, 17, 777 (1976).

55. E. Klesper, D. Strasilla and W. Regel, Makromol.
    Chem., 175/2, 523 (1974).

56. D. Strasilla and E. Klesper, Makromol. Chem.,
    175/2, 535 (1974).

57. E. Klesper and D. Strasilla, J. Polym. Sci.,
    Polym. Lett. Ed., 15, 23 (1977).

58. D. Strasilla and E. Klesper, J. Polym. Sci.,
    Polym. Lett. Ed., 15, 199 (1977).

59. N. A. Plate, L. B. Stroganov, T. Seifert and
    O. V. Noah, Doklad. Akad. Nauk. SSSR, 223, 396
    (1975); Chem. Abstr. 83, 164796d.

60. N. A. Plate, Pure and Appl. Chem., 46, 49 (1976)
    and references cited therein.

61. L. M. Landoll, T. Suzuki and H. J. Harwood, ACS
    Polymer Preprints, 15, 233 (1974).

62. T. K. Chen, Ph.D. Dissertation, University of
    Akron, 1979.
63. W. Ritter, K. H. Eichin and K. E. Elgert, Makromol.
    Chem., 178, 2837 (1977).
64. A. Tran and J. Prud'homme, Macromolecules, 10,
    149 (1977).
65. K. F. Elgert, B. Stutzel, P. Frenzl, H. J. Cantow
    and R. Streck, Makromol. Chem., 170, 257 (1973).
66. H. Hatada, Y. Tanaka, Y. Terawaki and H. Okuda,
    Polym. J., 5, 327 (1973).
67. Y. Tanaka and H. Sato, Polymer, 17, 113 (1976).
68. Y. Tanaka, H. Sato and T. Seimiya, Polym. J.,
    7, 264 (1975).
69. J. M. Thomassin, E. Walckiers, R. Warin and
    Ph. Teyssie,   J. Polym. Sci., Part B, 11,
    229 (1973).
70. Y. Tanaka, H. Sato, M. Ogawa, K. Hatada and
    Y. Terawaki, J. Polymer Sci., Part B, 12,
    369 (1974).
71. Y. Tanaka, H. Sato, K. Hatada, Y. Terawaki and
    H. Okuda, Makromol. Chem., 178, 1823 (1977).
72. A. Zambelli and C. Tosi, Adv. Polym. Sci.,
    15, 31 (1974).
73. A. Zambelli, G. Gatti, C. Sacchi, W. O. Crain, Jr.
    and J. D. Roberts, Macromolecules, 4, 475 (1971).
74. A. Zambelli, C. Bajo and E. Rigamonti, Makromol.
    Chem., 179, 1249 (1978).
75. G. J. Ray, P. E. Johnson and J. R. Knox,
    Macromolecules, 10, 773 (1977).
76. J. M. Sanders and R. A. Komoroski, Macromolecules,
    10, 1214 (1977).
77. A. E. Tonelli, Macromolecules, 11, 634 (1978).
78. M. H. Fisch and J. J. Dannenberg, Anal. Chem.,
    49, 1405 (1977).
79. D. Khlok, Y. Deslander and J. Prud'homme,
    Macromolecules, 9, 809 (1976).
80. M. I. Lobach, I. A. Poletayeva, A. S. Khatchaturov,
    N. N. Druz and V. A. Kormer, Polymer, 18, 1196
    (1977).
81. J. Furukawa, Angew. Makromol. Chem., 23, 189 (1972).
82. J. Furukawa, H. Amano and R. Hirai, J. Polym. Sci.,
    Part A1, 10, 681 (1972).
83. J. Furukawa and R. Hirai, J. Polym. Sci., Polym.
    Chem. Ed., 10, 3027 (1972).

84. T. Suzuki, Y. Takagami, J. Furukawa and R. Hirai, J. Polym. Sci., Part B, 9, 931 (1971).

85. C. J. Carman, Macromolecules, 7, 789 (1974).

86. M. Bruzzone, A. Carbonaro and C. Corno, Makromol. Chem., 179, 2173 (1978).

87. C. J. Carman and C. E. Wilkes, Macromolecules, 7, 40 (1974).

88. H. Y. Chen, J. Polym. Sci., Polym. Lett. Ed., 12, 85 (1974).

89. K. J. Ivin, D. T. Laverty and J. J. Rooney, Makromol. Chem., 178, 1545 (1977).

90. E. R. Santee, Jr., V. D. Mochel and M. Morton, J. Polym. Sci., Part B, 11, 453 (1973).

91. K. J. Ivin, T. Laverty and J. J. Rooney, Makromol. Chem., 179, 253 (1978).

92. H. Y. Chen, ACS Polymer Preprints, 17, 688 (1969).

93. S. J. Lee, J. McGinnis and T. W. Katz, J. Am. Chem. Soc., 98, 7818 (1976).

94. T. W. Katz, J. McGinnis and C. Altus, J. Am. Chem. Soc., 98, 606 (1976).

95. Y. Tanaka and K. Hatada, J. Polym. Sci., Part B, 11, 569 (1973).

96. M. Julemont, E. Walckiers, R. Warin and Ph. Teyssie, Makromol. Chem., 175, 1673 (1974).

97. F. Conti, A. Segre, P. Pini and L. Porri, Polymer, 15, 5 (1974).

98. B. Morese-Segula, M. St.-Jacques, J. M. Renaud and J. Prud'homme, Macromolecules, 10, 431 (1977).

99. D. H. Beebe, Polymer, 19, 231 (1978).

100. D. Blondin, J. Regis and J. Prud'homme, Macromolecules, 7, 187 (1974).

101. W. Ritter, K. F. Elgert and H. J. Cantow, Makromol. Chem., 178, 557 (1977).

102. T. Suzuki, Y. Tsuji and Y. Takegami, Macromolecules, 11, 639 (1978).

103. J. R. Ebdon, Polymer, 19, 1232 (1978).

104. F. Dawans and P. Teyssie, Ind. Eng. Chem. Prod. Res. Develop., 10, 261 (1971).

105. J. Furukawa, Pure and Appl. Chem., 42, 495 (1975).

106. K. F. Elgert, G. Quack and B. Stutzel, Polymer, 15, 612 (1974).

107. J. Furukawa, E. Kobayashi and T. Kawagoe, Polym. J., 5, 231 (1973).

108. W. Gronshi, N. Murayama, H. J. Cantow and
      T. Miyamoto, Polymer, 17, 358 (1976).
109. H. R. Kricheldorf and G. Schilling, Makromol.
      Chem., 179, 1175 (1978).
110. G. Schilling and H. R. Kricheldorf, Makromol.
      Chem., 178, 885 (1977).
111. H. R. Kricheldorf and W. E. Hull, Makromol.
      Chem., 180, 161 (1979).
112. H. R. Kricheldorf and W. E. Hull, J. Polym.
      Sci., Polym. Chem. Ed., 16, 583 (1978).
113. H. R. Kricheldorf, Makromol. Chem., 180, 147
      (1979).
114. T. Hiraoki, A. Tsutsumi, K. Kikichi and
      M. Kaneko, Polym. J., 8, 429 (1976).
115. C. Corno, A. Roggero and T. Salvatori, Eur.
      Polym. J., 10, 525 (1974).
116. C. Corno and A. Roggero, Eur. Polym. J., 12,
      159 (1976).
117. C. Corno, A. Roggero, T. Salvatori and
      A. Mazzei, Eur. Polym. J., 13, 77 (1977).

PRESENT STATUS OF THE CONFIGURATIONAL AND CONFORMATIONAL ANALYSIS OF STEREOREGULAR POLYMERS

P.Corradini,G.Guerra,B.Pirozzi

Istituto Chimico dell'Università di Napoli,
Via Mezzocannone 4, 80134 Napoli, Italy

CONSTITUTIONAL AND CONFIGURATIONAL ISOMERISM IN MACRO-MOLECULES(1)

In a molecule the constitution specifies which atoms are bonded to each other and with what kind of bonds, without considering their spatial dispositions; the configuration specifies the spatial disposition of the bonds, for an assigned constitution, without taking into account the molteplicities of the spatial dispositions, that may arise by rotation around single bonds. The spatial dispositions which arise from the specification of the internal rotation angles around single bonds represent possible conformations.

From any single monomer, different constitutional units may arise during the polimerization. Consider, for instance, the monomer isoprene; even in the case of a regular enchainment, the monomeric units may join the growing chain according to the three different constitutions which are indicated:

$$-CH_2-\overset{\overset{\textstyle CH_3}{|}}{C}=CH-CH_2- \quad ; \quad -CH_2-\overset{\overset{\textstyle CH_3}{|}}{\underset{\underset{\textstyle CH_2}{\overset{||}{CH}}}{C}}-$$

1,4   1,2

317

R. W. Lenz and F. Ciardelli (eds.), Preparation and Properties of Stereoregular Polymers, 317-352.
Copyright © 1979 by D. Reidel Publishing Company.

$$-CH_2-CH-$$
$$| $$
$$C-CH_3 \qquad (1)$$
$$||$$
$$CH_2$$

3,4

Consider now the 1,4 regular enchainment. Yet, a regular polymer having this constitution may have still two different configurations for the double bond along the chain:

CH$_3$ H   chain   H
   \\   /   \\   /
     C==C   (2)   C==C
   /   \\   /   \\
chain   chain   CH$_3$ chain

cis-configuration   *trans*-configuration

If all the units along the chain are *trans* or if all the units along the chain are *cis*, the polymer is called tactic: in the first case *trans*-tactic, in the second case *cis*-tactic. It may be noted that, in the case of 1,4 polyisoprene, these two possibilities correspond to guttapercha and natural rubber respectively; tactic polymers of isoprene are naturally occurring polymers.

From the previous example we have seen that a problem of configurational isomerism arises whenever we have a double bond along the chain. Another case in which a problem of configurational isomerism arises is the case in which we have along the chain a carbon atom which is further bonded to two different groups, S and L. Here are represented the two possible cases which may arise for two successive constitutionally equivalent carbon atoms —C(S)(L)— along the chain, that have a simmetrically constituted connecting group(if any):

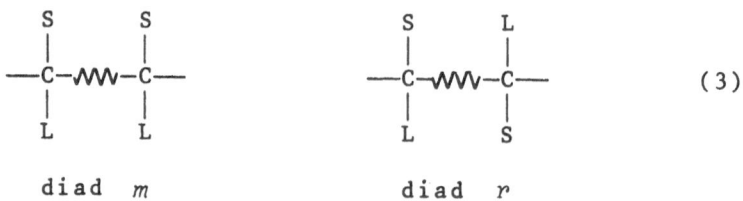

diad *m*   diad *r*

This representation of the space disposition of bonds, around these carbon atoms, is a modified Fisher projec-

tion, where the bonds at each carbon atom are seen as a
projection of a tetrahedral arrangement, so that the two
vertical bonds go in the direction of the observer, and
the two horizontal bonds go in the direction away from
the observer:

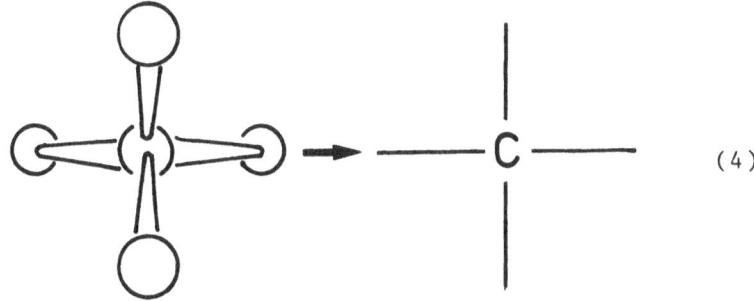

(4)

Stereosequences terminating in tetrahedral stereoisomeric centres at both ends, and which comprise two, three,
four, five etc. consecutive centres of that type, may be
called diads, triads, tetrads, pentads, etc., respectively. As indicated in the formulas (3), the diad m is defined as that with identical substituents on the same side in respect to the backbone in Fisher projection, and
diad r is defined as that with identical substituents in
the opposite sides in respect to the backbone, in the same projection.

All definitions in polymers refer to ideal situations, so an "ideal" isotactic vinyl polymer is a polymer
caracterized by a succession of all m diads; an "ideal"
syndiotactic vinyl polymer is a polymer which is caracterized by a succession of all r diads. According to these conventions, the isotactic polymer of propylene can
be indicated:

(Fisher projection)

(5)

(reference to a zig-zag
planar conformation)

while the syndiotactic polymer will be indicated:

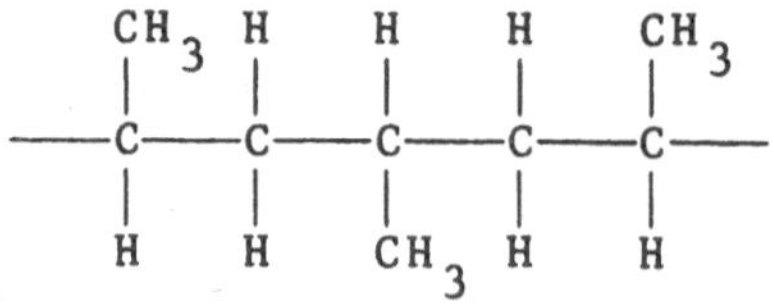

(Fisher projection)

(6)

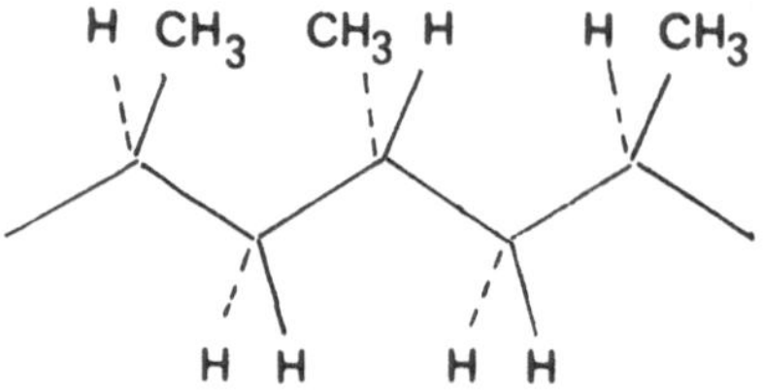

(reference to a zig-zag
planar conformation)

The first definition of an isotactic polymer was made with reference to a zig-zag planar conformation. Take, however, the case of polyethylidene; the isotactic polymer will be represented, in the modified projection, as follows:

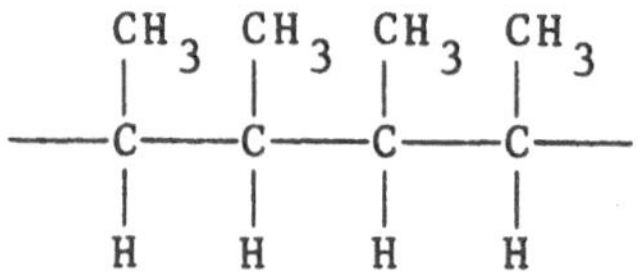

(7)

while the syndiotactic polymer in the same representation:

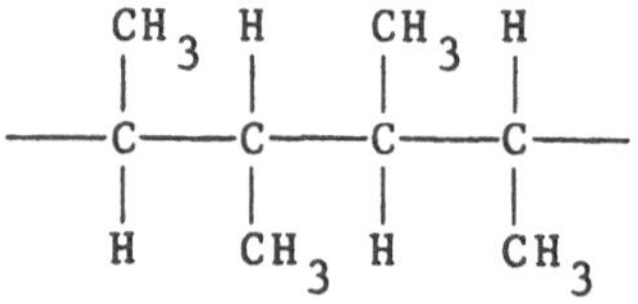

(8)

Note that, if you represent the chain in its zig-zag planar conformation, the result will be:

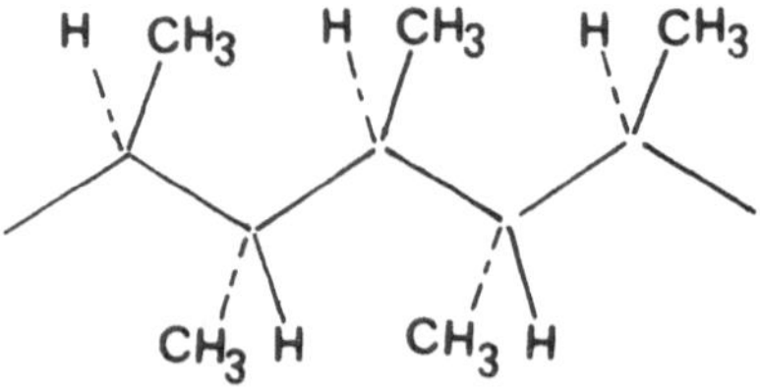

(9)

isotactic polymer

$$\text{(9)}$$

syndiotactic polymer

where the successive methyl groups are on opposite sides, in respect to the plane of the zig-zag, in the isotactic polymer, and on the same side in the syndiotactic
polymer; instead opposite conclusions, for analogue representations, are reached in the case of isotactic and
syndiotactic polypropylene.

Consider, now, the case of polypentadiene, in the 1,4
enchainment:

$$-\overset{*}{C}H(CH_3)-CH\overset{*}{=\!=}CH-CH_2- \tag{10}$$

and suppose that only one of the two sites of stereoisomerism * , in each constitutional unit in one sequence,
has defined stereochemistry. We can have the isotactic
polymer:

$$\left[\begin{array}{c} \overset{\displaystyle CH_3}{\underset{\displaystyle H}{\overset{|}{\underset{|}{C}}}}-CH\!=\!CH-CH_2-\overset{\displaystyle CH_3}{\underset{\displaystyle H}{\overset{|}{\underset{|}{C}}}}-CH\!=\!CH-CH_2 \end{array}\right]_n \tag{11}$$

(configuration of the double bond unknown or not defined)
the syndiotactic polymer:

$$\left[\begin{array}{c} \overset{\displaystyle CH_3}{\underset{\displaystyle H}{\overset{|}{\underset{|}{C}}}}-CH\!=\!CH-CH_2-\overset{\displaystyle H}{\underset{\displaystyle CH_3}{\overset{|}{\underset{|}{C}}}}-CH\!=\!CH-CH_2 \end{array}\right]_n \tag{12}$$

the *cis*-tactic polymer:

$$\text{(13)}$$

(configuration of the tertiary carbon atom unknown or not defined)
the *trans*-tactic polymer:

$$
\left[\begin{array}{c}
\text{--CH(CH}_3)\qquad\qquad\text{H} \\
\diagdown\qquad\diagup \\
\text{C}=\text{C} \\
\diagup\qquad\diagdown \\
\text{H}\qquad\qquad\text{CH}_2\text{---}
\end{array}\right]_n
\qquad(14)
$$

If both the sites of stereoisomerism have defined stereochemistry, the polymer is defined stereoregular. So stereoregular polymers are:

$$
\left[\begin{array}{c}
\overset{\displaystyle CH_3}{\underset{\displaystyle H}{\overset{|}{\underset{|}{C}}}}\text{---CH=CH---CH}_2\text{--}\overset{\displaystyle CH_3}{\underset{\displaystyle H}{\overset{|}{\underset{|}{C}}}}\text{---CH=CH---CH}_2\text{--} \\
\quad(trans)\qquad\qquad(trans)
\end{array}\right]_n
\qquad(15)
$$

isotranstactic

$$
\left[\begin{array}{c}
\overset{\displaystyle CH_3}{\underset{\displaystyle H}{\overset{|}{\underset{|}{C}}}}\text{---CH=CH---CH}_2\text{--}\overset{\displaystyle H}{\underset{\displaystyle CH_3}{\overset{|}{\underset{|}{C}}}}\text{---CH=CH---CH}_2\text{--} \\
\quad(trans)\qquad\qquad(trans)
\end{array}\right]_n
\qquad(16)
$$

syndiotranstactic

$$
\left[\begin{array}{c}
\overset{\displaystyle CH_3}{\underset{\displaystyle H}{\overset{|}{\underset{|}{C}}}}\text{---CH=CH---CH}_2\text{--}\overset{\displaystyle CH_3}{\underset{\displaystyle H}{\overset{|}{\underset{|}{C}}}}\text{---CH=CH---CH}_2\text{--} \\
\quad(\ cis\ )\qquad\qquad(\ cis\ )
\end{array}\right]_n
\qquad(17)
$$

isocistactic

$$
\left[\begin{array}{c}
\overset{\displaystyle CH_3}{\underset{\displaystyle H}{\overset{|}{\underset{|}{C}}}}\text{---CH=CH---CH}_2\text{--}\overset{\displaystyle H}{\underset{\displaystyle CH_3}{\overset{|}{\underset{|}{C}}}}\text{---CH=CH---CH}_2\text{--} \\
\quad(\ cis\ )\qquad\qquad(\ cis\ )
\end{array}\right]_n
\qquad(18)
$$

syndiocistactic

In general, according to the IUPAC definitions, a *regular* polymer is a polymer which is built up of identical constitutional units, which are called constitutional repeating units. A polymer is called *tactic* if at least

one site of stereoisomerism in each constitutional unit
has a regular stereochemistry. A polymer is called, instead, *stereoregular* when the molecules can be described
in terms of only one species of configurational unit,
having defined configuration at all the sites of stereoisomerism in the main chain, in a single sequential arrangement. Thus, a stereoregular polymer is always a tactic polymer, but a tactic polymer is not always stereoregular, because a tactic polymer need not have all sites of stereoisomerism with defined stereochemistry. The
polymers (11),(12),(13),(14) are tactic, the polymers
(5),(6),(7),(8),(15),(16),(17),(18) are both tactic and
stereoregular.

For the designation of relative configurations insi-
de of a given monomeric unit, with two non-constitutio-
nally equivalent carbon atoms of the main chain, bearing
substituents S_1,L_1 and S_2,L_2 respectively, a further convention (which is taken from the chemistry of carbohy-
drates) is used:

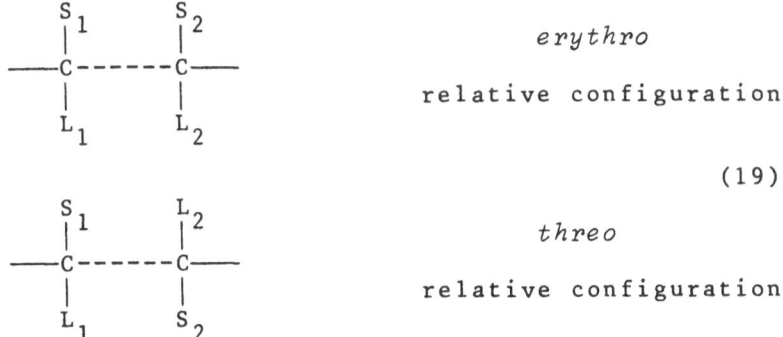

$$(19)$$

where $S_1 \neq S_2$ and/or $L_1 \neq L_2$ and L precedes S according
to the Cahn, Ingold, and Prelog rule of precedence.

The possible stereoregular polymers which may arise
from units of the previous kind are, in the case that S_1
$(=S_2)$ is a hydrogen atom and $L_1 = A \neq L_2 = B$:

$$(20)$$

erythro-diisotactic

```
   A   H   A   H   A   H
   |   |   |   |   |   |
——C———C———C———C———C———C——                          (21)
   |   |   |   |   |   |
   H   B   H   B   H   B
```

threo-diisotactic

```
   A   B   H   H   A   B
   |   |   |   |   |   |
——C———C———C———C———C———C——                          (22)
   |   |   |   |   |   |
   H   H   A   B   H   H
```

disyndiotactic

it is possible to get an *erythro* and a *threo* diisotactic
polymers whereas there is only one disyndiotactic polymer. But in the case, in which the substituents, A and
B, are joined in a ring it is possible to distinguish
two cases for the disyndiotactic polymer; the first case in which the rings join two atoms on the same side of
the Fisher projection (the polymer is then named *erythro*-disyndiotactic):

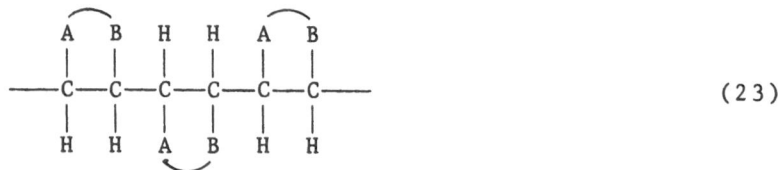

```
   A   B   H   H   A   B
   |   |   |   |   |   |
——C———C———C———C———C———C——                          (23)
   |   |   |   |   |   |
   H   H   A   B   H   H
```

erythro-disyndiotactic

the second case in which the rings join two atoms in the
opposite sides of the Fisher projection (the polymer is
then named *threo*-disyndiotactic):

```
   A   B   H   H   A   B
   |   |   |   |   |   |
——C———C———C———C———C———C——                          (24)
   |   |   |   |   |   |
   H   H   A   B   H   H
```

threo-disyndiotactic

THEORETICAL ASPECTS OF CONFORMATIONAL ANALYSIS

Internal coordinates(2)

The space form of the chain of a polymer depends on bond distances, on bond angles, and on dihedral angles: parameters which are called internal coordinates; the number of internal coordinates necessary to describe a chain with n atoms is $3n-6$. Fig. 1 shows as in a given chain bond lengths, bond angles, and internal rotation angles are most appropriately designated.

It is important to know which is the appropriate convention which is used to measure internal rotation angles. Take, for example, three successive bonds L_1, L_2, and L_3 (fig. 2). If you look in the direction of L_2 from the side of L_3, the dihedral angle is that from which we have to rotate the bond L_3 in order to superpose it to

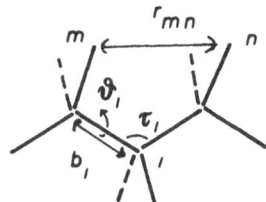

Fig. 1. Symbols used for: bond lengths, bond angles, internal rotation angles and distance between nonbonded atoms.

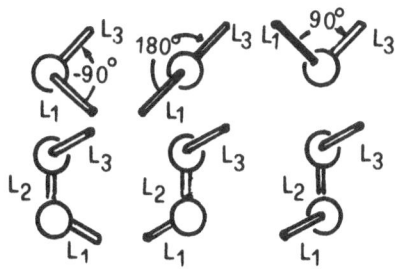

Fig. 2. Convention used to measure internal rotation angles.

L_1, describing the smaller angle possible. If the rotation is in the clockwise direction the angle is positive, viceversa it is negative; thus possible values of dihedral angles are included in the range $(-180°, 180°)$. It is easy to see that the same result would be obtained if, always looking in direction of L_2, but from the side of L_1, you measure the dihedral angle from which we have to rotate L_1 in order to superpose it to L_3 .

Special names and symbols are attributed to the following internal rotation angles:

trans or *antiperiplanar*   (T) for 180°
anticlinal   (A) for |120°|
gauche or *synclinal*   (G) for |60°|
cis or *symperiplanar*   (C) for 0°

A sign can be appended to the symbol, to indicate whether the internal rotation angle is plus or minus, while a prime may indicate an internal rotation angle which is slightly displaced from the exact corresponding value. The notation G , $\overline{G}$; A , $\overline{A}$ (and T , $\overline{T}$; C , $\overline{C}$ whenever the torsion angles are not exactly equal to 180° and 0°, respectively) are reserved for the designation of enantiomorph conformations, i.e. conformations of opposite but unspecified sign.

Internal potential energy(2)

The comprehension of the spatial relationships among the atoms of a molecule, which is the object of conformational analysis, is a universal prerequisite in the establishement of the connections between the graphic formula and the properties of a substance. The relevancy is even greater in the case of long chain molecules where the phemenomenon of rubber elasticity, the hydrodynamic and thermodynamic properties of the solutions, the rheology of the melts reflect the caracter of random coil of a single macromolecule, whereas many useful properties of polymers reflect their ability to crystallize.

The intramolecular potential energy is very important in determining the conformations of the macromolecules, both in the crystalline state and in the amorphous or solution state. In turn, the potential energy may be taken, in general, as a sum of terms of the kind: stretching, bending, torsion, nonbonded, electrostatic:

$$E = E_s + E_b + E_t + E_{nb} + E_{el} \qquad (25)$$

For small displacements from the minimum energy value, the stretching energy may be taken as:

$$E_s = \tfrac{1}{2} k_s (b-b_o)^2 \tag{26}$$

and an analogous formula may be used for the bending e-
nergy:

$$E_b = \tfrac{1}{2} k_b (\tau-\tau_o)^2 \tag{27}$$

where b_o and τ_o are the values of bond distances and angles chosen as energetic minima, and k_s and k_b constants depending on the particular kind of bonds. The k_s value in formula(26) is such as to prohibit displacements
of the bond distances from b_o greater than a few per ce-
nt. In fact, bond distances as determined from X-ray diffraction experiments, are generally almost constant in
going from one molecule to another, if we refer to atoms
in similar electronic environments.
    The bending energy parameters, which are used by Fl-
ory(3), in a consistent way with the non bonded parameters, which we shall indicate later on, are reported in
table 1. It is seen that a deviation of 5° from the minimum energy values does not imply very large energy differences; such differences are always lower than RT, at
room temperature (this is indicated in the third column
of the table).
    Most researchers, performing conformational analy-
sis, use also a torsion term, which for single C—C bonds
not adjacent to a double bond (for instance, ethane) is:

$$E_t = \tfrac{1}{2} E_o' (1 + \cos3\theta) \tag{28}$$

while for single bonds adjacent to a double bond (for

Table 1. Bending energy parameters used by Flory.

Bond angle	k_b (kcal mol^{-1}deg^{-2})	E ($\delta\tau=5°$)
<CCC	0.044	0.55
<CCH	0.029	0.36
<HCH	0.024	0.30

instance, propylene) is:

$$E_t = \tfrac{1}{2} E_0'' (1 - \cos 3\theta) \tag{29}$$

For his calculations, Flory takes $E_0' = 2.8$ kcal/mol and $E_0'' = 1.98$ kcal/mol. Take note that the relative positions of two hydrogen atoms in ethane are *trans* and *gauche* for the minimum energy value, while in the case of the methyl group of propylene the minimum energy relative positions of the hydrogen atoms of the methyl group, in respect to the carbon atom joined by a double bond, are *anticlinal* or *cis*, depending on the minus sign which appears in the formula(29):

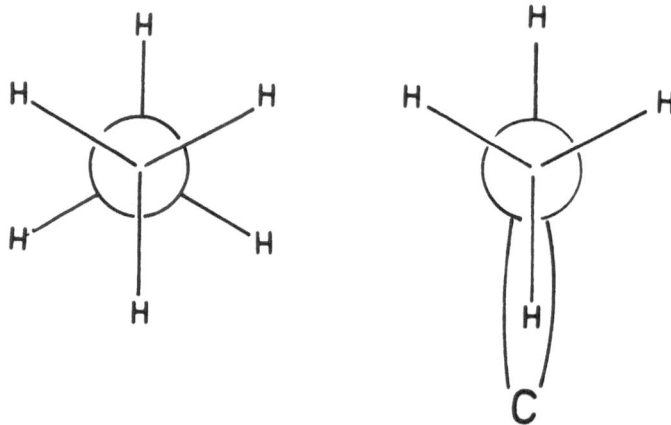

The nonbonded energies arise from the interactions between atoms which are not directly bonded and are taken to depend only on the distances r_{mn} (fig. 1) between each pair of atoms m and n, of species i and j. For the nonbonded energies two kinds of functions are generally used:

$$E_{nb} = \frac{d_{ij}}{r_{mn}^{12}} - \frac{e_{ij}}{r_{mn}^{6}} \qquad \text{Lennard-Jones}$$

$$\tag{31}$$

$$E_{nb} = a_{ij} \exp(-b_{ij} r_{mn}) - c_{ij}/r_{mn}^{6}$$

$$\text{Buckingham}$$

Some indicative data for contact types given by Flory are indicated in table 2. Similar data given by Scott and Scheraga(4) are indicated in table 3. A tabulation of the values of the function is given in the table 4.

When dipoles are present in the molecules (for instance, in the case of CO and NH groups in amides), Scott and Scheraga(4) include the electrostatic term of Eq.(25), by localizing partial charges on the atoms (table 5).

Table 2. Nonbonded parameters used by Flory.

Interacting pair	$d_{ij} \, 10^{-3}$ (kcal mol^{-1} Å^{12})	e_{ij} (kcal mol^{-1}Å^6)	r_{min} (Å)
C,C	398	366	3.6
C,H	57	128	3.1
H,H	7.3	47	2.6

Table 3. Nonbonded parameters used by Scott and Scheraga

Interacting pair	$d_{ij} \, 10^{-3}$ (kcal mol^{-1}Å^{12})	e_{ij} (kcal mol^{-1}Å^6)	r_{min} (Å)
C,C	286	370	3.4
C,H	38	128	2.9
H,H	4.46	46.7	2.4

Table 4. Values of the Lennard-Jones functions for different values of distances between the atoms. The energies are given in kcal mol^{-1}.

r_{mn} (Å)	with Flory parameters		
	E(C,C)	E(C,H)	E(H,H)
1.6	—	—	23.1
1.8	333.28	45.51	4.93
2.0	91.45	11.92	1.05
2.2	27.73	3.31	0.15
2.4	8.98	0.89	-0.05
2.6	2.99	0.18	-0.08
2.8	0.95	-0.02	-0.07
3.0	0.25	-0.07	-0.05
3.2	0.00	-0.07	-0.04
3.4	-0.07	-0.06	-0.03
3.6	-0.08	-0.05	-0.02
3.8	-0.08	-0.04	-0.01

r_{mn} (Å)	with Scott and Scheraga parameters		
	E(C,C)	E(C,H)	E(H,H)
1.6	—	—	13.06
1.8	236.35	29.08	2.48
2.0	64.04	7.28	0.36
2.2	18.98	1.83	-0.06
2.4	5.90	0.37	-0.12
2.6	1.80	-0.02	-0.10
2.8	0.46	-0.10	-0.08
3.0	0.03	-0.10	-0.05
3.2	-0.10	-0.09	-0.04
3.4	-0.12	-0.07	-0.03
3.6	-0.11	-0.05	-0.02
3.8	-0.09	-0.04	-0.02

Table 5. Partial charges on the atoms of amide group used by Scott and Scheraga.

Atom	Charge(in units of e)
H	+0.272
N	-0.305
C	+0.449
O	-0.416

Calculation of distances between the atoms(5)

The calculation of the potential energy of a disposition of atoms is possible if we know the coordinates
of the atoms in a cartesian system in order to get all
the distances between the atoms, which are essential in
the calculation of the nonbonded energy terms. One possible way to get the relevant cartesian coordinates of
a molecule, as a function of its internal coordinates,
is now indicated.

As specified before, the space form of a molecule
depends on $3n-6$ internal coordinates. Take as an example a succession of five atoms, as indicated in fig.3.
In this case the internal coordinates, which characterize the space form of such chain, correspond to four
bond distances (b_2, b_3, b_4, b_5), three bond angles $(\tau_2,
\tau_3, \tau_4)$, two dihedral angles (θ_3, θ_4).

We can put the first atom 1 at the origin of the cartesian system, the next atom 2 may be disposed with the
b_2 bond in the direction of the x-coordinate, and it is
also possible to fix the atom 3 in the x-y plane. The
coordinates of the first atom will be indicated as a column vector as follows:

$$\begin{vmatrix} x_1 \\ y_1 \\ z_1 \end{vmatrix} = \begin{vmatrix} 0 \\ 0 \\ 0 \end{vmatrix}$$

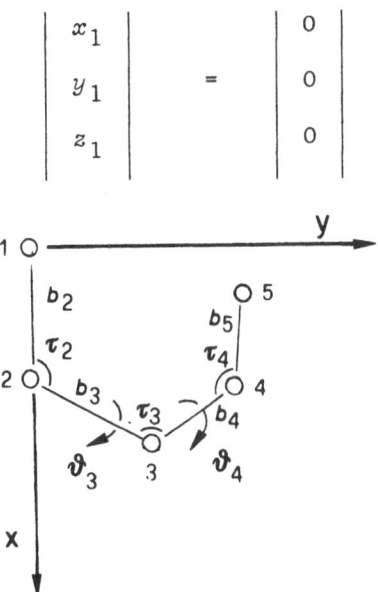

Fig. 3.A succession of five atoms in the cartesian system described in the text.

The coordinates of the second atom are indicated as:

$$
\begin{vmatrix} x_2 \\ y_2 \\ z_2 \end{vmatrix} = \begin{vmatrix} b_2 \\ 0 \\ 0 \end{vmatrix}
$$

The coordinates of the third atom may be obtained by summation of two vectors:

$$
\begin{vmatrix} x_3 \\ y_3 \\ z_3 \end{vmatrix} = \begin{vmatrix} -\cos\tau_2 & -\sin\tau_2 & 0 \\ \sin\tau_2 & -\cos\tau_2 & 0 \\ 0 & 0 & 1 \end{vmatrix} \begin{vmatrix} b_3 \\ 0 \\ 0 \end{vmatrix} + \begin{vmatrix} x_2 \\ y_2 \\ z_2 \end{vmatrix} =
$$

$$
= \begin{vmatrix} -b_3\cos\tau_2 + b_2 \\ b_3\sin\tau_2 \\ 0 \end{vmatrix}
$$

In order to get the coordinates of the atom 4, we have to sum the vector (x_3, y_3, z_3) to a vector whose coordinates depend on the internal rotation angle θ_3 and on the bond angle τ_3, so that we obtain:

$$
\begin{vmatrix} x_4 \\ y_4 \\ z_4 \end{vmatrix} = A_2^\tau \begin{vmatrix} 1 & 0 & 0 \\ 0 & \cos\tau_3 & -\sin\tau_3 \\ 0 & \sin\tau_3 & \cos\tau_3 \end{vmatrix} \begin{vmatrix} -\cos\theta_3 & -\sin\theta_3 & 0 \\ \sin\theta_3 & -\cos\theta_3 & 0 \\ 0 & 0 & 1 \end{vmatrix} \times
$$

$$
\times \begin{vmatrix} b_4 \\ 0 \\ 0 \end{vmatrix} + \begin{vmatrix} x_3 \\ y_3 \\ z_3 \end{vmatrix}
$$

where:

$$
A_2^\tau = \begin{vmatrix} -\cos\tau_2 & -\sin\tau_2 & 0 \\ \sin\tau_2 & -\cos\tau_2 & 0 \\ 0 & 0 & 1 \end{vmatrix}
$$

In the same way the coordinates of the atom 5 are obtained as follows:

$$
\begin{vmatrix} x_5 \\ y_5 \\ z_5 \end{vmatrix} = A_{\sim 2}^{\tau} \; A_{\sim 3}^{\tau\theta} \; A_{\sim 4}^{\tau\theta} \begin{vmatrix} b_5 \\ 0 \\ 0 \end{vmatrix} + \begin{vmatrix} x_4 \\ y_4 \\ z_4 \end{vmatrix}
$$

where:

$$
A_{\sim j}^{\tau,\theta} = \begin{vmatrix} 1 & 0 & 0 \\ 0 & \cos\tau_j & -\sin\tau_j \\ 0 & \sin\tau_j & \cos\tau_j \end{vmatrix} \begin{vmatrix} -\cos\theta_j & -\sin\theta_j & 0 \\ \sin\theta_j & -\cos\theta_j & 0 \\ 0 & 0 & 1 \end{vmatrix}
$$

For the general case of a chain of j atoms we get the formula (which can easily programmed for computer calculation):

$$
x_{\sim j} = A_{\sim 2}^{\tau} \; A_{\sim 3}^{\tau,\theta} \cdot \; \cdot A_{\sim j-1}^{\tau,\theta} \; b_{\sim j} + x_{\sim j-1} \tag{32}
$$

where

$$
x_{\sim j} = \begin{vmatrix} x_j \\ y_j \\ z_j \end{vmatrix} \qquad \text{and} \qquad b_{\sim j} = \begin{vmatrix} b_j \\ 0 \\ 0 \end{vmatrix}
$$

THE CONFORMATION OF POLYMERIC CHAINS IN THE CRYSTALLINE STATE

Equivalence principle(6)

In a system of polydisperse polymer molecules (as it is the case for synthetic polymers, where the molecules are never all alike, even in the case of $M_w/M_n \approx 1$), the crystalline state (which implies threedimensional long range order) may be conceived only in the approximation of not taking into account the terminals of the molecules (that is considering the molecules of infinite length) and implies in general, with exceptions which will be cited later on, the repetition of identical units along the chain axis.

The fig. 4 is a representation of the structure of cellulose, as given for the first time by Meyer(7).

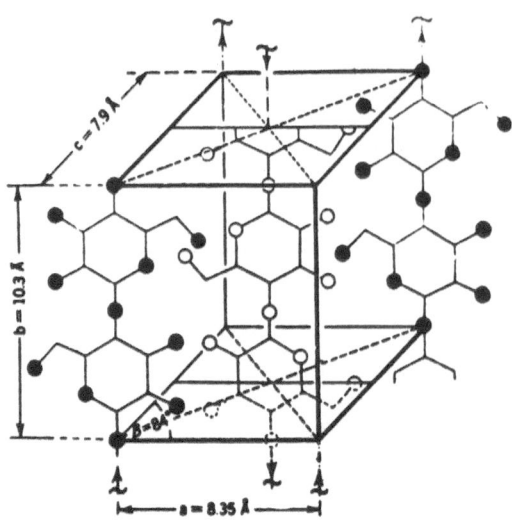

Fig. 4. Representation of the structure of cellulose as
given by Meyer.

The concept of macromolecule had been given a few
years before by Staudinger(8) in 1924 and was not accepted immediately. Many people continued to think of polymers as "colloidal" associations of small molecules.
The fact that the unit cell of cellulose was small could have been taken as an evidence of the non macromolecular character of that material and it was important,
at the time, the establishement of the idea that a molecule (in the case of chain molecules) needed not be confined to a unit cell, but could span many unit cells in
sequence along the chain axis.
       If we need to have repetition of identical units a-
long the chain of a polymer, it is clear that, at least
in principle, prerequisites for the crystallizability of
a polymer are:
1) regularity of chemical constitution
2) regularity of configuration (stereoregularity) for
   long sequences of monomeric units.
As we shall see, an exception may arise because of isomorphism of monomeric units having different constitutions or configurations and/or conformations, which we
shall discuss in the lecture "The crystalline structure
of addition polymers. Research problems".
       It may be seen that the repetition of identical u-
nits (repeating units) along the chain axis in the crystalline state implies the so called "equivalence princi-

ple" which can be formulated as follows:
The chain needs to be built up of structural units, which take geometrically equivalent positions in respect to
an axis.

As we shall see, such structural units are in general a fraction of the repeating units. While a repeating
unit may correspond even to a large number of monomeric
units (f.i., in helical polymers), the structural unit
corresponds very often to one monomeric unit, even though this is not necessarily so. For instance in the case of 1,4 *trans* or 1,4 *cis* polybutadiene, the structural
unit corresponds to one half a monomeric unit, while in
the case of polydimethyl-ketene in the ketonic enchainment, the structural unit corresponds to two monomeric
units. A similar occurrence may explain the conflicting
observations on crystalline gels of isotactic polystyrene, which are reported in(9) and we shall discuss in the
above cited lecture.

Line repetition groups(6)

The only simmetry operators which have a translational component and which are compatible with chain repetition are:
  t translation $\vec{c}$ along z (chain axis);
  c glide-plane (translation $\frac{1}{2}\vec{c}$ along z associated with a mirror on a plane containing z);
  s screw (helical) repetition of M units in N turns
(translation $\vec{c}/M$ along z plus rotation $2\pi N/M$ around z).
In the case of the helical repetition we use the terms:
unit height (h) for the translation $|\vec{c}/M|$, unit twist
(t) for the rotation $2\pi N/M$, number of residues per turn
(n) for the ratio M/N.
  Other symmetry operators which are compatible with
a chain repetition are: r_n ,$2\pi/n$ rotation around the chain
axis; i , center of symmetry; m , plane of symmetry perpendicular to the chain axis; d , plane of symmetry parallel to the chain axis; 2 , two-fold rotation perpendicular to the z axis.
  Some of them are just indicated for theoretical reasons. For instance rotations around the chain axis for
a single chain may be thought of only for very particular constitutional repeating units and very particular
values of $2\pi/n$. A rotation of 180° may occur if we have
two chains winding up on the same chain axis, but does
not refer to the conformation of one single chain. In
fact, not all the symmetry elements are compatible with
a given constitution and configuration of a polymer chain; for instance, whenever the unit has a directional

character, symmetry elements like 2 and *m* and *i* are ruled out automatically . The translational symmetry elements
and the further symmetry elements which we have indicated (excluding r_n) may be combined into the chain repe-
titions groups, indicated in table 6.

Table 6. Possible Chain Repetition Groups.

s(M/N)1	Isotactic polypropylene	*s*(3/1)1
particular case		
*t*1	1,4-*trans*-polyisoprene, Mod.α	*s*(1/1)=*t*1
s(M/N)2	Syndiotactic polypropylene	*s*(2/1)2
particular case		
*t*2	—	
tm	Nylon 77	
td	—	
tc	1,4-*cis*-polyisoprene	
ti	Ethylene-butene-2 isotactic alternating copolymer	
s(2/1)*m*	*trans*-polypentenamer	
s(2/1)*d*	Nylon 6 (planar chain conformation)	
tdm	—	
tid	Nylon 66 (planar chain conformation)	
tcm	Syndiotactic 1,2-polybutadiene	
tic	*cis*-1,4-polybutadiene	
s(2/1)*dm*	Polymethylene	

Thus, for example, the repetition group of the chain
of isotactic polypropylene (fig. 5), which is a threefold helix, may be indicated as *s*(3/1)1 , where the symbol *s* indicates the helical repetition, the symbol 3/1
the repetition of three units in one pitch and the further symbol 1 indicates that there is no further symmetry element but the identity which, according to the
crystallographyc rules, is indicated with 1.
    The symbol for syndiotactic polypropylene may be in-

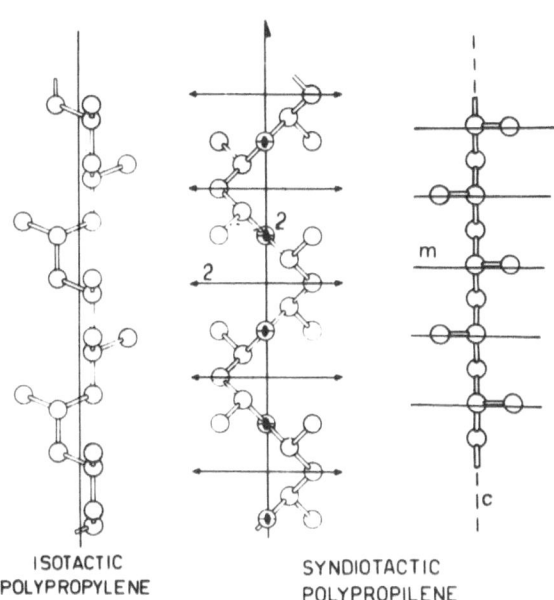

ISOTACTIC
POLYPROPYLENE

SYNDIOTACTIC
POLYPROPILENE

Fig. 5. Different conformations of polypropylene chains:
isotactic ($s(3/1)1$ group) and syndiotactic ($s(2/1)2$ and
tcm groups).

dicated with $s(2/1)2$ (fig. 5), and shows that neighbouring structural units are repeated through the operation
of twofold axes perpendicular to the chain axis and each
pair of units is repeated according to a helix containing two pairs in one pitch.
       The symbol ti applies for the   isotactic alternate
copolymer of ethylene and butene-2 (fig. 6a); in this
case the only symmetry element together with the translation is a center of symmetry.
       In the case of polymethylene almost all of the sym-
metry elements which have been indicated previously are
present; the appropiate symbol is $s(2/1)dm$ (fig. 6b) ,
but a center of symmetry and a glide plane are also present because they are generated by combination of the
symmetry elements indicated in the symbol (the screw a-
xis 2/1 and the mirror planes d and m).
       In the case of cis-1,4-polybutadiene the symbol is
tic. The center of symmetry and the glide plane are both
indicated, but these symmetry elements, combined together, generate also a twofold axis perpendicular to the
chain axis.
       The chain conformation of the four stereoregular po-
lymers, which may arise from the polymerization of 1,3

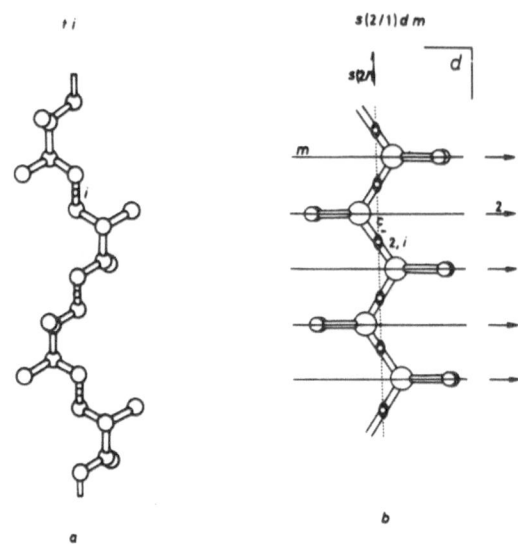

Fig. 6. Chain conformation and symmetry elements for:
a) Isotactic alternate copolymer of ethylene and bute-
ne-2.   b) Polymethylene.

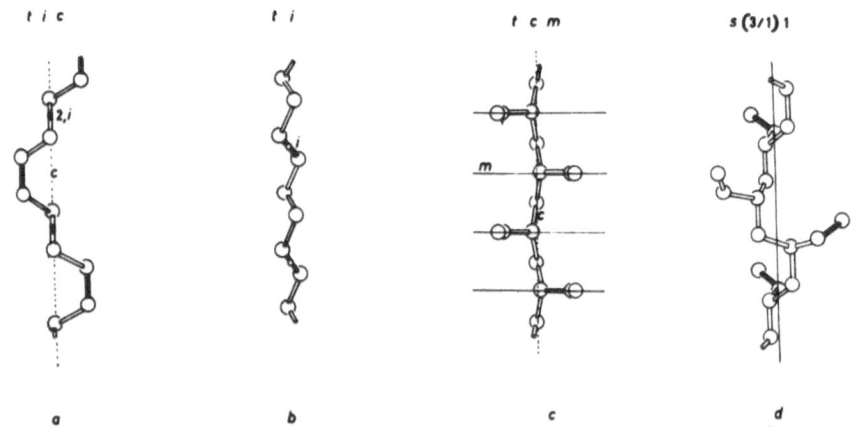

Fig. 7. Conformations of the four stereoregular polymers
of 1,3-butadiene with symmetry elements: a) *cis*-1,4
b)*trans*-1,4   c) syndiotactic 1,2   d) isotactic 1,2

butadiene, are indicated together with the appropriate chain repetition groups and the symmetry elements which are present in the chain in fig. 7. As said before, both in the case of 1,4-*cis* or 1,4-*trans* polybutadiene the indipendent structural unit is built up of only half a monomeric unit, because of the symmetry elements which are present along the chain.

Isotactic polymers get generally a helical structure with a number of monomeric units per pitch which ranges between 3 and 4; some examples are indicated in fig. 8.

Some selected examples of conformational energy calculations

Now we report the results of a conformational analysis which was performed on isotactic and syndiotactic polypropylene many years ago (more refined calculations on isotactic polymers performed by us lately, will be reported in the lecture "The crystalline structure of addition polymers. Research problems").

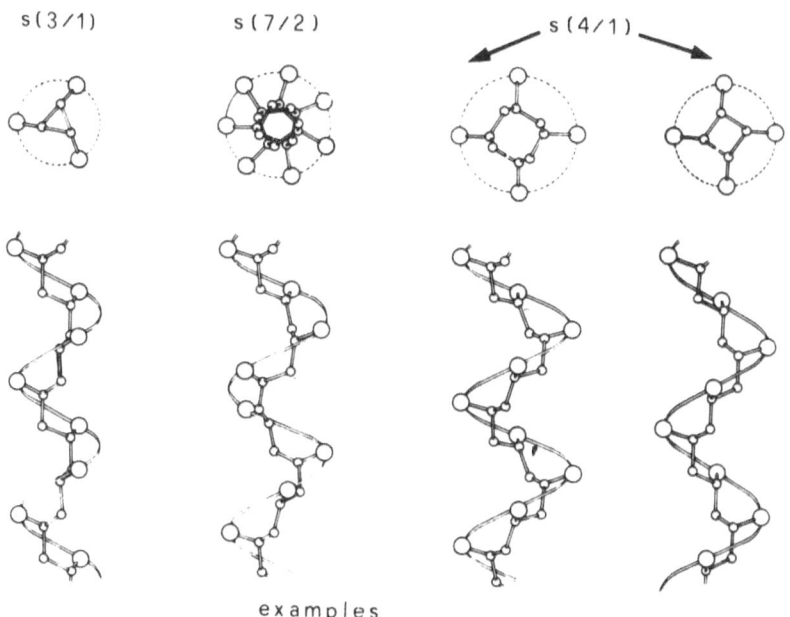

Fig. 8. Chain conformations of some isotactic polymers.

To study the conformations in the crystalline state, the calculations were performed on the basis of the e- quivalence principle, by taking into account the possi- ble variations of the internal rotation angles (bond an- gles and bond length being kept constant) and making the assumption that the structural unit was coincident with one monomeric unit. In such a case, it is easy to see that the isotactic polymer must be built by a succession of units in which a pair of different internal rotation angles θ_1 and θ_2 is repeated along the chain; whereas the chain of the syndiotactic polymer must be built up by a succession of the type θ_1, θ_1, θ_2, θ_2, θ_1, θ_1,... Con- sequently contour plots of the internal energy E as a function of two dihedral angles (θ_1 and θ_2), are suffi- cient to establish the conformation or conformations of minimum internal energy.

The fig. 9 shows, for isotactic polypropylene, two minima that correspond to the chain conformations $(TG_+)_3$ (left-handed helix) and $(G_-T)_3$ (right-handed helix) fo- und in the crystalline state.

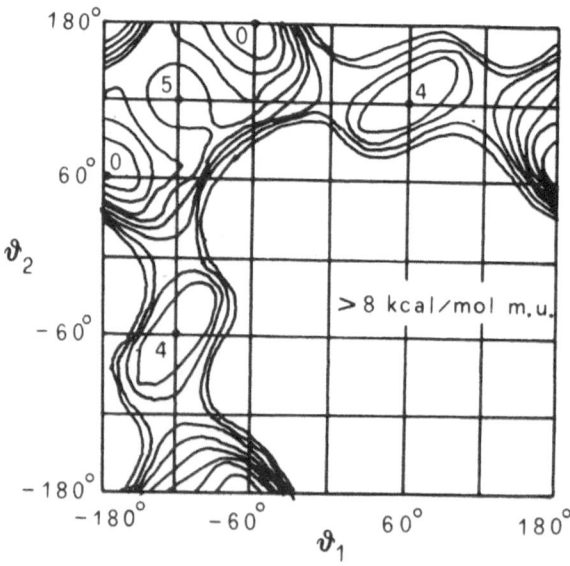

Fig. 9. Contour plot of internal energy E, as a function of two dihedral angles of the backbone for isotactic po- lypropylene.

The fig. 10 shows, for syndiotactic polypropylene, three minima in the energy map, two of which corresponding to a right-handed helix and a left-handed helix, while the third corresponding to a trans planar conformation.

This is in accordance with the fact that syndiotactic polypropylene is polymorphous. In fact, it can get two different crystalline forms, which differ because of the chain conformation (fig. 5).

This polymorphism is different from the case of isotactic polypropylene, which is also polymorphous, but in which, in the different crystalline forms, always a threefold helix is observed; or from the case of polybutene-1 which is also polymorphous, but the different helices observed correspond, however, to the same region of minimum of the conformational energy map. Instead in the case of syndiotactic polypropylene, as shown before, the two different crystalline forms correspond to chain conformations which are widely separated in the conformational energy map.

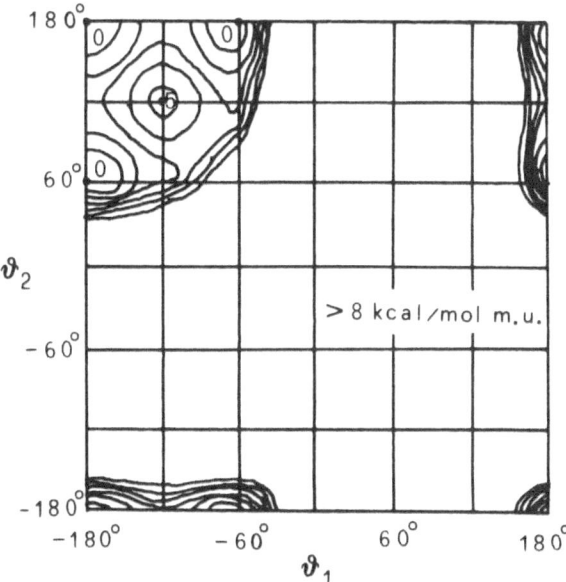

Fig. 10. Contour plot of internal energy E, as a function of the two dihedral angles of the backbone for syndiotactic polypropylene.

THE CONFORMATION OF POLYMERIC CHAINS IN SOLUTION AND IN
THE MELT(10)

The rotational isomeric model for liquid hydrocarbons

    The conformations of the macromolecules, in the ab-
sence of the constraints imposed by neighbouring molecules in an ordered crystalline state, need not be and
are not regular.
    The repartition of the bonds of a polymer, for an i-
solated chain with n carbon atoms, between different conformations may be evaluated by the methods of statistical mechanics. A complication arises because the energy
associated to a given conformational state of bond i may
not be assumed to depend only on its internal rotation
angle θ_i, but it depends in general also on the internal
rotation angles of all the neighbouring bonds. In the
approximation in which this dependence is restricted to
next-neighbour internal rotation angles only and in the
absence of flexible lateral groups, the treatment is simplified as follows. A statistical weight can be appended to bond i:

$$u_i = \exp\{-\left[E(\theta_i) + E(\theta_{i-1}, \theta_i)\right]/RT\} \tag{33}$$

where $E(\theta_i)$ represents the intrinsic torsional potential
of the bond and the nonbonded interactions which depend
exclusively on θ_i, while $E(\theta_{i-1}, \theta_i)$ includes the nonbonded interactions which depend jointly on the two in-
ternal rotation angles θ_{i-1} and θ_i. The dependence of e-
nergy on θ_{i+1} is included in the statistical weight relative to the bond $i+1$. The conformational partition function is, then, given formally by:

$$Z_{conf} = \sum_{\{\theta\}} \prod_{i=2}^{n-2} u_i$$

where the summation is taken over all the conformations
of the macromolecule.
    The problem can be further simplified taking into
account only a discrete number of rotational states (which are chosen in general to be coincident with conformational potential energy minima), in the calculation of
the partition function. This is the rotational isomeric
model, first proposed by Volkenstein and principally developped by Flory and his school.
    Firstly we consider the application of this model to
the molecule of n-butane. In fig. 11 the conformational
energy as a function of the dihedral angle $C_1-C_2-C_3-C_4$,

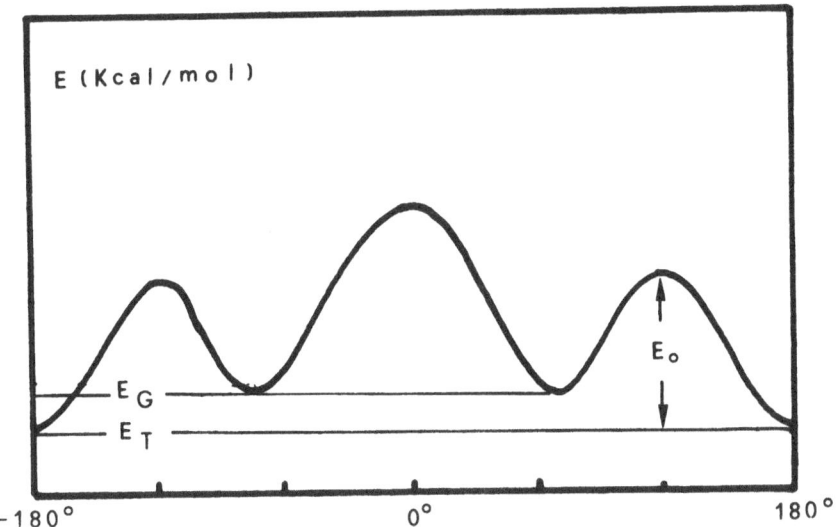

Fig. 11. Conformational energy of n-butane as function of the dihedral angle $C_1-C_2-C_3-C_4$.

is represented (this is a first approximation, in so far as the possible variation of the bond angles and of the torsion angles adjacent to the methyl groups is not considered).

The shape of the curve may be explained in terms of the previous considerations as arising from an intrinsic torsional potential and the interaction between the nonbonded methyl groups. There are three minima of the energy, corresponding to T, G_+, G_- conformations, the last two being energetically equivalent. For the evaluation of the partition function these are chosen as rotational isomeric states, and their statistical weights are:

$$u_T = \exp\ (-E_T/RT)\quad ;\quad u_{G_+} = u_{G_-} = \exp\ (-E_G/RT)$$

taking $E_T=0$ the partition function is:

$$Z_{conf} = 1+2\sigma\quad\text{with}\quad \sigma = \exp\ \{-(E_G-E_T)/RT\}$$

In the case of n-pentane, in an approximation similar to that used in the case of n-butane, the energy may be evaluated as a function of the two dihedral angles θ_{23} and θ_{34}. It is important to note that we cannot neglect the interactions between nonbonded atoms that depend on both θ_{23} and θ_{34} values jointly. In particular, if the two dihedral angles assume the values G_+ and G_-

(or G_- and G_+) in one sequence, the two terminal methyl groups approach each other to a distance much shorter than their Van der Waals radius ; the conformational energy is then widely larger than $2E_G$.

On the basis of the isomeric rotational approximation, we consider only the conformations generated by combining, in all possible ways the conformations T, G_+, G_- which refer to the first dihedral angle (θ_{23}) with the conformations T, G_+, G_- which refer to the second dihedral angle. We attribute to each resulting conformation a statistical weight and represent them with the table

$$
\begin{array}{c|ccc}
 & T & G_+ & G_- \\
\hline
T & 1 & \sigma & \sigma \\
G_+ & \sigma & \sigma^2 & \sigma^2\omega \\
G_- & \sigma & \sigma^2\omega & \sigma^2
\end{array}
$$

where the selected conformations for the two successive bonds are indicated with the relative symbols.

In this table we made the assumption that, excluding the sequences G_+G_- or G_-G_+, the interactions that depend on both dihedral angles jointly are negligible. The first element of the table is, then, 1 because the energy of a T conformation (and therefore of a TT conformation) is taken as equal to zero; moreover $\sigma = \exp(-\Delta E/RT)$, where ΔE is the energy difference between a *gauche* and a *trans* conformation. With the term ω we take into account the repulsive extra-energy E_ω of G_+G_- and G_-G_+ conformations (being $E_\omega = -RT\ln\omega$), that we discussed before.

The conformational partition function, being the sum of all the terms appearing in the table, may be written for n-pentane, in the rotational isomeric approximation, as:

$$Z_{conf} = 1 + 4\sigma + 2\sigma^2 + 2\sigma^2\omega$$

For a linear hydrocarbon of n atoms, the number of terms to be summed, in the approximation of three rotational isomeric states for each bond, is 3^{n-3}, a number that becomes high very rapidly; therefore the use of a matrix formulation of the partition function is necessary.

We note that, for the n-butane, it is possible to write:

$$Z_{conf} = 1+2\sigma = \overline{\begin{vmatrix} 1 & \sigma & \sigma \end{vmatrix}} \begin{vmatrix} 1 \\ 1 \\ 1 \end{vmatrix}$$

Moreover the partition function of n-pentane which is the sum of all the elements of the table reported may be written as the following product:

$$Z_{conf} = \overline{\begin{array}{ccc} 1 & \sigma & \sigma \end{array}} \begin{vmatrix} 1 & \sigma & \sigma \\ 1 & \sigma & \sigma\omega \\ 1 & \sigma\omega & \sigma \end{vmatrix} \begin{vmatrix} 1 \\ 1 \\ 1 \end{vmatrix}$$

In general for a linear hydrocarbon of n atoms, it is possible to see that the partition function is:

$$Z_{conf} = \overline{\begin{array}{ccc} 1 & \sigma & \sigma \end{array}} \begin{vmatrix} 1 & \sigma & \sigma \\ 1 & \sigma & \sigma\omega \\ 1 & \sigma\omega & \sigma \end{vmatrix}^{n-4} \begin{vmatrix} 1 \\ 1 \\ 1 \end{vmatrix} \qquad (35)$$

Polyethylene and isotactic polymers(11)

The conformational partition function of a polyethylenic chain of n atoms may be written, according to the results of the previous paragraph:

$$Z_{conf} = \overline{\begin{array}{ccc} 1 & \sigma & \sigma \end{array}} \begin{vmatrix} 1 & \sigma & \sigma \\ 1 & \sigma & \sigma\omega \\ 1 & \sigma\omega & \sigma \end{vmatrix}^{n-4} \begin{vmatrix} 1 \\ 1 \\ 1 \end{vmatrix}$$

Matrix methods, in the case of n large, bring to the conclusion that:

$$Z_{conf} \simeq \lambda_1^n \qquad (36)$$

where λ_1 is the largest root of the secular equation(in other words the largest eigenvalue) for the matrix of the statistical weights:

$$\begin{vmatrix} 1 & \sigma & \sigma \\ 1 & \sigma & \sigma\omega \\ 1 & \sigma\omega & \sigma \end{vmatrix} \qquad (37)$$

For instance, taking $\omega = 0$, the secular equation is:

$$\det \begin{vmatrix} 1-\lambda & \sigma & \sigma \\ 1 & \sigma-\lambda & 0 \\ 1 & 0 & \sigma-\lambda \end{vmatrix} = 0 \tag{38}$$

and the largest root is: $\lambda_1 = \frac{1}{2}(1 + \sigma + \sqrt{1+6\sigma+\sigma^2})$.

Matrix methods of the kind illustrated are the basis for the evaluation of various thermodynamical properties. For instance, the frequency of occurrence of *gauche* states in a polyethylenic chain is given by:

$$f_G = \frac{1}{n} \frac{\partial \ln Z}{\partial \ln \sigma} \simeq \frac{\partial \ln \lambda_1}{\partial \ln \sigma} \tag{39}$$

If $E_G-E_T = 500$ cal/mol and $T = 400$ K (about the melting temperature of the polyethylene), that is $\sigma = 0.5, f_G \approx 40\%$. Assumption of independence among bond rotations would have given, as for the case of n-butane, $\lambda_1 = 1+2\sigma$ and the fraction of *gauche* bonds, with the same σ value is $f_G \simeq 2\sigma/(1+2\sigma) \simeq 50\%$.

A simplified treatment of the conformation of isotactic polymers in solution can be made in the following way, that we shall discuss for two extreme cases: the case of polypropylene and that of polystyrene. We start from the identification of the minimum internal energy conformations available for a piece of a chain of the kind:

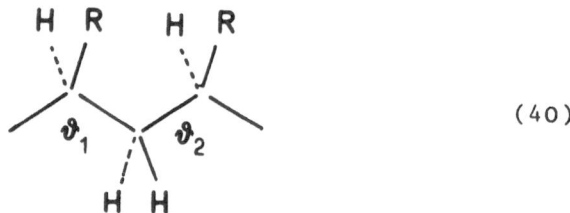

$$(40)$$

In the case of polypropylene, the conformational map shows only two minimum internal energy conformations : TG_+ and G_-T. If these conformations are present in a long sequence, they produce a left-handed helix and a right-handed helix respectively (fig. 9). The perpetuation of such helicoidal sequences can be interrupted by a pair of internal rotation angles (θ_x and θ_y), in two different ways, if we go in the sense from the left to the right in the chains below indicated. If we go from a left-handed to a right-handed helix: (fig. 12)

$$\ldots\ldots|TG_+|TG_+|TG_+|T\theta_x|\theta_y T|C_-T|C_-T|G_-T|\ldots\ldots$$

the possible minimum energy pairs for the angles θ_x and θ_y may be taken in the rotational isomeric approximation as corresponding approximately to A_+G_- and G_+A_- conformations, with energy of the order $E = 2.5$ kcal/mol, the corresponding statistical weight being $\omega = \exp(-E_\omega/RT)$. If we go from a right-handed to a left-handed helix:

$$\ldots\ldots|G_-T|G_-T|G_-T|G_-\theta_x|\theta_y G_+|TG_+|TG_+|TG_+|\ldots\ldots$$

the pair θ_x θ_y may assume the low energy conformation TT. There is no increase of the energy at the inversion of the spiralization sense and the corresponding statistical weight may be taken as 1 (fig. 13). We can write now a simplified matrix of statistical weights analogous to (37), written for polyethylene, but relative to pairs of bonds. The compacted matrix has the form:

$$
\begin{array}{c|cc}
 & G_-T & TG_+ \\
\hline
G_-T & 1 & 1 \\
TG_+ & 2\omega & 1
\end{array}
$$

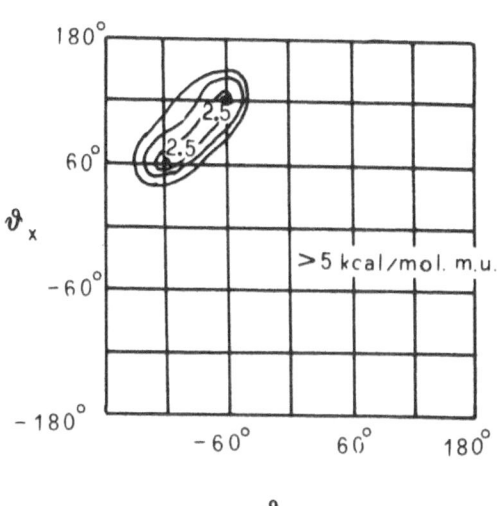

Fig. 12. Conformational energy map for the sequence:
$$\ldots\ldots|TG_+|TG_+|T\theta_x|\theta_y T|G_-T|G_-T|\ldots\ldots$$

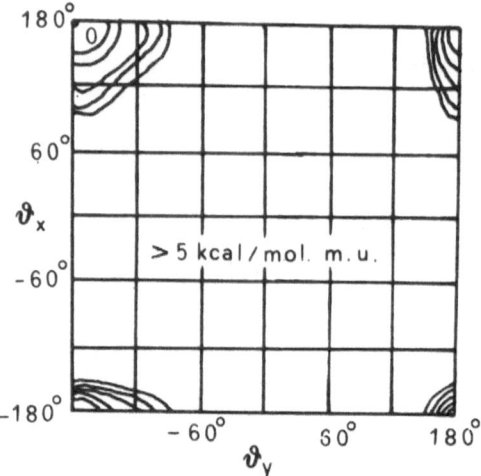

Fig. 13. Conformational energy map for the sequence:
.... |G_T|G_T|G_θ$_x$|θ$_y$G$_+$|TG$_+$|TG$_+$| ....

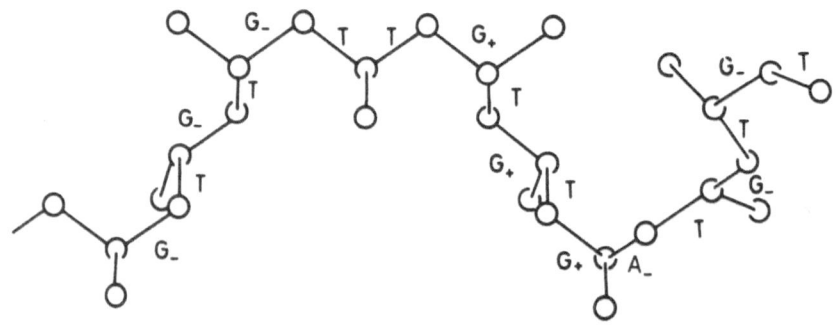

Fig. 14. A model for the chain of isotactic polypropylene in solution θ or in the melt.

and for a chain of n monomeric units with n large:

$$Z \approx \mathrm{spur} \begin{vmatrix} 1 & 1 \\ 2\omega & 1 \end{vmatrix}^n \approx (1 + \sqrt{2\omega})^n$$

where spur is the sum of diagonal elements.
The fraction of the inversions in the spiralization sen-
se is $f_i \approx \sqrt{2\omega}/(1+\sqrt{2\omega})$, and, if T = 450 K, $f_i \approx 25$ %.
A resulting model for the chain in solution is indica-
ted in fig. 14.
    In the case of isotactic polystyrene or of polyacry-
lates, other effects must be taken into account. In particular, together with the conformations TG_+ and G_-T ,
for a piece of chain of the kind (40), also conformations near to TT planar are available. Without discussing
the more complicated rotational isomeric model which result, it is interesting to explain why conformations near to the TT planar are possible for polystyrene and more unlikely for a polymer such as polypropylene.
    Consider a piece of chain in the conformation TT ,
for isotactic polypropylene and for isotactic polystyrene, with the bond angles and the relevant distances
indicated in fig. 15; in both we can see that the distances and hence the interactions energy between late-

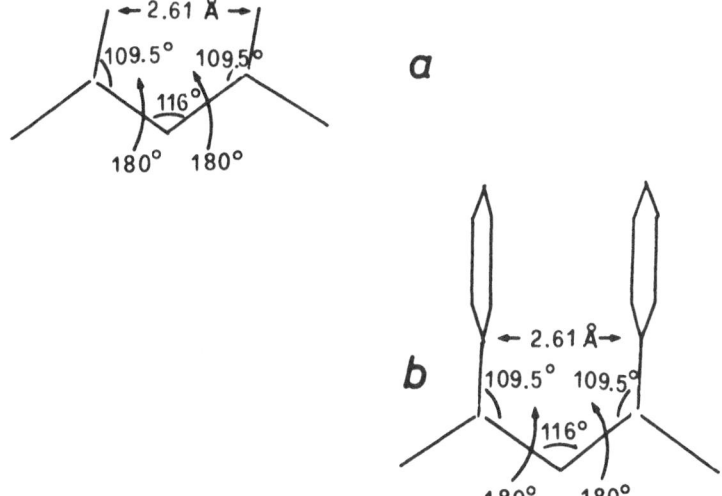

Fig. 15. Pieces of chains in the conformation TT for:
a) isotactic polypropylene b) isotactic polystyrene

ral groups are prohibitive. If we change the two internal rotation angles by as much as 10°, as illustrated in
fig. 16, the increase of the torsion term of the energy
is very small. But, while in the case of polypropylene
the distance between methyl groups is still energetically prohibitive, in the case of polystyrene such distance refers to interactions between carbon atoms which are
"nude"; moreover a good part of the 36 distances between
the carbon atoms of the phenyl groups (if we take them
staggered in respect to the chain) is attractive.

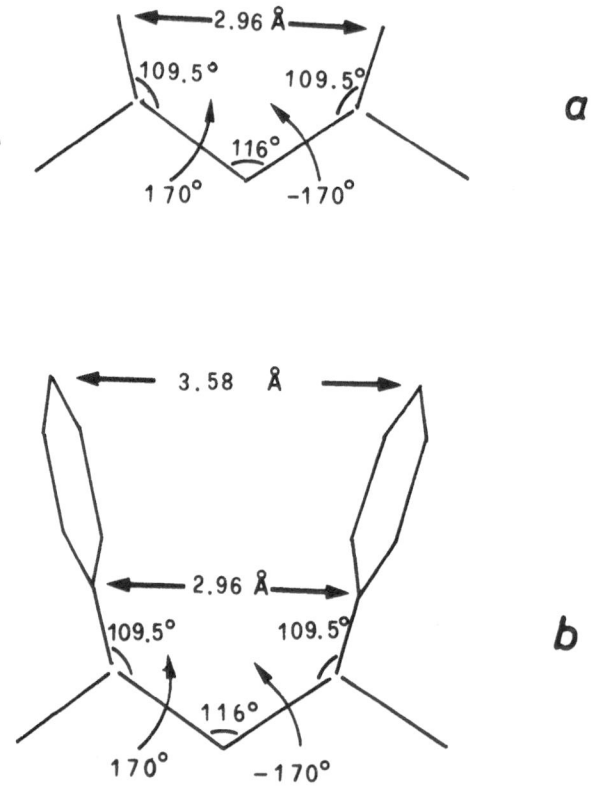

Fig. 16. Pieces of chains in the nearly TT (170°,-170°)
conformation for:
 a) isotactic polypropylene b) isotactic polystyrene

Finally we want to point out that an application of the rotational isomeric model to a polymer is the evaluation of the portion of the entropy of melting, which depends on the conformational freedom which the chains acquire in the melt. The conformational contribution to the difference in entropy may be evaluated, again by standard methods of statistical mechanics, as:

$$S_{conf} = k \ln Z_{conf} + \frac{T}{Z_{conf}} \frac{dZ_{conf}}{dT} \tag{41}$$

where Z_{conf} is the conformational partition function, easily evaluated in the rotational isomeric approximation. Some data calculated by Tonelli(12) for various polymers are reported in table 7.

Table 7. Comparison between the experimental entropy of fusion at constant volume and calculated conformational contribution to the entropy of fusion.

Polymer	$(\Delta S_m)_v$ (e.u./mole of monomer)	S_{conf}
Polyethylene	1.77	1.76
Polyoxymethylene	2.8	3.00
Polyoxyethylene	4.22	5.10
1,4-*cis*-polyisoprene	1,7	5.41
1,4-*trans*-polyisoprene	5.1	5.47
1,4-*cis*-polybutadiene	5.96	5.52
Polyethyleneterephthalate	8.2	7.51
Polytetrafluoroethylene	0.76	1.6
Nylon-6	11.5	11.91

References

1) The nomenclature of polymer configurations reported in the chapter are based on the IUPAC reports:
   a) IUPAC. Commission on Macromolecular Nomenclature (1974). *Pure and Applied Chemistry*, 40, 479.
   b) IUPAC. Commission on Macromolecular Nomenclature (1975). *Pure and Applied Chemistry*, 48, 373.
   c) IUPAC. Commission on Macromolecular Nomenclature.*Pure and Applied Chemistry*. In press.

2) P. Corradini, *Recent Advances in Fiber Science*, F. Happey ed., 1977, vol I°.

3) U. W. Suter and P. J. Flory, *Macromolecules*, 8, 765 (1975)

4) R. A. Scott and H. A. Scheraga, *J. Chem. Phis.*, 45, 2091 (1966)

5) M. Yokouchi, H. Tadokoro and Y. Chatani, *Macromolecules*, 7, 769 (1974)

6) P. Corradini, *The Stereochemistry of Macromolecules*,(A. D. Ketley, ed.), Dekker, New York, Part III, pp. 1-60.

7) K. H. Meyer and H. Mark, *Ber.*, 61, 593(1928).

8) H. Staudinger, *Ber.*, 57, 1203 (1924).

9) E. D. T. Atkins, D. H. Isaac, A. Keller and K. Miyasaka, *J. Polymer Sci.* , *Phis. Ed.* , 15, 211 (1977).

10) A good account of rotational isomeric model is given in:
    a) M. V. Volkenstein, *Configurational Statistics of Polymeric Chains*, Interscience, New York, 1963.
    b) P. J. Flory, *Statistical Mechanics of Chain Molecules*, Interscience, New York, 1969.

11) P. Corradini, *J. Polymer Sci.: Symposium* , 50, 327 (1975).

12) A. E. Tonelli, *J. Chem. Phis.*, 52, 4749 (1970).

APPLICATION OF CHIROPTICAL PROPERTIES TO CONFORMATIONAL ANALYSIS
OF STEREOREGULAR POLYMERS

F.Ciardelli, C.Carlini, E.Chiellini, P.Salvadori,
L.Lardicci, R.Menicagli, and C.Bertucci

Centro CNR Macromolecole Stereordinate ed Otticamente
Attive, Istituti di Chimica Organica e di Chimica Organica Industriale, University of Pisa, Italy.

1. INTRODUCTION

    Determination of secondary structure (conformation) in solu-
tion of macromolecules is a rather difficult task even for stereoregular polymers. Indeed rotation of both main chain and side
chains is possible and the conformational equilibrium is characterized by the fast interconversion of an extremely large
number of conformers. This is particularly true for synthetic
polymers having a hydrocarbon backbone expecially if they do not
possess highly polar groups in the side chains. The situation
is therefore in principle very different with respect to
polypeptides which can maintain in solution ordered conformations
[1] thanks to intramolecular hydrogen bonding involving peptide
groups of different units. Accordingly polypeptides give rise to
a sharp helix-coil transition with varying temperature, solvent
and pH, whereas in vinyl polymers conformational changes occur
linearly. This behavior does not exclude the existence in solution of an appreciable extent of conformational order in terms
of chain sections. In other words this means that a large amount
of monomer units can be in a well definite conformation giving
rise to chain sections with ordered secondary structure, linked
by shorter sections containing units in conformations with
higher energy. This would also allow the chains to assume a
substantially coiled conformation which is the most probable
one. The problem of the determination of this type of conformational order can be conveniently approached by chiroptical
techniques provided such an arrangement displays optical
activity. In the present chapter we will limit ourselves to some
significant examples in the field of stereoregular vinyl polymers and methacrylic derivatives which have been the most

R. W. Lenz and F. Ciardelli (eds.), Preparation and Properties of Stereoregular Polymers, 353-368.
Copyright © 1979 by D. Reidel Publishing Company.

largely investigated and the ones having given the most
strightforward indications among synthetic polymers (with the
exception of poly-α-aminoacids which are better related to
polypeptides and proteins). Moreover their hydrocarbon main
chain allows simple conformational analysis and gives the
opportunity to relate in a more definite way conformation to
chiroptical properties. Thus, after reconsidering the basic
results obtained in the 60s for poly(α-olefin)s and for some
vinyl polymers with side chain chromophores, we shall discuss
the stereoregular copolymers between chiral vinyl monomers and
achiral comonomers bearing a chromophoric group.

2. POLYMERS OF OPTICALLY ACTIVE α-OLEFINS

       Since the discovery of stereospecific polymerization of
vinyl monomers leading to highly isotactic crystalline polymers,
it was shown by X-ray methods that the macromolecules assume in
the crystalline state a helical conformation [2]. In the polymers derived from prochiral monomers, like propylene, both screw
senses are equally probable; accordingly left-handed and right-
handed helices have been found in the crystalline lattices in
equal amount. The macromolecules conformation in the crystalline
state is, in general, that having the minimum of intramolecular
energy and therefore the same conformation is also the most
probable in the solution, particularly with hydrocarbon polymers
or with polymers containing no highly polar groups. Statistical
calculations have given indications supporting this hypothesis,
even if the conformation in solution must contain helical blocks
with lower average length than in the crystals and connected by
conformational reversals [3-5].
       The isotactic polymers derived from optically active α-
olefins assume, in general, helical conformation of a single
screw sense in the crystalline state. Because of the chirality
in the side chains, in fact the two screw senses have no longer
the same thermodynamic stability. The prevalence of one screw
sense can be detected by measuring optical rotation of polymer
solutions and by comparison with suitable low molecular weight
models. The optical activity, per monomeric residue, of highly
isotactic polymers derived from α-olefins having the asymmetric
carbon atom in the α or β position with respect to the main
chain is much larger, in absolute value, than for low molecular
weight structural models strictly resembling the structure of
the repeating unit (Table 1). This result can be explained by
taking into account the larger conformational homogeneity in the
tactic polymer where the monomer residue is allowed to assume
only few conformations all implying a helical structure and
with high optical rotation (Figure 1). Good agreement between
observed and calculated [6] optical rotation is found only
assuming the presence of conformations leading to a single screw

Table 1

Experimental and calculated rotatory power of isotactic poly(α-olefin)s and of their structural low molecular weight models

Polymer	Number of[a] conformations	$[\Phi]_D^{25}$ exp.	$[\Phi]_D^{25}$ calc.[a]	Structural model	Number of conformations	$[\Phi]_D^{25}$ exp.	$[\Phi]_D^{25}$ calc.
---CH$_2$-CH--- CH$_3$-C*-H C$_2$H$_5$	2	+161	+180	H$_3$C-CH-CH$_3$ H$_3$C-C*-H C$_2$H$_5$	4	-11.4	-15
---CH$_2$-CH--- CH$_2$ H$_3$C-C*-H C$_2$H$_5$	2	+288	+240	H$_3$C-CH-CH$_3$ CH$_2$ H$_3$C-C*-H C$_2$H$_5$	3	+22.9	+20

a) For the favoured screw sense.

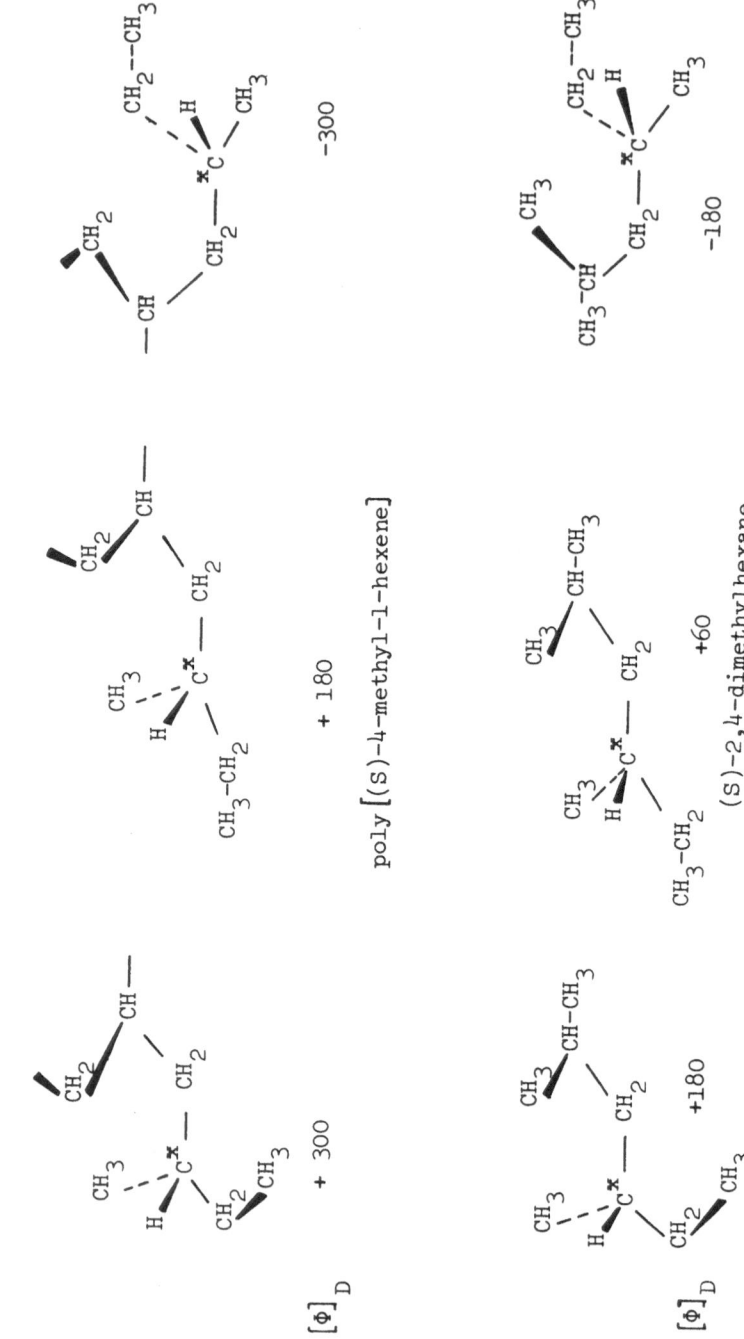

FIGURE 1. Allowed conformations of isotactic poly[(S)-4-methyl-1-hexene] and of (S)-2,4-dimethyl-hexane and their calculated rotatory power.

sense (Table 1). Similar optical rotation as in the polymer is calculated for specific conformations of the structural low molecular weight models (Figure 1). An experimental support to this statement arises from conformationally homogeneous chiral paraffins which can show rotatory power of the same order of magnitude as the above polymers (Table 2).

A further proof of the substantially one screw sense helical conformation in solution of isotactic polymers from chiral α-olefins is given by the optical rotation in the crystalline state. In fact a good agreement between rotatory power in solution and in the crystalline state has been observed when the macromolecules have been found by X-ray examination to assume in the crystals one screw sense helical conformations [7,8]. Clearly, for entropic reasons [8], the macromolecules in solution have to be considered as consisting of a succession of chain sections spiraled in both screw senses, the average length of the sections in a single screw sense being much larger than that of opposite sense at least in the cases reported in Tables 1 and 2, where the asymmetric carbon atom is close to the main chain.

The different internal energy between the two senses has been proposed to be due to the cooperative interaction among side chains disposed along isotactic blocks of the backbone [6,9]. Accordingly the optical rotation decreases with lowering isotacticity degree[6,10] and, with comparable tacticity, by average shortening of isotactic blocks by random insertion of ethylene units in the chain[11] (Table 3). The cooperative interaction survives in copolymers between a chiral α-olefin and a substituted non chiral α-olefin. Thus the coisotactic copolymers of (S)-4-methyl-1-hexene with 4-methyl-1-pentene display[12] larger specific rotatory power than the corresponding homopolymers mixture whatever the composition (Table 4). Such a result is consistent with the inclusion of 4-methyl-1-pentene units in helical sections of a single screw sense jointly with the chiral units from (S)-4-methyl-1-hexene. Isotactic macromolecules from 4-methyl-1-pentene can exist with equal probability in right or left-handed helical conformation where the monomer units are in enantiomeric relationship (Figure 2). For these two conformations a molar rotatory power (per monomer residue) of +240 and -240 can be calculated, respectively. As appears from Table 4 the former value is approached by increasing the content in the copolymer of units from (S)-4-methyl-1-hexene which induce a left handed screw sense.

3. HOMOPOLYMERS WITH CHROMOPHORES IN THE SIDE CHAINS

Additional information about conformation in solution of vinyl polymers can be obtained by chiroptical techniques when a chromophore absorbing in the accessible spectral region

Table 2

Experimental and calculated rotatory power for isotactic poly(α-olefin)s and their conformational low molecular weight models

Polymer[a]	Number of conformations[b]	$[\Phi]_D^{25}$ exp.	calc.[b]	Conformational model[c]	Number of conformations	$[\Phi]_D^{25}$ exp.	calc.
---CH$_2$-CH--- H$_3$C-C*-H C$_2$H$_5$	2	+161	+180	CH$_3$ H$_3$C-C-CH$_3$ H$_3$C-C*-H C$_2$H$_5$	1	-150	-180
---CH$_2$-CH--- CH$_2$ H$_3$C-C*-H C$_2$H$_5$	2	+288	+240	CH$_3$ H$_3$C-C-CH$_3$ H-C*-CH$_3$ CH$_2$ H$_3$C-C*-H C$_2$H$_5$	2	+120	+120

a) See ref.6. b) For the favoured screw sense. c) S.Pucci, M.Aglietto, P.L.Luisi, and P.Pino, "Conformational Equilibria in Low and High Molecular Weight Paraffins" in Conformational Analysis, Academic Press Inc., 1971, 203-218.

Table 3

Optical rotation of the homopolymers of (S)-4-methyl-1-hexene
and of its random copolymer with ethylene

Structure	Stereoregularity	$[\phi]_D^{25}$ [a]
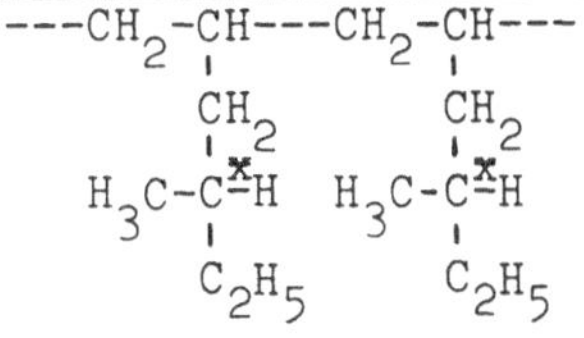 ---CH$_2$-CH---CH$_2$-CH--- CH$_2$ CH$_2$ H$_3$C-C$\overset{x}{}$H H$_3$C-C$\overset{x}{}$H C$_2$H$_5$ C$_2$H$_5$	isotactic	+288
	atactic	+174
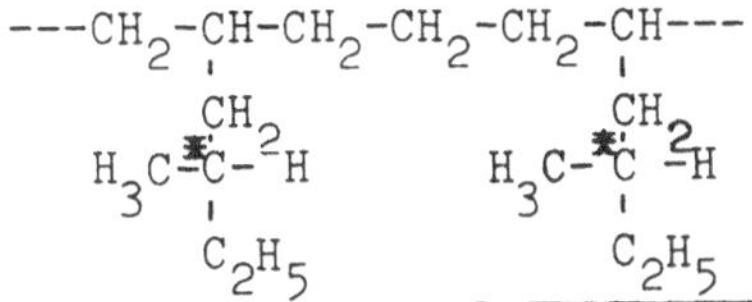 ---CH$_2$-CH-CH$_2$-CH$_2$-CH$_2$-CH--- CH$_2$ CH$_2$ H$_3$C-C-H H$_3$C-C-H C$_2$H$_5$ C$_2$H$_5$	isotactic[b]	+133

a) Based on one (S)-4-methyl-1-hexene unit.
b) Moles of (S)-4-methyl-1-hexene: 51%.

Table 4

Optical rotation of units from 4-methyl-1-pentene (4MP) in coisotactic copolymers with (S)-4-methyl-1-hexene (4MH)

Sample	Units from 4MP (mole-%)	$[\alpha]_D^{25}$ copolymer	$[\alpha]_D^{25}$ homopolymers mixture	$[\phi]_D^{25}$ [a] of units from 4MP
1	30	+260	+196	+160
2	56	+233	+123	+155
3	77	+161	+ 64	+ 98

a) Calculated assuming that 4MH units have in the copolymer the
same optical rotation as in the homopolymer [12].

FIGURE 2. Allowed conformations for isotactic poly(4-methyl-
            1-pentene)

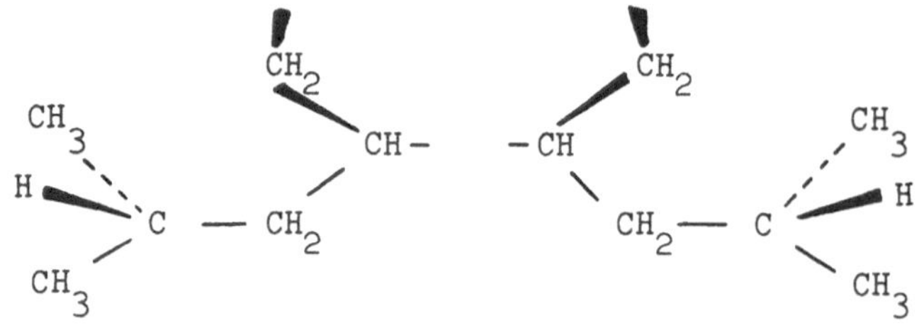

$$[\Phi]_D + 240 \qquad\qquad\qquad [\Phi]_D - 240$$

Left-handed helix   Right-handed helix

is present in the side chains. In polymers of optically active
alkylvinylethers the only chromophoric system observable is
the ethereal oxygen in the side chain which shows two bands related to the electronic transitions connected with non-bonding
electrons of the oxygen atom itself [13]. These two transitions
are located below 200 nm and are very close each other making
detection and assignement very difficult. The data concerning
the Cotton effects related to the ethereal chromophore are in
most cases uncertain even combining information derived from
ORD and CD spectra and it is not possible to extablish if and
to what extent ethereal oxygens can mutually interact in polymers
having different stereoregularity. The existence, however, of a
Cotton effect strong enough to affect appreciably the optical
rotation down to 589 nm has been shown in polymers and low
molecular weight alkyl ethers taken as models [13].

    Moreover the optical rotation is clearly dependent on
stereoregularity indicating that this last affects remarkably
the position of the conformational equilibrium [13].

    In conclusion, in spite of the difficulties arising from
the peculiar electronic structure of the side chain chromophore,
all collected data are consistent with the presence in solution
of macromolecules having a predominant one screw sense helical
conformation as proposed for poly(α-olefin)s.

    A dependence of optical rotatory properties on stereoregulari-
ty was observed also in poly(L-menthylacrylate) [14], poly(L-
bornylacrylate) [14] , poly{[(S)-2-methylbutyl]-methacrylate}
[15], poly(L-menthylmethacrylate)[15,16] and poly{[(S)-α-methyl-
benzyl]-methacrylate}[16-18]. A detailed examination of the ORD
curves of the last two polymers demonstrated that in the
former the longest wavelength optically active electronic
transition is centered between 235 and 208 nm and is due to the

$n \rightarrow \Pi^x$ electronic transition of the ester group, while in the latter an additional optically active transition is present at longer wavelength, associated with the lowest energy $\Pi \rightarrow \Pi^x$ transition (^1L_b) of benzene [17]. The position of some bands at the limit of the instrumental penetration and the large distance between main chain and asymmetric carbon atom which, on the other side, is connected by a conformationally very free ester group, limited to obtain definite results.

Polymers of optically active alkyl vinyl ketones appeared very suitable as they show a well distinct dichroic band of the $n \rightarrow \Pi^x$ transition of carbonyl chromophore at about 280-290 nm. The ORD spectra of poly{[(S)-1-methylpropyl]vinylketone}, poly{[(S)-2-methylbutyl]-vinylketone} and poly{[(S)-3-methylpentyl]-vinylketone}evidenced a negative Cotton effect between 292 and 288 nm, the position of which was not affected by stereoregularity, whereas its amplitude was markedly increasing with this last [19]. The successive CD study [20]together with tacticity determination by 220 MHz H-NMR confirmed that the amplitude of the Cotton effect corresponding to $n \rightarrow \Pi^x$ electronic transition of the keto group is strongly dependent on stereoregularity, which clearly affects the position of the conformational equilibrium. However as this transition is electronically forbidden, and then very weak, poly(vinylketone)s failed to give additional information with respect to the previously discussed polymers as far as direct CD evidence of helical conformations in solution is concerned.

4. COPOLYMERS OF CHIRAL VINYL MONOMERS WITH AROMATIC MONOMERS

A substantial improvement in analysis of macromolecular conformation in solution by chiroptical techniques, in particular CD, has been achieved by studying copolymers of optically active "transparent" monomers with comonomers containing an aromatic chromophore, which acts as spectroscopic probe .

In copolymers of styrene with α-olefins the only absorbing moiety in a very accessible spectral region is the benzene chromophore, the electronic transitions related to its Π-electron system being located over 180 nm. In the coisotactic copolymer of styrene with optically active 3,7-dimethyl-1-octene, where the asymmetric carbon atom is in the α position to the main chain, the benzene chromophore is optically active as shown by the presence of a dichroic band around 265 nm where its lowest energy electronic $\Pi \rightarrow \Pi^x$ transition (^1L_b or $^1A_{1g} \rightarrow {}^1B_{2u}$) is located. The ellipticity of this dichroic band is considerably larger than in low molecular weight structural models [21] (Figure 3). Successive examination down to 185nm showed [22] that also the ^1L_a ($^1A_{1g} \rightarrow {}^1B_{1u}$) and $^1B_1$ ($^1A_{1g} \rightarrow {}^1E_{1u}$) electronic transitions gave rise to CD bands, the last one assuming the

typical exciton splitting into two dichroic bands with
similar ellipticity and opposite sign.

FIGURE 3. Schematic structure of the styrene/(R)-3,7-dimethyl-
          1-octene random copolymer (A) and of the low molecular
          weight structural model (3R:9R)-3,9-dimethyl-6-
          phenylundecane (B)

--(-CH_2-CH-)-(-CH_2-CH-)--   CH_2-CH_2-CH-CH_2 - CH_2
                    |* |* |*
                   H-C*-CH_3   H-C*-CH_3   H -C*-CH_3
                    | | |
                   (CH_2)_3   C_2H_5   C_2H_5
                    |
                   CH(CH_3)_2

              A   B

$\Delta \epsilon_{262}$ = -0.27 $\Delta \epsilon_{262}$ = -0.017

The "conformational" low molecular weight model (S)-2,2-
dimethyl-3-phenylbutane, for which the conformational analysis
predicts a similar situation as in the macromolecules (Figure 4),
displays comparable optical rotation and ellipticity as the
isotactic copolymer [23]. However this conformational model of
styrene copolymers, even if showing dichroic bands between 270
and 185 nm of comparable intensity as the polymer, does not show
the above splitting, only one dichroic band being observed in
the region of the ^1B electronic transition.

FIGURE 4. Preferred conformation around the phenyl ring in
          isotactic polystyrene and in a low molecular weight
          conformational model.

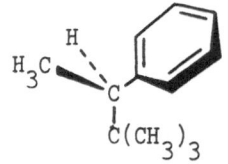

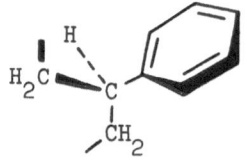

(S)-2,2-dimethyl-3-phenylbutane   Styrene unit in a left
                                         handed helical chain

Thus while conformational homogeneity gives ellipticity increase
of all $\Pi \rightarrow \Pi^x$ electronic transitions, the occurrence of the
exciton splitting of the ^1B band must be related to electrostatic
dipole-dipole interactions between aromatic chromophores disposed

FIGURE 5. Coisotactic copolymer of styrene with optically active
α-olefins

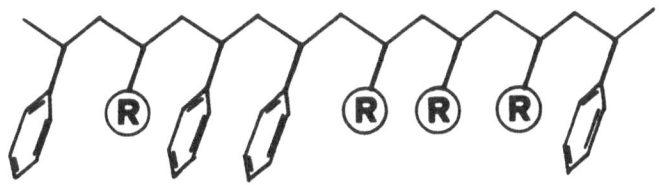

zig-zag planar conformation

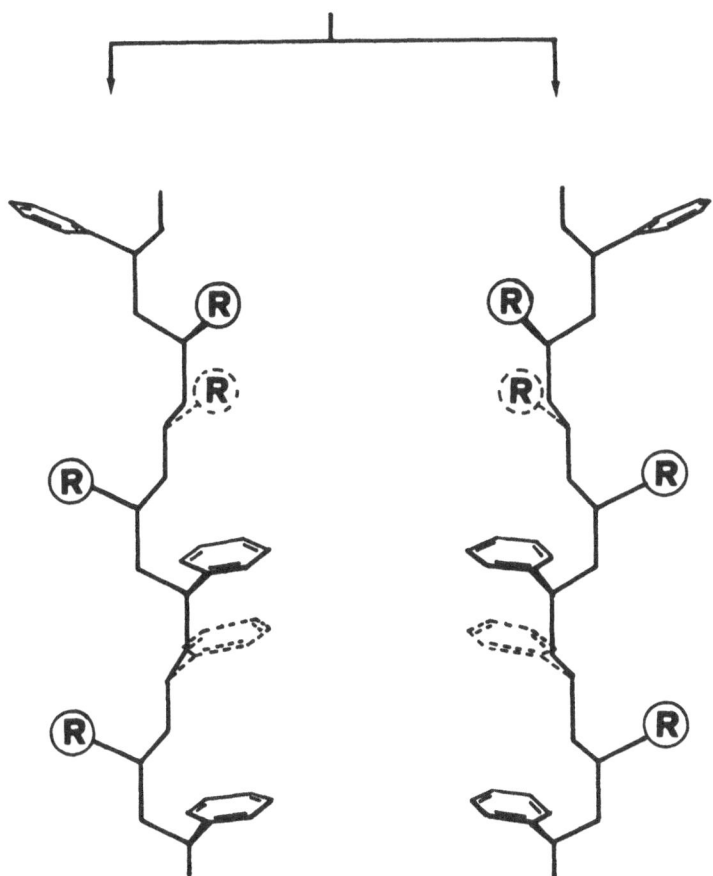

left-handed helix   right-handed helix

(R) : chiral alkyl group of a single configuration

in a mutual dissymmetric disposition in the same molecule. This situation can be achieved in one screw sense helical sections of the copolymer macromolecules (Figure 5).

A confirmation of this interpretation comes out from semiempirical calculations of CD spectra by the De Voe theory [24], applied to the model shown in Figure 5. The agreement between experimental and calculated CD spectra was excellent, particularly in the region of the ^1B electronic transition where a couplet comparable to that observed is found [25]. More detailed structural information can be obtained by looking at the influence of composition on the chiroptical properties and the different correspondence between calculated and experimental spectra for the different dichroic bands. The stronger dependence of ellipticity on aromatic units content, observed in the calculations performed assuming statistical distribution of the counits, can be explained by taking into account a substantially block distribution experimentally detected [26]. Extension of these concepts to analogous copolymers of 1-vinylnaphthalene [27] and 2-vinylnaphthalene [28] (Figure 6) with the same optically active α-olefin brought to similar conclusion even if the situation appears more complicated because of larger complexity of naphthalene chromophore with respect to benzene. In fact, even if exciton splitting occurs also in these cases in the ^1B transition range, the sign of the couplet is opposite in the two cases; moreover in the latter one a second splitting seems to be present at a slightly shorter wavelength.

FIGURE 6.   Schematic structures of copolymers of (R)-3,7-
dimethyl-1-octene with 1-vinylnaphthalene (A) and
2-vinylnaphthalene (B).

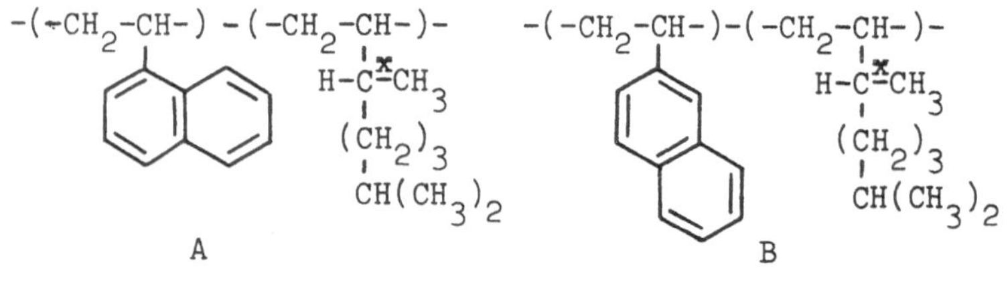

A   B

The concept of the optically active aromatic chromophore as "conformational probe" in stereoregular copolymers can be further extended by employing optically active aromatic monomers. Thus the coisotactic copolymers of (R)-4-phenyl-1-hexene with 4-methyl-1-pentene [29] (Figure 7) can be used for correlating the sign of CD of the aromatic Π-electrons system with the macromolecular conformation. Both the above copolymer and poly[(R)-4-phenyl-1-hexene] show negative CD between 270 and 250 nm (^1L_b-band) comparable in absolute value ($\Delta\epsilon_{max} \sim 0.15$) and with the same negative sign as the styrene/ (R)-3,7-dimethyl-1-octene copolymer for which the right-handed helical conformation is more probable. On the other side the units from non chiral 4-methyl-1-pentene display molar optical activity up to -208 as expected from their insertion in a chain with a largely prevailing (86%) right-handed screw sense [12].

FIGURE 7. Schematic structures of poly [(R)-4-phenyl-1-hexene] (A) and of (R)-4-phenyl-1-hexene/4-methyl-1-pentene random copolymer (B).

A   B

An analogous case is offered by the copolymers between optically active [(S)-α-methylbenzyl]-methacrylate and trityl-methacrylate [30] (Figure 8), the optical activity of which varies with composition turning from negative to positive when the content of trytilmethacrylate units overtakes 80% mole.

FIGURE 8. Schematic structure of [(S)-α-methylbenzyl]-methacrylate/ tritylmethacrylate random copolymer

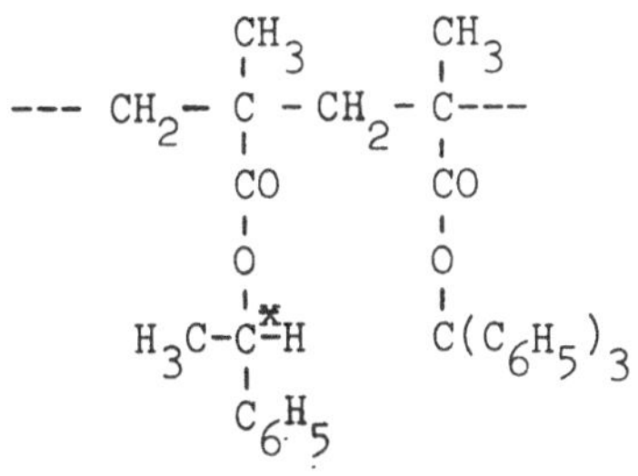

Analogous variation and change of sign occur for all the CD bands in the investigated region (270-210 nm). A very reasonable explanation of these results is that with increasing the content of tritylmethacrylate units, these last are increasingly included in isotactic helical blocks with a largely

predominant screw sense, induced by even small amounts of
units from [(S)-α-methylbenzyl]-methacrylate. In this situation
the tritylmethacrylate units give a contribution to the optical
activity with opposite sign with respect to the chiral comonomer.

5. CONCLUSIONS

From the examples discussed in this chapter it results
clearly that chiroptical techniques supply a powerfull method
for investigating the molecular conformation of macromolecules
in solution. The evident limitation of the method is that
optically active polymers only display chiroptical properties.
While this occurs in the most naturally occurring polymers, it
is not true for the most common stereoregular synthetic polymers.
In these last the asymmetric carbon atoms have either one or the
other absolute configuration with the same probability and
inter-and intra-molecular compensations cancel any optical
rotation.

The optically active homopolymers synthesized therefore
must be regarded as sophisticated models for obtaining information
about macromolecular arrangement in solution. It would be
certainly dangerous to extend simply these conclusions to
analogous systems from non chiral monomers. Rather the results
presented demonstrate that, even in the absence of hydrogen
bonding, steric interactions among bulky side chains along
an isotactic backbone can provide ordered conformation in
solution. It is likely that with opportune modifications this
is valid also for non optically active isotactic macromolecules.
On the other side these simple systems provide unique information about relationship between chiroptical properties of suitable chromophores and geometry of the macromolecules. Such
information is of great help for interpreting more complicated
data of poly-α-aminoacids and proteins [31].

A different situation is found for coisotactic copolymers
of non chiral monomers and optically active comonomers. In these
cases the non chiral monomer units can be forced to assume a well
definite conformation as shown by the induced optical activity.

This cooperative effect, depending on comonomer structure
and polymer stereoregularity, can be transmitted for long chain
sections, a small amount of chiral comonomer being sufficient
to induce chirality of one single type in the whole macromolecule.
This effect can be of great interest for the simple and economical
synthesis from commercial achiral monomers and a small amount of
chiral comonomers of functional optically active copolymers with
an ordered chiral secondary structure to be used as reagents or
catalysts for asymmetric synthesis.

REFERENCES

1.   E.R.Blout in"Fundamental Aspects and Recent Developments in ORD and CD", Chapter 4.5 F.Ciardelli and P.Salvadori Eds, Heyden, London, 1973.

2.   G.Natta, Makromol.Chem., 35, 93 (1960).

3.   T.M.Birshtein and P.L.Luisi, Vysokomol.Soed., 6, 1238 (1964).

4.   G.Allegra, P.Corradini and P.Ganis, Makromol.Chem., 90, 60 (1966).

5.   A.Abe, J.Am.Chem.Soc., 90, 2205 (1968).

6.   P.Pino, F.Ciardelli, G.P.Lorenzi and G.Montagnoli, Makromol.Chem., 61, 207 (1963).

7.   V.Petraccone, P.Ganis, P.Corradini and G.Montagnoli, Eur.Polymer J., 8, 99 (1972).

8.   P.Pino, F.Ciardelli and M.Zandomeneghi, Ann.Rev.Phys.Chem., 21, 561 (1970).

9.   P.L.Luisi, Polymer, 13, 232 (1972).

10. P.Pino, P.Salvadori, E.Chiellini and P.L.Luisi, Pure Appl.Chem., 16, 469 (1968).

11. O.Pieroni, F.Ciardelli and G.Stigliani, Chim.Ind.(Milan) 52, 289 (1970).

12. C.Carlini, F.Ciardelli and P.Pino, Makromol.Chem., 119, 244 (1968).

13. P.Pino, P.Salvadori, G.P.Lorenzi, E.Chiellini, L.Lardicci, G.Consiglio, O.Bonsignori and L.Lepri, Chim.Ind.(Milan) 55, 182 (1973).

14. R.C.Schulz and H.Hilpert, Makromol.Chem., 55, 132 (1962).

15. E.I.Klabunovskii, M.I.Schvartsman and Yu.I.Petrov, Vysokomol.Soed., 6, 1579 (1964); Izv.Akad.Nauk SSSR, 223 (1966).

16. K.Matsuzaki, A.Ishida and N.Tateno, J.Polymer Sci. C 16, 2111 (1967).

17. K.J.Liu, J.S.Lignowski and R.Ullman, Makromol.Chem., 105, 18 (1967).

18. H.Yuki, K.Ohta, K.Uno and S.Murahashi, J.Polymer Sci. A1, 6, 829 (1968).

19. O.Pieroni, F.Ciardelli, C.Botteghi, L.Lardicci, P.Salvadori and P.Pino, J.Polymer Sci. Part C, 22, 993 (1969).

20. A.Allio and P.Pino, Helv.Chim.Acta, 57, 616 (1974).

21. P.Pino, C.Carlini, E.Chiellini, F.Ciardelli and P.Salvadori, J.Am.Chem.Soc., 90, 5025 (1968).

22. F.Ciardelli, P.Salvadori, C.Carlini and E.Chiellini, J.Am.Chem.Soc., 94, 6536 (1972).

23. P.Salvadori, L.Lardicci, R.Menicagli and C.Bertucci, J.Am.Chem.Soc., 94, 8598 (1972).

24. H.De Voe, J.Chem.Phys., 43, 3199 (1965).

25. W.Hug, F.Ciardelli and I.Tinoco Jr., J.Am.Chem.Soc., 96, 3407 (1974).

26.   E.Chiellini, F.Maestrini, P.Vergamini and G.Ceccarelli,
       Chim.Ind.(Milan), 57, 131 (1975).
27.   F.Ciardelli, P.Salvadori, C.Carlini, R.Menicagli and
       L.Lardicci, Tetrahedron Letters, 1779 (1975).
28.   F.Ciardelli, C.Righini, M.Zandomeneghi and W.Hug, J.
       Phys.Chem., 81, 1948 (1977).
29.   C.Carlini, F.Ciardelli, L.Lardicci and R.Menicagli,
       Makromol.Chem., 174, 27 (1973).
30.   H.Yuki, K.Ohta, Y.Okamoto and K.Hatada, J.Polymer Sci.
       Letters Ed., 15, 589 (1977).
31.   F.Ciardelli, E.Chiellini, C.Carlini, O.Pieroni, P.Salvado-
       ri and R.Menicagli, J.Polymer Sci., Polymer Symposia,
       (Dublin),62, 143 (1978).

RECENT DEVELOPMENTS OF VIBRATIONAL SPECTROSCOPY OF POLYMERS: DYNAMICS AND SPECTRA OF DISORDERED POLYMERS

Giuseppe Zerbi

Istituto di Chimica, Università di Trieste, Italy.

In nature many molecular systems can be found which can be described as "chain molecules". The main structural property which characterizes these systems is generally the fact that intramolecular forces are at least one order of magnitude larger that intermolecular forces.

The majority of synthetic or natural polymers can be described as chain molecules where each molecular sub-unit or chemical repeat unit is linked to the other by covalent forces and form a long molecular chain with very large molecular weight. [1]

In addition to organic macromolecules a few other classes of chemical compounds can be described as chain systems. In many organic crystals molecules crystallize as chains formed by intramolecular hydrogen bonds. Solid formic and acetic acid, succinic acid, N methyl acetamide, solid methanol, solid HX acids (X = F, Cl, Br I) phenol, napthol etc. are typical simple examples of classes of chain systems in the solid where intramolecular forces are certainly stronger than the intermolecular ones. Very many inorganic molecules can also crystallize as chains (e.g. HgO, Te, ...etc.).

The shape of these molecules are primarily determined by atom-atom interactions which force the chain to take up in space a structure of minimum energy which can be described with a set of conformational angles. The prediction of the minimum energy conformation is at present possible with a fair reliability using semiempirical atom-atom potentials as proposed by

R. W. Lenz and F. Ciardelli (eds.), Preparation and Properties of Stereoregular Polymers, 369-385.
Copyright © 1979 by D. Reidel Publishing Company.

various authors[2,3]. More than one minimum in conforma
tional energy can and has been predicted for many
molecules by such a type of calculations thus indicating that these molecules may undergo structural
phase transitions from one molecular shape into another
one following certain energetical paths[4]. Generally,
interchain packing forces in a crystal are weaker
contributions to the total stability of the chain and
do not greatly change the structure which the chain
takes up "in vacuo" because of intramolecular
interactions.

     The rigorous picture just described of the struc-
ture of a macromolecule would imply that chains should
be considered as one dimensional highly anisotropic
infinite perfect crystals for which a one-dimensional
rototranslational simmetry operator can be defined in
order to generate the infinite chain. In reality
organic chain molecules never posess a perfect structure but contain several types of chemical and structural disorder. The rigorous concepts just discussed
must be "relaxed" and one has to take into account
chemical defects, kinks, jogs and folds in the polymer
chain. The most typical case of folded molecules can
be found in the single crystals of organic polymers
such as polyethylene, polytetrafluoroethylene,
polypropylene etc.[5] In order to form a small single
crystal with a thickness of, say, a few hundreds A, the
long ribbon-like molecule has to fold into itself in
order to meet the packing requirements of a certain
crystal space group. The energetics and kinetics of
kink formation is the subject of extensive studies by
many authors.[6,7] The importance of the knowledge of
the structure and concentration of kinks or fold lies
in the fact that the physical properties of these materials depend on their microstructure at the molecular
level.

DYNAMICS OF PERFECT POLYMERS

     Vibrational spectroscopy has been known since
many years to be one of the useful sources of experimental information for the determination of the
overall structure of chain molecules.[8,9] The detailed
interpretation of the dynamics of these systems aimes
at determining the phonon frequencies and the shape of
the phonon waves which travel along the one-dimensional, infinite crystal.[10,12]

     Space (line) group selection rules determined by the
symmetry of the polymer chain allow the predict the
activity of the $\underline{k} = 0$ phonon modes in the infrared or

Raman and their state of polarisation for stretch-
-oriented polymers. Coherent neutron scattering experiments provide the detailed shape of the phonon
dispersion curves $\omega(\underline{k})$[13] and incoherent scattering
experiments provide the amplitude weighted density of
vibrational states $g(\omega)$[14]. Phonon dispersion curves
can also be constructed from the study of the optical
spectra of molecular models with increasing chain
length. The contributions in the field of polymer
structure and dynamics of vibrational spectroscopy in
the last twenty years are discussed in great length
by several review articles and books.[10,15-19] The
techniques generally adopted for the calculation of
dispersion curves for one or three dimensional organic
crystals in terms of Wilson's internal coordinates are
fully discussed in the literature [10,11]. It is generally convenient for polymers and crystals to solve
the problem in cartesian coordinates while all force
constants are generally given in terms of internal
coordinates. If $\underline{\underline{B}}(\underline{k})$ is the $\underline{k}$ dependent linear
transformation from cartesian to internal phonon
coordinates [11] the $\underline{k}$ dependent potential energy matrix
in cartesian coordinates $\underline{\underline{F}}_X(k)$ can be obtained from the
$\underline{k}$ dependent potential energy matrix in internal
coordinates $\underline{\underline{F}}_R(\underline{k})$ by the following transformation:

$$\underline{\underline{F}}_X(\underline{k}) = \widetilde{\underline{\underline{B}}}(\underline{k})\ \underline{\underline{F}}_R(\underline{k})\ \underline{\underline{B}}(\underline{k}) \qquad (1)$$

The dispersion relation for an organic solid can then
be written as

$$\left[\underline{\underline{D}}(\underline{k}) - \omega^2(\underline{k})\ \underline{\underline{E}}\right]\underline{L}(\underline{k}) = 0 \qquad (2)$$

where $\underline{\underline{D}}(\underline{k}) = \underline{\underline{M}}^{-\frac{1}{2}}\ \underline{\underline{F}}_X(k)\ \underline{\underline{M}}^{-\frac{1}{2}}$, $\underline{\underline{M}}$ the diagonal matrix of
the masses, ω the vibrational frequencies and $\underline{L}(\underline{k})$ the
corresponding vibrational displacements or polarisation
vectors. The geometry of the polymer chain enters
through the $\underline{\underline{B}}(\underline{k})$ matrix while the atomic masses appear
in the matrix $\underline{\underline{M}}$ defined above.
   The size of the dynamical matrix, hence the order
of the secular equation, to be solved for each $\underline{k}$, is
tractable for simple polymers (Polyethylene). But
the size becomes soon untractable for just slightly
more complicates systems. For instance, the crystallographic repeat unit of isotactic polypropylene contains
three monomer units and the resulting secular determinant has dimension 81. For polytetrafluoroethylene in
the most stable 15/7 structure the dispersion relation
contains 135 branches; 144 are the branches for
isotactic polystyrene. Each of the secular equation

has to be solved as many times as many $\underline{k}$ values one decides to use (normally at least 10,20). The introduction of symmetry factoring is not of great help in polymer calculations since the line group at the Γ point ($\underline{k}$ = 0) is generally already low and becomes even lower throughout the Brillouin zone for $\underline{k} \neq 0$.

2. DYNAMICS OF PARTIALLY ORDERED POLYMERS.

The problem of the calculation of eigenvalues and eigenvectors becomes even more complex when the translational symmetry is destroyed by the introduction of defects. The concept of phonon waves is lost and the $\underline{k}$ dependency in equation 1 and 2 is lost. One has then to calculate eigenvalues and eigenvectors of very large dynamical matrices corresponding to the size of the piece of crystal or segment of polymer chain one has chosen. Even if drastic structural simplifications are introduced, the problem still remains untractable by the standard numerical procedures.

The dynamics of imperfect lattices has been already treated in several text books of solid state physics,[20] but most of the treatments presented deal with mass or force constants impurities in simple crystalline solids, with a small concentration of generally non interacting defects. The case of organic solids we are dealing with becomes more difficult since we have to account, with great accuracy, for the dynamics of rather complex chemical units made up by many atoms and containing a large concentration of interacting defects. Most of the disorder is due to structural defects which require the simultaneous treatment of geometry and force constant defects. Mass defects occur only in particular cases, such as in isotopic mixtures[21] or isotopic copolymers.[22]

In this paper we wish mainly to focus our attention at the methods recently developped for the understanding of the real structure of these materials which is very far from the ideal model generally considered in most of the dynamical treatments.

In fig. 1 (as an example in our discussion) we report a comparison of the calculated unweighted density of vibrational states of a perfect single chain of polyethylene with the actual infrared absorption spectrum. For a perfect trans planar chain of polyethylene with D_{2h} symmetry the structure of the irreducible representation for $\underline{k}$ = 0 modes and the corresponding infrared and Raman activities for a stretch oriented rod in the one dimensional model are the following:

$$T_{vibr} = 3A_g(R) + A_u + 2B_{1g}(R) + B_{1u}(IR, /\!/) + 2B_{2g}(R) + 2B_{2u}(IR, \perp) +$$

$$+ B_{3g}(R) + 2B_{3u}(IR, \perp)$$

If we neglect the C-H stretching region, the infrared spectrum should show only three active modes non coincident with 6 Raman active modes. In figure 1 we notice instead the following facts: i) several singularities in the density of states seem to find some coincidence with infrared bands which do not correspond to $k = 0$ modes; ii) some extra peaks are observed which do not find any correspondence with singularities originating from the perfect one dimensional crystal.[12] The fact that some of these infrared peaks are related to conformational defects in the polymer chain is experimentally verified when the spectrum of high density extended chain polyethylene is compared with the infrared spectrum of the same material after normal crystallisation from the melt.[23,24] In fig. 2 we clearly identify the C-H out-of-plane deformation modes of the vynil groups which terminate the polymer chain. Their concentration is shown not to change when a sample of extended chain PE is melt and re-crystallized under normal conditions. If we take these two bands as internal references one can see that the intensities of the peaks at $\sim$1365 - 1350 cm^{-1} are strongly increased in the melt crystalli zed material thus showing that the concentration of the molecular species (or local conformational isomers) which originate these absorptions is greatly increased.

The existence of the extra bands such as those just discussed is a common fact in polymer spectroscopy and has suggested the existence of an "amorphous" part of the polymer substance. These kinds of bands have been generally called "amorphous" bands and their intensity has been taken as measure of the amorphous or non crystalline part of the sample.[25] Processes aimed at the purification of polymer samples have been developped and the so called amorphous bands in the infrared used as an indication of the "amorphicity" of the samples. The content of syndiotactic structure in syndiotactic polypropylene was generally determined using the so called syndiotacticity index measured from the intensity of some infrared bands.[26,27]

At present, three methods have been proposed for the treatment of the dynamics of disordered organic polymers each one having advantages and limitations.

Following the traditional approach of solid state physics to ionic crystals [20], Schmid and collaborators have developped, in the past few years, a Green's function method for the calculation of the density of states of polyethylene chains containing various types of conformational defects.[28] In spite of its elegance and completeness, the Green function method is limited by the enormous amount of algebra which must be developped even for the simplest cases. Such an algebraic complexity has forced the authors to simplify the molecular models treated to point mass models thus reducing the analysis to the motions along the ω_5 and ω_9 branches. We have however already shown that several interesting features occur in the higher energy of the spectrum. A Green's function approach has also been discussed by the russian school of Kozyrenko et al.[29]

The second method is that which bases the interpretation of the vibrational spectrum of a disordered material on the knowledge of the dynamics of small model molecules. The classical example is that of Snyder on liquid alkanes[30]. While such a treatment is well justified when highly localized modes are considered and the vibrational motions of adjoining sections of the polymer chain can be neglected, the problem becomes more difficult when we deal with highly delocalized resonance modes. The shape of these resonance modes will never we reproduced by short chain models.

3. NUMERICAL METHODS.

In our laboratory we have been extensively using a numerical method which allows to overcome the complexity of the dynamical problem of atomic or molecular systems even with complex molecular structure, containing any type of defects with any desidered concentration. Basically, the method allows to calculate directly the number of vibrational states in a given frequency interval thus reaching, if necessary, a single isolated eigenvalue. The numerical metod, commonly called NET (Negative Eigenvalue Theorem), has been first introduced by Dean[31] who also treated the vibrations of several disordered systems, namely monoatomic chains with a random distribution of masses, silica glass and ice.

The procedure is as follows.[32] Let p be the number of atoms in the chemical repeat unit and N the number of units which make up our systems. We wish to compute the number $n(\omega_2 - \omega_1)$ of eigenvalues of the $3Np \times 3Np$ dynamical matrix $\underline{\underline{D}}$ which lie in the interval (ω_1, ω_2) where ω_1 and ω_2 are positive real numbers such that $\omega_2 > \omega_1$. The number $n(\omega_2 - \omega_1)$ is given by

$$n(\omega_2 - \omega_1) = \eta(\underline{D} - \omega_2\underline{E}) - \eta(\underline{D} - \omega_1\underline{E}) \qquad (3)$$

where $\underline{E}$ is the 3Np x 3Np unit matrix and $\eta(\underline{D} - \omega\underline{E})$ is the number of negative eigenvalues of the matrix

$$\underline{D}_i = \underline{D} - \omega_i\underline{E} \qquad (4)$$

The computation of the negative eigenvalues of $\underline{D}_i$ is performed by a particular partitioning of $\underline{D}_i$ which applies to any symmetrical matrix. $\underline{D}$ is symmetrical and for chain molecules has also a codiagonal form, the number of codiagonals depending on the complexity of the chemical repeat unit and on the extent of dynamical coupling. The NET states that, given a symmetrical matrix $\underline{P}$ of dimensions r x r, partitioned as follows:

$$\underline{P} = \begin{bmatrix} \underline{A}_1 & \underline{B}_2 & & \\ \tilde{\underline{B}}_2 & \underline{A}_2 & \underline{B}_3 & \\ & & \ddots & \underline{0} \\ \underline{0} & & & \\ & & \tilde{\underline{B}}_k & \underline{A}_k \end{bmatrix} \qquad (5)$$

where $\underline{A}_i$ has dimensions r_i x r_i, $\underline{B}_i$ has dimensions r_{i-1} x r_i and $\sum_{i=1,k} r_i = r$, then number $\eta(\underline{P} -x \underline{E})$ of negative eigenvalues of the matrix $\underline{P} -x \underline{E}$ is given by

$$\eta(\underline{P}-x\ \underline{E}) = \sum_{i=1}^{k}\eta(\underline{U}_i) \qquad (6)$$

where

$$\underline{U}_i = \underline{A}_i - x\underline{E}_i - \tilde{\underline{B}}_i\underline{U}_i^{-1}\underline{B}_i$$

$$\underline{U}_1 = \underline{A}_1 - x\underline{E}_i \qquad (7)$$

The partitioning on $\underline{D}_i$ is the following. Let $\underline{D}_i^1$ denote the matrix $\underline{D}_i$ partitioned as

$$\underline{D}_i \begin{bmatrix} \underline{X}_1 & \underline{Y}_1 \\ \tilde{\underline{Y}}_1 & \underline{Z}_1 \end{bmatrix} , \qquad (8)$$

where $\underline{X}_1$ is a 1 x 1 matrix, $\underline{Y}_1$ is a 1 x (3Np -1) matrix and $\underline{Z}_1$ is a (3Np $-$ 1) x (3Np -1) matrix. Then

$$\eta(\underline{D}_i) = \eta(\underline{D}_i^{(1)}) = \eta(\underline{X}_1) + \eta(\underline{D}_i^{(2)})$$

$$\underline{\underline{D}}_i^{(2)} = \underline{\underline{Z}}_i - \underline{\underline{\tilde{Y}}}_1 \underline{\underline{X}}_1^{-1} \underline{\underline{Y}}_1 \tag{9}$$

One can continue the process as indicated in eq. (9)
for $3Np - 1$ times until one reaches the result that

$$\eta(\underline{\underline{D}}_i) = \sum_{j=i}^{3Np} \eta(X_j) \tag{10}$$

The partitioning just described avoids the inversion
of matrices since $\underline{X}_i$ is a 1 x 1 matrix. The time for
computing each $\underline{\underline{D}}_i^{(k)}$ is greatly reduced by the fact
that the row matrix $\underline{Y}_{k-1}$ has only $C - 1$ non zero
elements where, C is the number of codiagonals of $\underline{\underline{D}}$.

The intervals (ω_1, ω_2) can be restricted to any
desidered accuracy. The density of states for each
interval can then be plotted as hystogram with any
desidered mesh. By narrowing the interval (ω_1, ω_2)
one can also reach the single isolated eigenvalues
which could in principle be computed by the traditional solution of the secular equation. For the
computation of the eigenvectors one needs a more
precise knowledge of the approximate eigenvalue, thus
one needs to narrow the interval $d\omega = \omega_2 - \omega_1$
until only one eigenvalue is contained. Let $\bar{\omega}_i$ be
the value $(\omega_2 - \omega_1)/2$ and ω_i the (unknown) exact
eigenvalue of $\underline{\underline{D}}$ which occurs in d. The procedure
applied to compute $\bar{\omega}_i$ ensures that

$$|\omega_i - \bar{\omega}_i| < |\omega_k - \bar{\omega}_i|$$

where ω_k is any other exact eigenvalue of $\underline{\underline{D}}$.

The time required to compute a step of the
hystogram for $g(\omega)$ for a matrix $\underline{\underline{D}}$ with dimensions
3600 and 30 non zero codiagonals is at present of~10
seconds on a UNIVAC 1108. Having reached such short
computational times one has a great flexibility and
freedom in the numerical calculations. The calcula-
tion of the approximate eigenvectors $\underline{L}_i$ corresponding
to the approximate eigenvalue $\bar{\omega}_i$ can be done using the
"inverse iteration method"[23] originally proposed by
Wilkinson.[21] The time required to compute one
eigenvector of a matrix 900 with 30 codiagonals on
UNIVAC 1108 is of ~50 seconds. It has also to be
added that the construction of the dynamical matrix $\underline{\underline{D}}$
can now be made by suitable programs which automati-
cally generate the large matrix from smaller building
blocks which contain all the informations on the
geometry, masses, force constants, type of defects,
randomness, length and end groups of the chain.[22,34,35]
The application of NET to several organic

materials has allowed to improve our understanding of
their dynamics and spectra. Some attempts of generali-
sations can at present be attempted. The introduction
of defects in the polymer gives rise to the following
facts. We give as examples the most recent results
while refer to the literature for the discussion of
previous cases.

a) - Localized or gap modes.

If one of the vibrational levels of these very
complex systems happens to occur in a energy gap
which is not spanned by the frequency band of the
perfect host linear lattice, the corresponding vibrational motion cannot be transmitted by the surrounding
medium and the motion remains localized in space at
the defects.[21] The vibrational amplitudes decay
rapidly along the polymer chain and generally should
give rise to sharp narrow bands.

Two kinds of localized modes have been found in
real case of organic molecules.

i) isotopic substitution - The introduction of
deuterium in a host lattice of hydrogen atoms in
n-alkane chains generates very clear gap modes in the
C-D stretching region in an energy range completely
free from absorptions of any other fundamentals. The
corresponding absorptions have been observed but are
of no use for structural studies since the vibrations
are uncoupled from the neighouring units. Much more
interesting are the gap modes corresponding to the
CD_2 rocking motion which jump out from the CH_2
twisting-rocking branch ($1100 - 720$ cm^{-1}) and generate
characteristic peaks. Because of the coupling with
some of the neighbouring CH_2 units these gap modes are
conformationally dependent and the corresponding
frequencies become a very useful probe of the
conformational microstructure of the chain molecule.
Snyder has proposed the method for the study of
polyethylene.[36] The study of conformational dependent
gap modes in selectively deuterated n-hydrocarbons [37]
and n-fatty acids has allowed to map the conformational geography of some of these molecules and to derive
informations on the conformational transitions.[38]

ii) localisation because of geometrical distorsions -
For particular geometries the vibrations of groups of
atoms or clusters of atoms along the chain generate
modes whose frequencies occur near the edge of a
frequency branch just at the beginning of the
frequency gap. Examination of the eigenvectors
corresponding to such modes generally show that a
large number of atoms is involved in such a normal
mode. Cases have been observed for head-to-head defects

of Polyvinylchloride [39] and for the bending motion of
n-alkane chains just above the ν_5 branch.[28]
Sometimes gap modes due to the motions of groups of
atoms become characteristic of particular geometrical
defects. In the study of the dynamics of tight folds
in polyethylene, a gap mode calculated at 715 cm^{-1}
seems to be characteristic of a group of atoms
organized in a tight fold (with conformational
sequence GGTGG). This mode has been observed in the
infrared spectrum of the crystalline cyclic hydrocarbon
molecule $C_{34}H_{68}$. These predictions should be useful as
a tool for the study of the folding surface of single
crystal of polyethylene.[41]

b) - Resonance modes.
        Some of the eigenstates of a disordered organic
polymer may occur within the band spanned by the
perfect host lattice giving rise to pseudo localized
resonance modes. This is a very common case predicted
in theory and experimentally observed in the spectra
of several polymeric materials studied in our laboratory. The location of some of these resonance modes
in the spectra has a great diagnostic value in the
study of the microstructure of polymers. It may well
happen that, even if the eigenfrequencies of these
modes occur within the band spanned by the host lattice, the actual vibrational displacements show that the
motion is localized at the defects and is relatively
little coupled with the host lattice. These modes beco
me characteristic of the defects. Characteristic
resonance modes have been calculated and observed for [23,30]
distorted chain in the wagging region of polyethylene,
resonance modes in the ν_4 branch of polyethylene are
indicated as "gauche bands" and are commonly adopted
as a probe for the structure of phospholypids,
membranes and model compounds.[18] Resonance modes in
the low energy region of the spectrum have been
calculated by Schmid et al. for the understanding of
the structure of disordered Polyethylene.[28]

4. ACTIVATION OF THE DENSITY OF STATES OF THE HOST

   LATTICE.

        In principle the removal of the translational
periodicity because of disorder relaxes the selection
rules imposed by group theory and all phonons of the
host lattice may become active. In principle, one can
then expect to observe the activation of the whole
density of states g( ) of the host lattice because of
the perturbations by the end groups. The extent of

activation depends upon the perturbations of the end groups and by the electrical characteristics of the medium which determine the infrared and Raman transition moments.

The asymmetric wing at the higher energy side of the CH_2 rocking mode in Polyethylene can be taken as an activation of the whole frequency branch which starts from the cutoff at 720 cm^{-1} and ends at 1000 cm^{-1}. Analogoulsy the temperature dependent infrared spectrum with maximum at 500 cm^{-1} [23] may correspond to the highest singularity at the top of the ν_5 acoustic branch of polyethylene. The precise identification of the activation of density of states for organic materials is hindered by the fact that many defect modes in resonance with the host lattice may occur and the observed absorption spectrum or scattering may be the envelope of such transitions. A typical example is the infrared peak at 348 cm^{-1} for polytetrafluoroethylene which so far has been justified as a disorder induced activation of the density of state of the host lattice for which a singularity is calculated at that frequency.

5. COMPARISON OF CALCULATED $g(\omega)$ WITH EXPERIMENTAL DATA; CALCULATION OF THE TRANSITION MOMENTS.

In the previous section we have reported the present state of the art the calculation of the eigenvalue and eigenvectors of the very large dynamical matrices which must be handled in a detailed study of the dynamics of disordered or partially ordered organic materials. Even if these new techniques are of great help in the study of frequencies, one soon realizes that frequency fitting of vibrational data from Infrared, Raman and Neutron scattering, is not any more enough for a reliable structural analysis. The same limitations are commonly found in normal calculations when only frequency fitting of the fundamentals is required while the intensities of each individual bands are neglected.

An improvement in this kind of studies could be obtained if the density of states $g(\omega)$ could be properly weighted by a dipole factor in the infrared or polarizability factor in the Raman. Amplitude weighted $g(\omega)$ from neutron scattering can be easily evaluated [13] since vibrational amplitudes can be routinely calculated for crystal and polymers.

The weighting of $g(\omega)$ by a proper transition moment may completely change the spectral pattern, namely, a few phonons or localized modes may gain large intensity while strong singularities may be damped almost to zero by small transition moments.

The actual calculation of the transition moments, or in other words of the absorption coefficient in the infrared and scattering activity in the Raman, requires the knowledge of the electrical properties of polymers at the equilibrium and during the vibrational motion. Unlike ionic crystal, effective point charges give unsatisfactory results for covalent molecules such as all those treated in this paper.

The only possibility to overcome this difficulty is to start from the study of the vibrational intensities in the infrared and Raman of small molecules chemically similar to the polymer of interest and to parametrize them according to a suitable model. The study of infrared and Raman intensities had received some attention several years ago but both the experimental and computational difficulties have limited the development of these studies. Only recently intensity studies have been revived and measurements and calculations are being made in several laboratories.[1,42]

Our approach to the problem is precisely that of being able to use a certain model of dipole moment changes and polarizability changes which may be transferred between chemically similar molecules[43] in order to predict the intensity of infinite polymers[44] and in particular of partially disordered polymers. The main object of our work is to be able to recognize a given structural defect from its absorption frequency and to calculate its concentration using the predicted absorption coefficient or scattering activity.

The model of bond moments for infrared and bond polarizabilities for the Raman (electrooptical parameters, eop) proposed by the russian school proves to be a very good model for the parametrisation of vibrational intensities.[41]

The main object of our work in this field is to use experimental intensity data from small model molecules to calculate by least squares procedures sets of electrooptical parameters which may be transferable between chemically similar molecules as well to polymers. A simultaneous "overlay" least squares calculation of the infrared intensities of methane, deuteromethane, ethane, deuteroethanes,[43] propane, deuteropropane and several n-paraffins[45] up to heptane has provided a set of eop which can very satisfactorily reproduce the spectrum of Polyethylene. In analogous way calculations have been made on the

Raman intensities of methane, deuteromethane,[46] ethane, deuteroethane[47] and cyclohexane [48] for the determination of the Raman eop's for the calculations of the intensities of polymers. Experiments have also been made on the integrated relative Raman intensities of stretch oriented polyethylene and perdeuteropolyethylene in different scattering geometries and polarization conditions such as to be able to excite selectively the various components of the derived polarizability tensor.[49] The calculated eop's nicely correspond with those obtained from the series of smaller hydrocarbons just mentioned. The number of experimental data in the case of infrared is much larger than the case of the Raman and has allowed a more systematic process of debugging of the various possible sign choices in order to provide more exact informations on the equilibrium bond dipole moments and their changes during vibrations [41]. The lack of experimental data does not yet allow a similar work in the case of the Raman.

From the evidence so far collected, at least in the case of hydrocarbons, it can be stated that a careful use of the intensity data yelds sets of eop's which are satisfactorily transferable within a [42] class of chemically and structurally similar molecules.

It has been recently shown, however, that the infrared intensities of rotational isomers of hydrocarbons cannot be very satisfactorily reproduced with the set of eop's derived from all trans molecules. One is then faced with the problem of the determinations of the electrical interactions between adjacent bonds in gauche and trans positions. This problem is presently being analyzed in our laboratory.[50]

The possibility of a reliable transition moment weighting of g( ) of organic polymers is at present at a very early stage of development. Many more experimental data both in infrared and Raman on smaller molecules are needed for a generalisation of the method.

From our experience a positive step in this field may come from a combined effort of "ab initio" quantum mechanical calculations and spectroscopic analysis.

Aknowledgments.

Most of the work discussed in this chapter has been made possible by the constant collaboration of Dr. M. Gussoni.

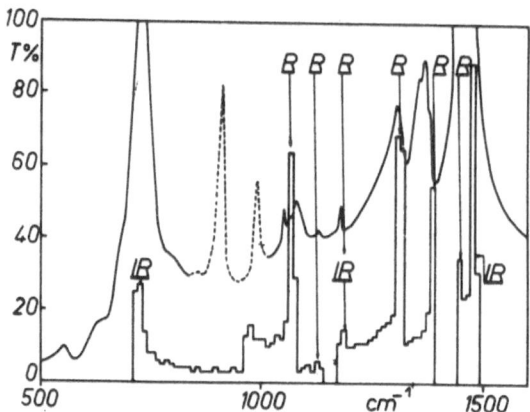

Fig. 1 - Comparison between the infrared absorption
spectrum of polyethylene with the density of vibrational states calculated for a perfect polymethylene
single chain. The spectroscopic activity (Raman and
Infrared) of the k = 0 critical points are indicated.

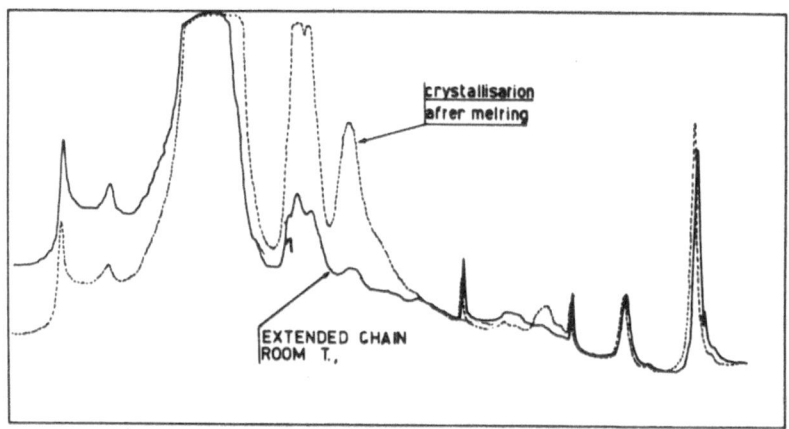

Fig. 2 - Section from 1700 to 900 cm^{-1} of the infrared
spectrum of a sample of extended chain polyethylene
before and after heat treatment.

REFERENCES

1. *Stereoregular Polymers and Stereospecific Polymerisation*, Vols 1 and 2, (G. Natta and F. Danusso, es.) Pergamon Press, New York (1967).
2. P. De Sanctis, E. Giglio, A.M. Liquori and A. Ripamonti, *J. Polymer Sci*. A1, 1383 (1963).
3. G.N. Ramachandran and V. Sasisekharan, Conformation of Polypeptides and Proteins, *in Advanced in Protein Chem*. 23, 284 (1968).
4. G. Masetti, F. Cabassi, G. Morelli and G. Zerbi, *Macromolecules* 6, 700 (1973).
5. B. Wunderlich, *Macromolecular Physics*, vol. 1, Academic Press, New York (1973).
6. P.E. McMahon, R.L. McCullogh and A.A. Schlegel, *J. Appl. Phys*. 38, 4123 (1967).
7. V. Petraccone, G. Allegra and P. Corradini, *J. Polym. Sci*, C38, 419 (1972).
8. R. Zbinden, *Infrared Spectroscopy of High Polymers*, Academic Press, New York (1964).
9. S. Krimm, *Advances in Polymer Science*, 2, 51 (1960)
10. G. Zerbi, *Molecular Vibrations of High Polymers*, Applied Spectr. Reviews (E. Brame ed.) vol. 2, 193 (1969).
11. L. Piseri and G. Zerbi, *J. Mol. Spectr*., 26, 254 (1968).
12. L. Piseri and G. Zerbi, *J. Chem. Phys*., 48, 3561 (1968); ibid. 49, 3840 (1968).
13. S. Trevino and H. Boutin, *J. Macromol. Sci*., A1, 723 (1967).
14. T. Kitagawa and T. Miyazawa, *J. Chem. Phys*., 47, 337 (1967).
15. J.H. Schachtschneider and R.G. Snyder, *Spectrochim. Acta*, 19, 17 (1963).
16. R.G. Snyder and J.H. Scachtschneider, *Spectrochim. Acta*, 21, 169 (1965).
17. R.G. Snyder and G. Zerbi, *Spectrochim. Acta*, 23A, 391 (1967).
18. G. Zerbi and G. Masetti "Raman Spectra of Polymers" in *"Analytical Raman Spectroscopy"* (W. Kiefer ed.) J. Wiley, in Press.
19. G. Zerbi "Dynamics and Spectra of Polymers" in *"Modern Methods in Vibrational Spectroscopy"* (Orville-Thomas, Barnes Eds) Elsevier, Amsterdam, (1977).
20. P.G. Dawber and R.J. Elliott, *Proc. Roy. Soc*., A273, 222 (1963); P.G. Dawber and R.J. Elliott, *Proc. Phys. Soc*., 81, 453 (1963).
21. M. Gussoni and G. Zerbi, *J. Chem. Phys*., 60, 4862 (1974).

22. M. Tasumi and G. Zerbi, _J. Chem. Phys._, _48_, 3813 (1968).
23. G. Zerbi, L. Piseri and F. Cabassi, _Mol. Phys._, _22_, 241 (1971).
24. G. Zerbi, in "Phonons" (M.A. Nusimovici ed.) Flammarion, Paris (1971).
25. G. Zerbi, F. Ciampelli and V. Zamboni, _J. Polymer. Sci._, _C_, 141, (1964).
26. J. Boor and E.A. Youngmann, _J. Polym. Sci._, _A1_, 1861, (1966).
27. G. Masetti, F. Cabassi and G. Zerbi, _Polymer_, in press.
28. K. Hölzl, C. Schmid and P.C. Hägele, _J. Phys. C, Solid State Phys._, _11_, 9 (1978).
29. V.N. Kozyrenko, I.V. Kumpanenko and I.D. Mikhailov, _J. Polym. Sci., Polym. Phys. Ed._ _15_, 1721 (1977).
30. R.G. Snyder, _J. Chem. Phys._, _47_, 1316 (1967).
31. P. Dean, _Rev. Mod. Phys._, _44_, 127 (1972).
32. G. Zerbi, _Pure and Applied Chemistry_, _26_, 501 (1971); ibid. _36_, 35 (1973).
33. G. Zerbi, "Defect in Organic Crystal, Numerical Methods" in "_Lattice Dynamics and Intermolecular Forces_" (S. Califano ed.) Academic Press (1975).
34. A. Rubcic and G. Zerbi, _Macromolecules_, _7_, 754 (1954); ibid. _7_, 759 (1974).
35. G. Zerbi and M. Sacchi, _Macromolecules_, _6_, 692 (1973).
36. R.G. Snyder and M.W. Poore, _Macromolecules_, _6_, 708 (1973).
37. R. Magni, Thesis, University of Milano, (1979).
38. G. Zerbi, Unpublished.
39. A. Rubcic and G. Zerbi, _Chem. Phys. Letters_, _34_, 343 (1975).
40. G. Zerbi and M. Gussoni, to be published.
41. M. Gussoni, in "_Advances in Infrared and Raman Spectroscopy_" (eds. R.J.H. Clark and R.E. Hester), vol. 6, Heyden, London (1979).
42. G. Zerbi, "_Vibrational Intensities in Infrared and Raman Spectroscopy_" (W.B. Person and G. Zerbi eds.) Elsevier, Amsterdam (1979).
43. M. Gussoni, S. Abbate and G. Zerbi, _J. Chem. Phys._, in press.
44. S. Abbate, M. Gussoni, G. Masetti and G. Zerbi, _J. Chem. Phys._, _67_, 1519 (1977).
45. M. Gussoni, S. Abbate and G. Zerbi, to be published.
46. S. Abbate, M. Gussoni and G. Zerbi, _J. Mol. Spectry_, _73_, 415 (1978).
47. S. Abbate, M. Gussoni and G. Zerbi, _Indian J. Pure Appl. Phys._, _16_, 199 (1978) _Raman Memorial Volume_.

48. M. Gussoni, S. Abbate and G. Zerbi, <u>J. Raman</u>
    <u>Spectroscopy</u>, 6, 289 (1977).
49. S. Abbate, G. Masetti and G. Zerbi, <u>J. Chem. Phys.</u>,
    ibid. in press; S. Abbate, M. Gussoni and G. Zerbi,
    ibid. in press.
50. L. Colombo, Thesis, University of Milano, (1979).

THE CRYSTALLINE STRUCTURE OF ADDITION POLYMERS. RECENT RESEARCH RESULTS

P. Corradini, G. Guerra, B. Pirozzi

Istituto Chimico dell'Università di Napoli,
Via Mezzocannone 4, 80134 Napoli, Italy.

INTRODUCTION

As we pointed out in the lecture "Present status of configurational and conformational analysis in stereoregular polymers", the structure of macromolecules in the crystalline and a morphous states can be advantageously rationalized through the methods of conformational analysis.

In the first paragraph of this chapter we shall report on some recent energetic calculations in order to point out their predictive character.

In the second paragraph we shall discuss about disordered macromolecular conformations compatible with long range order (crystallinity).

Both in ordered and in disordered structures the importance of taking into account the variability of valence angles, be sides the variability of dihedral angles, will be stressed. Intriguing cases of polymer structures, resisting the effort of finding a solution, have been solved by appropriate consideration of the variability of valence angles. Enphasis on results obtained in our laboratories will be given.

In the third paragraph we shall discuss the results of some Monte-Carlo calculations related to the structure of "liquid" polyethylene; preliminary evidence will be given that bundles of chains are not present at equilibrium in the melt of polyethylene and their existence is not required to explain the X-ray diffraction data with its characteristic 4.5 Å halo.

R. W. Lenz and F. Ciardelli (eds.), Preparation and Properties of Stereoregular Polymers, 387–405.
Copyright © 1979 by D. Reidel Publishing Company.

CONFORMATIONAL ANALYSIS OF THE CHAIN OF SOME ISOTACTIC POLYMERS

In the crystalline state, the conformation of the chain of an isotactic vinyl polymer is generally helical and corresponds to a succession of nearly trans and nearly gauche internal rotation angles. Slightly different internal rotation angles and bond angles in the backbone give rise to different unit heights ($h = c/M$) and unit twists ($t = 2\pi N/M$ generally found between 3 and 4).

In the various polymers, the chain conformation is related to the bulkiness of the side groups in the vicinity of the backbone.

As we stated in the introduction, a good way to examine and foresee the conformations of the polymers in the crystalline state is provided by the methods of conformational analysis (energetic calculations). To enphasize this point of view, we will show in this paragraph a series of results of this kind of calculations.

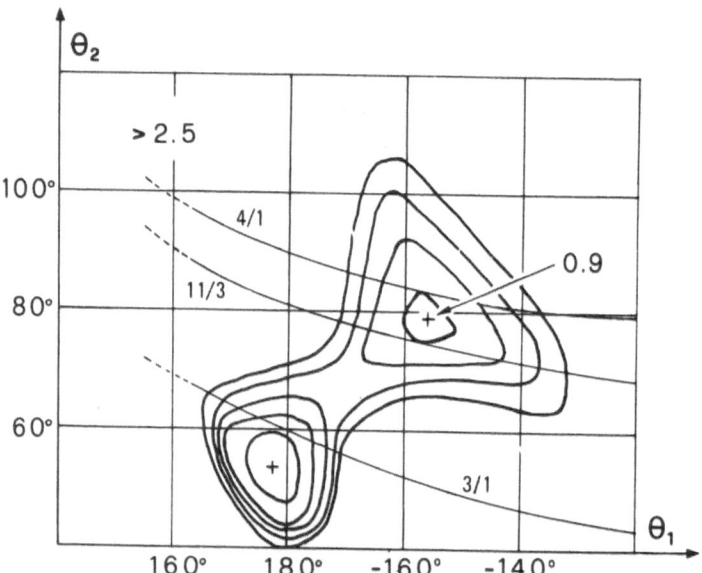

Fig. 1 – Internal energy of isotactic poly-α-butene as a function of the two internal rotation angles in the backbone θ_1 and θ_2. For each pair, both the bond angles and the internal rotation angles of the lateral group assume the values that minimize the internal energy. The curves are reported at intervals of 0.5 Kcal (mol of CRU)$^{-1}$. The open curves are the loci of points which correspond to the unit twist of the three conformations experimentally observed in the crystalline state.

We have performed recently some energetic calculations on the isolated chains of isotactic polypropylene (PP), poly-α-butene (PB), poly-3-methylbutene (P3MB), poly-(S)-3-methylpentene-1 (P(S)3MP) and polystyrene (PS), under the restriction of a periodic repetition and taking as variables both the internal rotation angles and the bond angles of the main chain and of the lateral group. The results (1) have been reported as maps of energy versus θ_1 and θ_2, the two internal rotation angles of the backbone that determine the helical conformation. The values of all the other parameters which have been varied in the calculation are those which minimize the energy for each pair θ_1, θ_2.

The calculations for poly-α-butene are reported, as an example, in fig. 1. The $E(\theta_1, \theta_2)$ map represented is that one relevant to the possible conformations of left-handed helices. Differently from the case of PP (above cited lecture) the energetic minimum is split into two. Correspondingly, chain polymorphism is experimentally observed for PB(2). The different crystalline modifications have $s(3/1)1$, $s(11/3)1$, $s(4/1)1$ chain conformations; the loci of points corresponding to such symmetries are also represented in the figure.

In the table 1, we compact some results of the calculations

Table 1 - A comparison between the values of the internal rotation angles and the bond angle on the CH_2 of the backbone corresponding to the minima of potential energy. The comparison is between the data obtained without any previous assumption and under the restriction that $h = h_{exp}$ and $t = t_{exp}$ (if they are known). The unreported parameters are practically the same. The ΔE values are in K_{cal}/mol of CRU.

Polymer	Parameters found without any previous assumption			Parameters found under the restriction $h = h_{exp}$, $t = t_{exp}$			ΔE
	θ_1	θ_2	τ_{CH_2}	θ_1	θ_2	τ_{CH_2}	
PP	176°	59.5°	116°	177°	62°	116°	0.05
PB	177.5°	54°	116°	177°	62°		
PB	204°	79°	113°	200° (form II)	75.5°	115°	0.20
				199.5° (form III)	83.5°	116°	0.30
P3MB	205°	78°	113°	208°	82.5°	113°	0.35
P3MB	175°	57.5°	116°	–	–	–	–
P(S)3MP	176° *	58° *	116°	–	–	–	–
P(S)3MP	205° *	77.5° *	113°	208°	82.5°	113°	0.40
P(S)3MP	-205°	-78°	113°	–	–	–	–
P(S)3MP	-174°	-59°	116°	–	–	–	–
P(S)3MP	205° **	79.5° **	113°	208°	82.5°	113°	0.30
P(S)3MP	177° **	53.5° **	116°	–	–	–	–
PS	176°	64°	114°	169°	66°	116°	0.50

* Minimum for $\vartheta_4 = 60°$ ** Minimum for $\vartheta_4 = 180°$ (see fig. 2)

performed on the polymers cited above. We compare in this table
the conformational parameters that we foresee without any previous assumption, with those obtained under the constraint that
the unit height $h = h_{exp}$ and the unit twist $t = t_{exp}$.

The accordance is good, as shown by the small differences
in energy between the two respective calculated conformations
(see the last column of the table). The ΔE values are representative of the order of magnitude of the extent to which intermolecular forces are able to modify the energetic minima calculated for the isolated chain.

Let us consider now the case of an optically active polymer
(P(S)3MP). The conformation of the chain is influenced not only
by the hindrance but also by the configuration of the lateral
group. As you can see from fig. 2, the conformations of minimum
potential energy are two for the left-handed helix and only one
for the right-handed helix. It is experimentally observed that
in the solid polymer, the conformation of the lateral group may
take statistically both conformations of minimum energy found
with these calculations, the helices being only of one sense
(left-handed). A requirement for good packing is that the two
conformations be represented in almost equal amounts in the cry
stalline lattice (3).

Since in vinyl polymers having optically active side groups
the optical activity may keep itself high also in solution, an
interesting problem is whether there may be a prevailing spiralization sense of the chains in the solutions of such polymers.
The problem has a positive answer and is tackled in the Appendix to this lecture.

Going back to the energetic calculations on isotactic vinyl
polymers, a point to be noted is the presence of two minima not
only for PB but also for P3MB and P(S)3MP, one corresponding to
the region of a threefold helix and the other to the region of a
fourfold helix. In the region of a threefold helix, the values
of the torsional energy are at a minimum, but the bulkiness of
the side groups forces the bond angles at the CH_2's to open up.
As a result, at the minimum of the nonbonded energies, the defor
mation of that angle is higher ($\sim$116°) than in the region of a
fourfold helix ($\sim$113°). (1)

The deformation of the bond angles assume a role which is
particularly important for some crowded polymers, such as poly-i-
sobutylene. For this polymer the chain conformation has $\tau_{CH_2} = 124°$
and corresponds to a succession of nearly trans and nearly gauche bonds (4).

Energetic calculations did not disclose the corresponding
energy minimum, when the bond angles at the CH_2 were taken as
fixed and nearer to tetrahedral.

Another case of a structure that was solved only through the
assumption of values of the bond angles at the CH_2 as large as
those found for polyisobutylene is provided by isotactic poly(me
thylmetacrylate) (PMMA). The energetic map of this polymer is

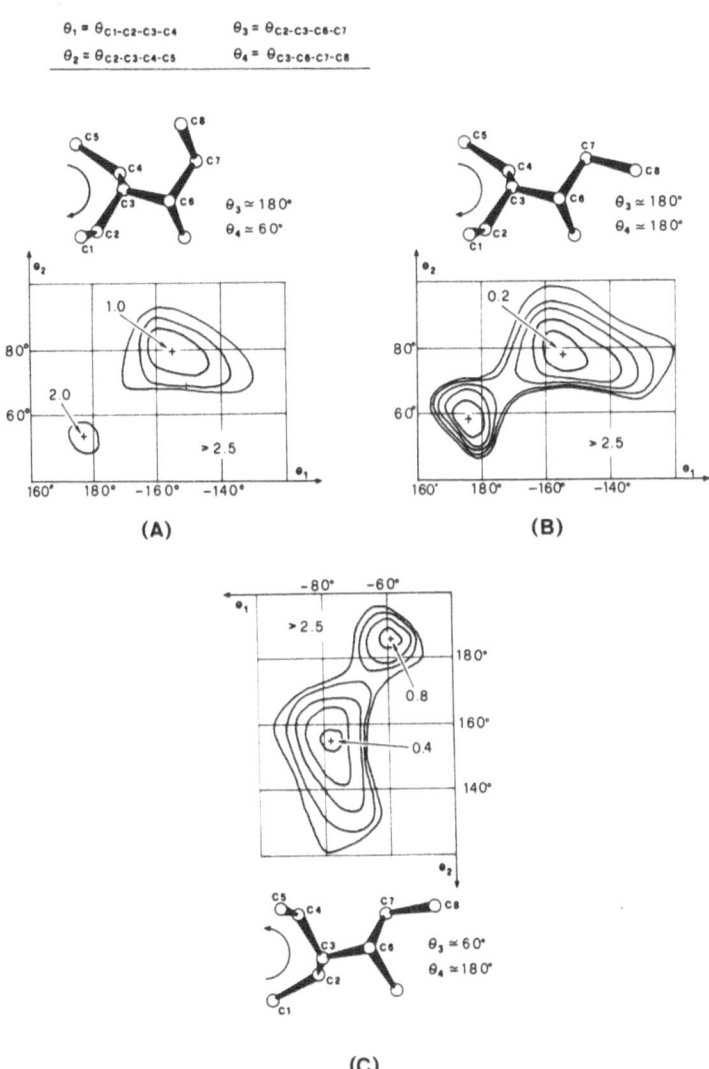

Fig. 2 - Internal potential energy of isotactic P(S)3MP. In the parts A and B are reported the regions of the multidimensional energy surface corresponding both to a left-handed helix for the two different conformations of the lateral group observed in the crystalline state. In the part C the minimum energy region corresponding to a right-handed helix is reported. The energies, referred to the absolute minimum (part B $\theta_1 \backsim 180°$ and $\theta_2 \thicksim 60°$), are in Kcal(mol of CRU)$^{-1}$.

shown in fig. 3. The bond angle at the CH_2 is taken as 122° (5).
The absolute minimum appears for the pair of internal rotation
angles of the backbone, that corresponds to a helix of large radius comprising ten configurational units in one pitch. The X-
-ray data indicates instead an identity period along the chain
axis c = 10.4 Å, the repetition comprising five configurational
repeating units.

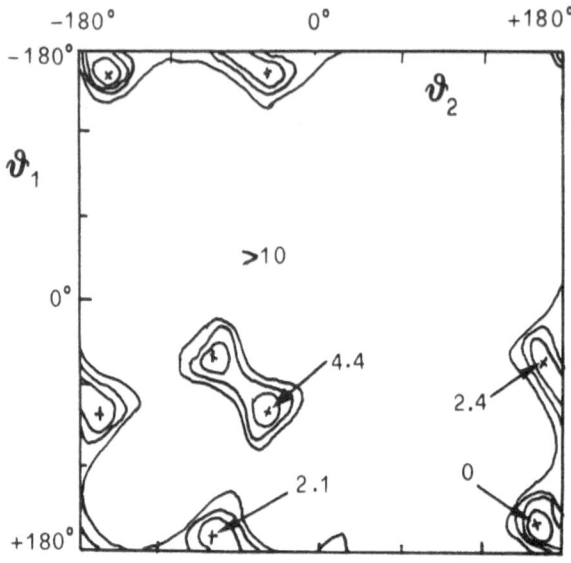

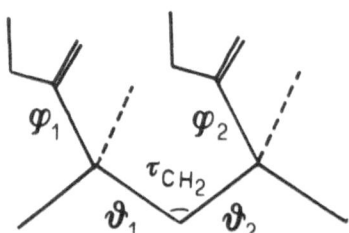

Fig. 3 - Potential energy map of the meso dyad of PMMA (for
φ_1 = 180° and φ_2 = 180° and τ_{CH_2} = 122°).
The energies are in Kcal/mol and relative to the lowest minimum, labeled 0.

The reconciliation of the contrasting results of the X-ray determination of the identity period and of the conformational analysis of a simple chain was brilliantly achieved by Tadokoro (6): the chains (10/1 helices) are interwinned in pairs in such a way that the identity period is half of that of a 10/1 helix (fig. 4).

It is very interesting to note that the 10/1 helix is locally nearly zig-zag planar. This conformation has been foreeseen by the conformational analysis only when an "unusual" bond angle at the CH$_2$ was considered, and proved to be energetically better than those wich less distorted angles.

A second evidence of a nearly zig-zag planar chain conformation for an isotactic polymer is now provided by the crystalline gels of isotactic polystyrene, first studied by Keller et al.(7).

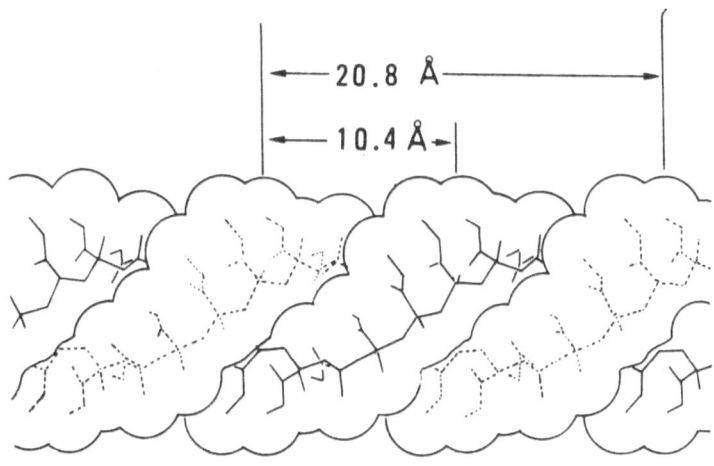

Fig. 4 - Model of the double helix of isotactic PMMA as given by Tadokoro et al. (6).

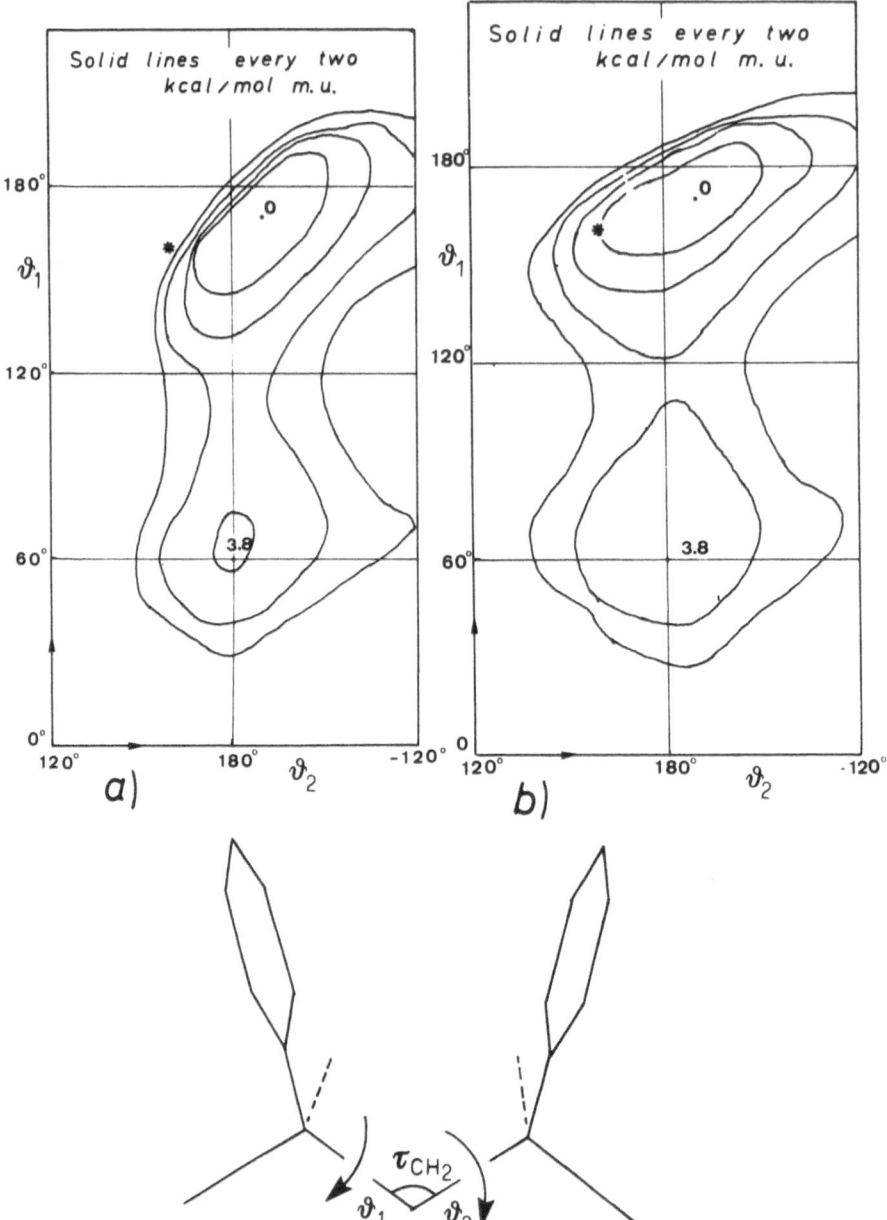

Fig. 5-Conformational energy map of the represented meso dyad of
PS. In the part a $\tau_{CH_2}=112°$ and the benzene rings are taken
in the staggered positions. In the part b $\tau_{CH_2}=116°$ and the
benzene rings are rotated 10° from the staggered positions.
The energies are in Kcal(mol of CRU)$^{-1}$.

Fig. 5a shows a portion of the conformational energy map of the indicated meso dyad of polystyrene.

The map is one of many calculated by us and corresponds to taking the angle at the CH_2 as 112° and the benzene rings staggered.

It may be seen that the minimum of energy in the region of the pair of internal rotation angles (170° +δ, -170° + δ, with /δ/<15°) is even lower than that one corresponding to the "classical" pairs 60°, 180° or 180°, -60° (this last one not appearing on the map).

However, it may be shown easily (as it was done for instance by Hägele (8)) that the repetition of pairs of internal rotation angle θ, -θ (in our case 170°, -170°) corresponds to repetition around a ring and not along a chain.

Let us now go back to the experimental X-ray observations on the crystalline gels of isotactic polystyrene: they indicate a helical repetition of structural units, with a corresponding unit height of 5.1 Å. It is difficult to think of a unit height of 5.1 Å far from a nearly planar zig-zag conformation.

If the structural unit which repeats along the helix comprises two monomeric units, it may be seen that an elongated chain, satisfying the requirement of having a 5.1 Å meridional reflection, is achieved by a repetition of the kind: ...170°+δ, -170°+δ, θ_3, θ_4....with θ_3 and θ_4 displaced from 180°, but with the same sign (fig. 6).

The availability of this region on the θ_1, θ_2 map of fig. 5a is not evident; however, slight adjustments of the bond angles and of the internal rotation angles which define the orientation of the benzene ring may permit previously unavailable pairs of in ternal rotation angles along the chain (f.i. compare the pair indicated with a star in fig. 5a and 5b).

While the exact solution of the chain structure of polystyrene in the crystalline gels would require more accurate energy minimizations in respect to all the internal coordinates (f.i. preliminary calculation show that the benzene rings are better energetically when not exactly staggered) and a comparison of calculated and observed X-ray data, the discussion above shows that a very elongated chain conformation can be built up for isotactic polystyrene.

Note that a structure such as the one above would be the first one, according to our knowledge, in which the helical repetition refers, in an isotactic vinyl polymer, to pairs of monomeric units.

CRYSTALLINE STRUCTURE WITH DISORDERED MACROMOLECULAR CONFORMATIONS

Different conformations of configurational repeating units may succeed each other along the chain of a polymer while some kind

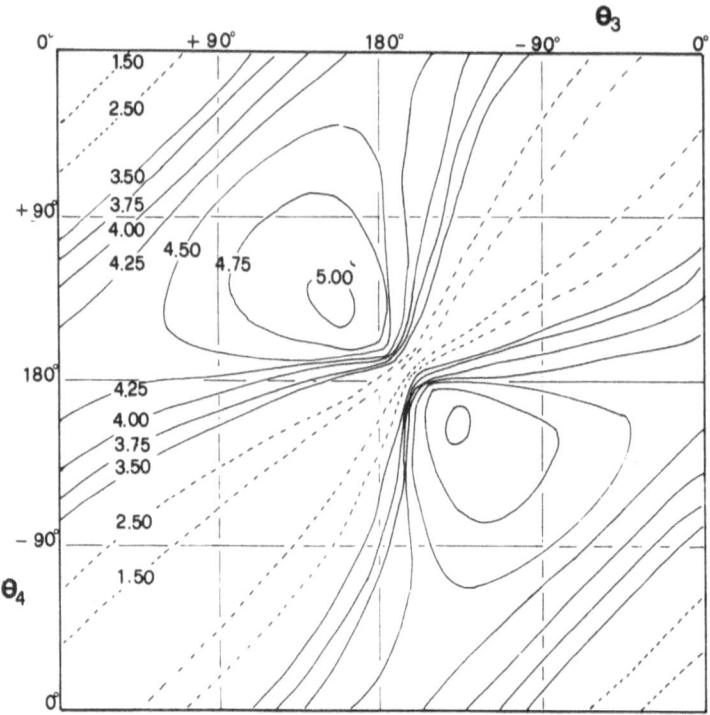

Fig. 6 – Unit heights, in Å, of helices of an isotactic polymer

The helical repetition, refers to two CRU's in the conformation 165°, -175° θ_3, θ_4. The map is not symmetrical
in respect to the line $\theta_3 = \theta_4$ because $\delta = -5° \neq 0°$.

of threedimensional order, and hence of crystallinity, is maintained. The crystallinity is, indeed, possible even if the long
range order is referred only to some of the atoms and even to
some feature only of the structure (9).

Here we shall discuss the two cases of polytetrafluoroethyle
ne (PTFE) and 1,4-cis polyisoprene, studied in our laboratory. In
both cases the main-chain atoms may assume in the same site of
the crystal different conformations, more or less at random.
A prerequisite for this occurrence is that the different conformations give rise to similar overall shapes of the macromole-

cules and the disordered succession maintains the chain axis
straight.

   In the case of the chain conformation of the high temperature modification of PTFE that we have lately studied (10), let
us confine our examination to a statistical succession of rotational isomeric states (see the above cited lecture) T_+, T, T_-
for which the internal rotation angles are 163.5°, 180°, 163.5°
respectively. The ordered repetition of T_+ bonds or T_- bonds
corresponds to the helix as found in the low temperature modification.

   We note that, if we do not permit that two successive bonds
are in the states T_+ T_- (or T_- T_+), the chain remains nearly unaltered in respect to the ordered one; moreover the chain is
maintained exactly straight (that is the fluorine atoms maintain
a fixed distance from an axis) with extremely small, energetical
ly unimportant, deformations of the internal rotation angles
(always lower than 5°) and of the bond angles (always lower than
0.4°) (fig. 7).

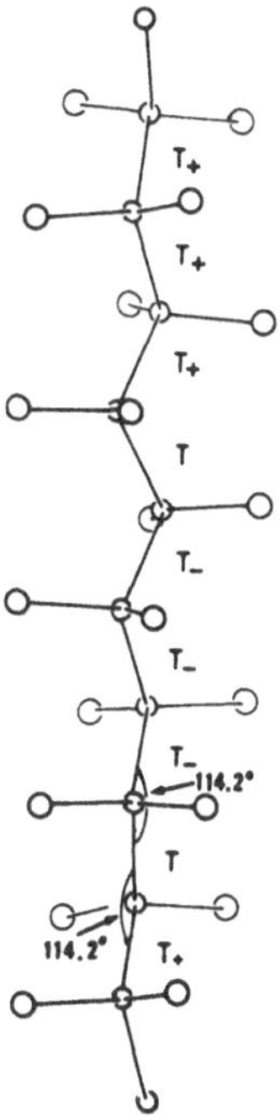

Fig. 7 – The chain axis of
the PTFE is maintained exactly straight with extremely
small deformations of the
bond angles at the inversion
in respect to the value observed in the low temperature form (C-Ĉ-C = 114.6°).
The succession of the dihedral angles at the inversion
is 163.5°, 167.8°, 180°,
−167.8°, −163.5.

   From the calculations of the energies for different sequences of internal rotation angles, we have obtained at 303 K for
an approximately straight chain axis, the matrix of statistical
weights (see the above cited lecture):

$$
\begin{array}{c|cc}
 & T_+(\text{or } T_-) & T \\
\hline
T_+(\text{or } T_-) & 1 & 0.22 \\
T & 1 & 0.68
\end{array}
$$

Through standard methods of statistical mechanics (11), it is possible to get the average lengths of the sequences in the three states (in the first approximation that the packing energy is indipendent of the number of inversions of the helical sense):

$$<y_{T_+}> = <y_{T_-}> \simeq 3 \qquad\qquad <y_T> \simeq 2$$

The X-ray diffraction data above 30°C are in fairly good agreement with the Fourier transform calculated according to this model.

These results confirm, hence, a model in which different senses of spiralization succeed each other frequently along the chain (for instance in fig. 7 the statistical succession of dihedral angles ... T_+ T T_- T_- T_- T T_+ T_+ T_+... is represented). But, in spite of the disorder, the threedimensional order is assured because the atoms are able to maintain a fixed distance from the chain axis, so that the fluorine atoms remain confined in cylindrical envelops having nearly the same size as in the ordered modification.

An analogous case of conformational disorder is provided by 1,4-cis polyisoprene:

$$
\begin{array}{ccccccccccc}
 & & & CH_3 & & & & & & CH_3 & \\
 & & & | & & & & & & | & \\
- CH_2 & - & CH & = & C & - CH_2 & - & CH_2 & - & CH & = & C & - CH_2 -
\end{array}
$$

The conformations of minimum internal energy for each triplet of single bonds are (12)

120°/ 180°/ -120° (A_+, T, A_-) or -120°/ 180°/ 120° (A_-, T, A_+)

The "disordered" chain conformation is built up by a random succession of such units; for instance:

....(A_+,T,A_-) cis (A_-,T,A_-) cis (A_-,T,A_+) cis (A_+,T,A_-)....

In the crystal, the structure of each chain may be thought of as a succession of the units represented in fig. 8.

With the symbol D are indicated units with "down" double bonds and with the symbol U units with "up" double bonds; the subscripts refer to the two possible orientations of the $CH_2 - CH_2$ bonds.

In the model proposed (13), after units of the D group may follow only units of the U group and "viceversa", moreover the second subscript of all the units in a sequence must be equal to the first subscript of the successive unit in order to match the orientation of the $CH_2 - CH_2$ bonds.

In such a way, whichever is the succession of the two possi-

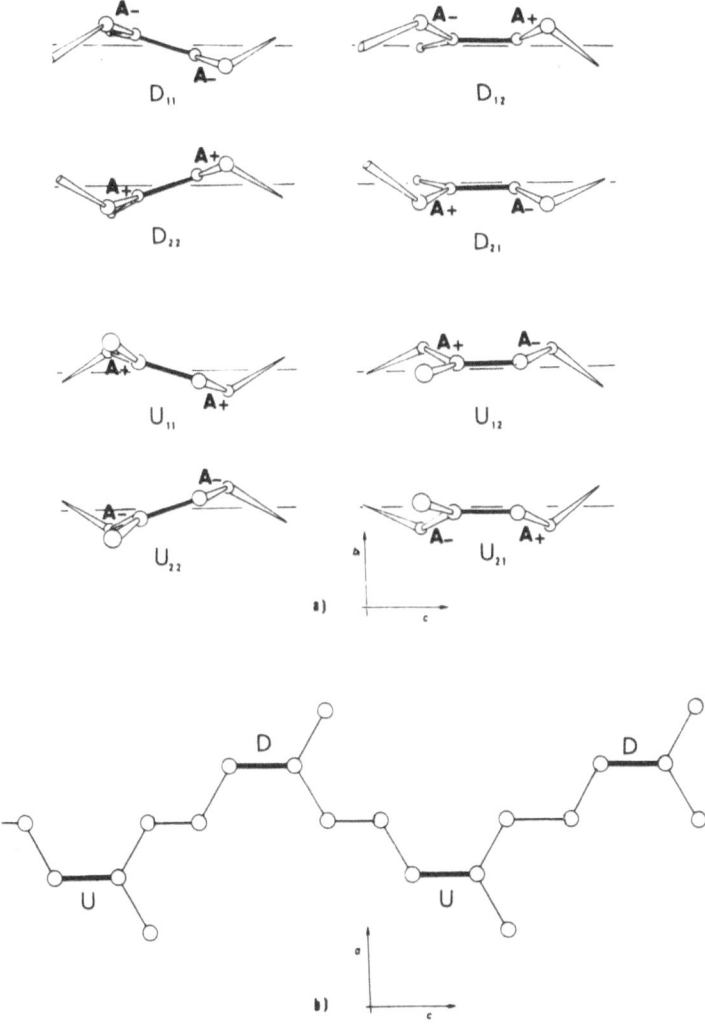

Fig. 8 - a) All the possible orientations in the crystal of the
units of 1,4-cis polyisoprene, when the single bonds a-
djacent to the double bond assume the conformations A_+
or A_-. The symbols D and U indicate whether the double
bond is "down" or "up" respectively; the subscripts refer to the two possible orientations of the CH_2-CH_2 bonds.
b) The overall shape of the chain of 1,4-cis-polyisopre-
ne in the crystallographic plane y if after units D_{ij}
may follow only units U_{j1} and analogously after units
U_{ij} may follow only units D_{j1}.

ble, above cited, conformations ($A_+ T A_-$ or $A_- T A_+$) the overall
shape of the chain remains unchanged in the ac plane of the crystal; the chain has the shape of a ribbon with hollows and bul-
ges provided by the methyl groups (Fig. 8b). As we can easily
see, the four atoms of the four units within each group (D or U)
have x and z coordinates identical (indeed the coordinates are
related by partial or total mirroring at the crystallographic
plane y = 1/8), and a threedimensional order is maintained.

Provided that small deformations of bond angles and slight
deviations of the internal rotation angles from A_+ and A_- are allowed, this model presenting a statistical sequence of conformations maintains also the chain axis straight.

MONTE-CARLO CALCUALTIONS ON "LIQUID" POLYETHYLENIC CHAINS

From the study of disordered chain conformations, while
threedimensional order of some feature of the structure is maintained, we went into the consideration of model building of the
"structure" of amorphous polymers.

The understanding of how the molecules of a polymer are organized in the amorphous state is one of the most exciting in
polymer science (14).

We started from the consideration that the X-ray diffraction spectrum of a liquid normal alkane is practically identical
to the X-ray spectrum of polyethylene in the melt (15).

Therefore, we started a treatment of normal alkane molecules (C_{10} at the beginning, and C_{30} more recently) by Monte Carlo
methods.

Common physical systems comprise a number of particles of
the order of 10^{23}; the treatment with a computer of systems of
the order of 10^3 particles already complicates notably the mat-
ters from the point of view of calculation times, and does not
avoid a series of inconveniences, substantially due to "surface"
effects.

In order to avoid such effects, we applied the widely applied device of considering a periodic structure, which is "liquid like" only locally.

It consists in considering a cell (fundamental cell) of volume V and containing N atoms (with N of the order of 10^3) sur-
rounded on all sides by identical cells (image-cells) with the
same number of particles, which reproduce identically the configuration of the fundamental cell. When an atom is moved in the
fundamental cell, the corresponding atoms of the "image cells"
are moved concertedly.

This permits the elimination of "surface" effects.

A bidimensional example is shown in figure 9.

We started the movements both from ordered and disordered
starting dispositions of the n-alkane molecules.

In a typical example, a cell having dimensions of 30 Å is

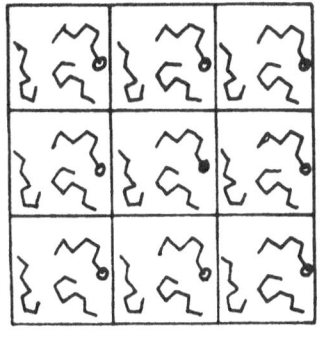 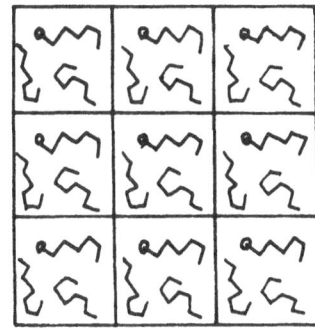

PART (A)   PART (B)

Fig. 9 - Bidimensional model of a "periodic" assembly of disorde-
         red hydrocarbons molecules (A). Part B illustrates the ef
         fect of a movement: the atom encircled in (A), at the
         head of a molecule, has been cancelled, and substituted
         by the atom encircled in (B) at the tail of the same mo-
         lecule.

filled with the carbon atoms of C_{30} molecules up to the experimental density.

        A Taylor potential is assumed for torsions and a Mason and Kreevoy
potential is assumed for the non-bonded interactions of CH_2 groups.
Bond angles and bond lengths are assumed to be constant.

        The strategy to move the molecules from their starting po-
sition is to add at chance an atom to the tail of a molecule if
appropriate according to the statistical weight; and if so, to
cancel an atom from the head of the molecule.

        We go on, starting with strategies which permit easier move-
ments at the very beginning, until we reach a sort of equilibrium
state. After a number of movements of the order of 10^5 there is
no more any recollection of the starting model, but to be sure we
continued many times as much with the movements. We felt then
that the molecules could be considered as being in their equilibrium arrangement, subject to the potential used.

        All the coordinates of the carbon atoms are then available
for calculation of properties.

        Some of the results are as follows:

1) The distribution of TG, GT, TT and GG pairs is in accordance
   with the distribution provided by the rotational isomeric model, as applied to isolated molecules (1);

2) The molecular size is of the right order of magnitude
   ($<r^2>/nl^2 = 5.1$, $<s^2>/nl^2 = 0.71$) for a random coil

conformation (according to Flory's rotational isomeric state
scheme, ($<r^2>/nl^2 = 5.6$ $<s^2> /nl^2 = 0.75$);

3) The calculated X-ray spectrum is not far from the experimental one (Fig. 10) for polyethylene (note however that the
halo at $2\frac{sen\vartheta}{\lambda} = 4.5$ Å is somewhat sharper in this second
case);

4) The orientation function between "meta" vectors (connecting
C_i with C_1 in any $C_i - C_k \backsim C_1$ moyety) of different molecules, as a functions of the distance between C_k's even at distances as low as 5 Å does not show any evident indication
of correlation of orientations of neighbouring molecules.

   Thus, from our model calculations, it seems that, contrary
to previous objections to the random coil model of liquid hydrocarbons, the correct density may be achieved by Monte-Carlo fil-
ling of the space and the X-ray amorphous halo (-which many of
us considered as an evidence for the presence of bundles of
chain molecules) may be fairly well reproduced in intensity, while no evidence of chain parallelism is present.

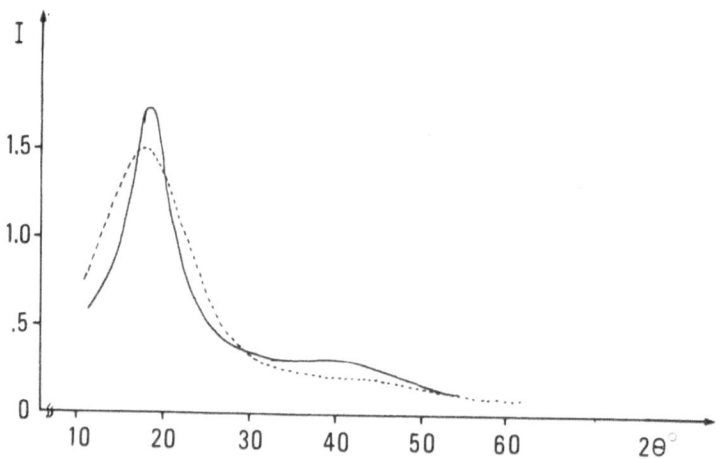

Fig. 10 - X-ray intensity calculated from the Monte-Carlo model
          (dashed line) compared with the experimental intensi-
          ty (heavy line) of polyethylene (Cuk_α).

APPENDIX

   Starting from the simplified model of the chain of isotactic
polypropylene in solution or in the melt reported in the above

cited lecture, we have to consider that the probability of occurrence of (TG_+) or (G_-T) conformations of bonds for each single
monomeric unit may not be any more the same if the side groups
are chiral (see fig. 2 as an example) (16).

Let us consider a model, for which there are m conformations
accessible to the side groups with energies $e'_1, e'_2 \ldots e'_m$, for
the chain conformation $(TG_+)_n$ (left-handed helix) and q different conformations accessible to the side groups, with energies
$e''_1, e''_2, \ldots e''_q$, for the chain conformation $(G_-T)_n$ (right-handed helix).

So, it is possible to define a ratio ε between the internal
partition functions, relative to the conformational states of the
side groups in a left-handed and in a right-handed helix:

$$\varepsilon = \frac{\sum_{i=1}^{m} exp \ (-e'_i/RT)}{\sum_{j=1}^{q} exp \ (-e''_j/RT)}$$

This ratio ε will be in general different from unity; it will approach unity only when the distance of the optically active carbon atom from the main chain is high.

To derive the frequency of occurrence of (TG_+) units and the
frequency of occurrence of (G_-T) units, we can apply again standard methods of statistical mechanics (11). If the same simplifications implied in the previous treatment of polypropylene can
be applied, we can write the matrix of statistical weights:

$$
\begin{array}{c c}
 & \begin{array}{c c} G_-T & TG_+ \end{array} \\
\begin{array}{c} G_-T \\ TG_+ \end{array} & \left| \begin{array}{c c} \varepsilon & 1 \\ Z\omega & 1 \end{array} \right|
\end{array}
$$

with $\omega = exp \ (-E_\omega/RT)$.

The largest eigenvalue of this matrix is

$$\lambda_1 = \tfrac{1}{2} \ (1 + \varepsilon + \sqrt{(1 - \varepsilon)^2 + 4Z\omega} \ )$$

and the elements of the relative eigenvector and eigenrow are as
follows:

$$A_{21} = (\lambda_1 - 1) \ A_{11}; \ B_{12} = \frac{(\lambda_1-1) \ B_{11}}{Z\omega}$$

$$A_{11}B_{11} + B_{12}A_{21} = 1$$

For sufficiently long chains the frequency of occurrence of a sta
te η is

$$f_\eta = A_{\eta 1} B_{1\eta}$$

hence we obtain:

$$f_1 = f(TG_+) = A_{11}\,B_{11} = 1 - A_{21}\,B_{12} = \frac{Z\omega}{Z\omega + (\lambda_1 - 1)}$$

and

$$f_2 = f(G_-T) = 1 - f(TG_+) = \frac{(\lambda_1-1)^2}{Z\omega + (\lambda_1-1)^2}$$

In the fig. 11 we report the values of the percent excess of units spiralized in the sense (TG_+) with respect to those spiralized in the sense (G_-T):

$$p = f(TG_+) - f(G_-T) = \frac{Z\omega - (\lambda_1 - 1)^2}{Z\omega + (\lambda_1 - 1)^2}$$

for the simplified model considered above, with $\omega = 0.015$, $(E_\omega/RT = 4.2)$ and $Z = 2$ as in polypropylene when the ratio ε varies between 0.4 and 1.

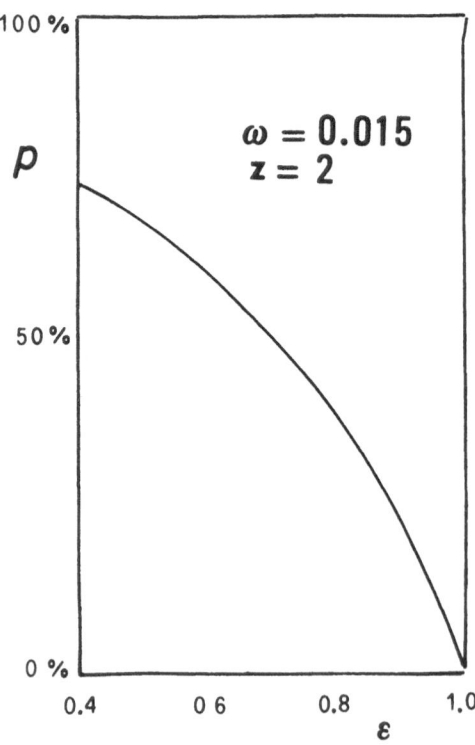

$\omega = 0.015$
$z = 2$

Fig. 11 - The values of the percent excess of u-nits spiralized in one sen se, with respect to those spiralized in the other sense, (p) are plotted a-gainst ε if $Z = 2$, $\omega = 0.015$. (See text)

REFERENCES

1. P. Corradini, V. Petraccone and B. Pirozzi
   Europ. Polym. J., 12, 831 (1976).

2. V. Petraccone, B. Pirozzi, A. Frasci and P. Corradini
   Europ. Polym. J., 12, 323 (1976).

3. V. Petraccone, P. Ganis, P. Corradini and G. Montagnoli
   Europ. Polym. J., 8, 99 (1972).

4. G. Allegra, E. Benedetti and C. Pedone
   Macromolecules, 3, 727 (1970).

5. P.R. Sundarajan, P.J. Flory
   J. Am. Chem. Soc. 96, 5025 (1974).

6. H. Hiroshi, H. Tadokoro and Y. Chatani
   Macromolecules, 9, 531 (1976).

7. E.D.T. Atkins, D.H. Isaac, A. Keller and K. Miyasaka
   J. Polym. Sci.: Physics Ed., 15, 211 (1977).

8. L. Beck and P.C. Hägele
   Colloid & Polym. Sci., 254, 228 (1976).

9. P. Corradini
   J. Polym. Sci., Symp. N. 51, 1 (1975).

10. P. Corradini and G. Guerra
    Macromolecules, 10, 1410 (1977).

11. P.J. Flory
    "Statistical Mechanics of Chain Molecules" Interscience,
    New York, 1969.

12. J.E. Mark
    J. Am. Chem. Soc. 88, 4354 (1966); ibid. 89, 6829 (1967).

13. E. Benedetti, P. Corradini and C. Pedone
    Europ. Polym. J., 11, 585 (1975).

14. P. Corradini
    J. Polym. Sci., Symp. N. 50, 327 (1975).

15. M. Vacatello, A. Tuzi, G. Avitabile and P. Corradini
    Rend. Acc. Sci., Napoli, Vol. 46 (in press).

16. G. Allegra, P. Corradini, P. Ganis
    Die Makr. Chemie 90, 60 (1966).

NEW INSIGHTS AND UNSOLVED PROBLEMS IN THE CRYSTALLINE STRUCTURE OF POLYESTERS

F. Brisse, R. H. Marchessault and S. Pérez

Department of Chemistry, Universite de Montreal
C. P. 6210, Succ. A., Montreal, Canada

INTRODUCTION

In the early distinction by Carrothers (1) between condensation and addition polymers, polyesters were uniformly
considered as examples of the former. However it was soon
appreciated that a large family of polymers could be made which
were functionally "polyesters" but did not fall into the synthesis
based definition of condensation polymers. This family of
polymers are usually prepared by ionic routes, generally a
ring opening polymerization, and are structurally different
from the condensation polyesters by having a chain sense i.e.
the crystalline chain conformation possesses an intrinsic dipole
moment. Accordingly, a structurally based definition of polyesters in terms of "polar and non-polar" crystalline conformations will be used in this report. The non-polar molecules
considered will be aromatic polyesters and the polar molecules
will be polyalkanoates.

This study reviews experiments at the Universite de Montréal
over the past decade. The use of x-ray fiber diffraction coupled
with computer model building, based on minimum energy considerations, has allowed substantial progress in understanding
the crystal structures of polyesters. Of fundamental importance
in such analyses are certain a priori structural assumptions:
a) absolute certainty of the chemical structure and geometry
b) narrow limits on the torsional angles available to
the ester group.
The latter feature is a theorem which has been thoroughly considered by molecular physicists (2).

R. W. Lenz and F. Ciardelli (eds.), Preparation and Properties of Stereoregular Polymers, 407–430.

The implications of this theorem are in keeping with the
experimental observations (3) on crystalline esters which prefer
the planar trans conformation. This means, according to the
schematics of ester dihedral angles

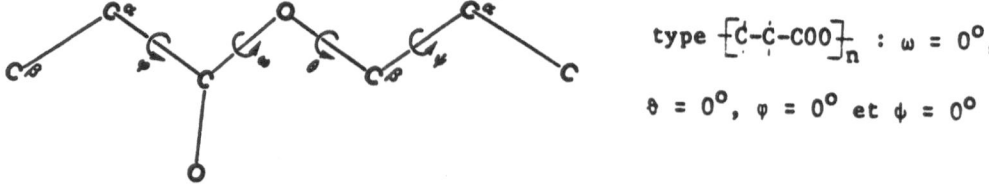

type $\left[\dot{C}-\dot{C}-COO\right]_n$: $\omega = 0°$,

$\vartheta = 0°$, $\varphi = 0°$ et $\psi = 0°$

that angle ω, which has partial double bond character, can
depart only slightly (a few degrees) from the planar trans
position before a large internal energy is added to the molecular
conformational state. In a non-dynamic situation, such as
occurs in the crystalline state, symmetry and energy considerations
will greatly favor those states where $\omega \to o$, that is the "trans"
conformation of the ester group(4).

CONFORMATION AND CRYSTALLINE STRUCTURE OF AROMATIC POLYESTERS

Although polymer diffractionists refer to small molecule(5,6)
structures (determined with high reliability) and use standard
bond angles and bond lengths in their structure determinations,
it would appear that this approach is not rigorous enough.
In the case of poly (oligomethylene terephthalates) the small
energy differences between trans and gauche conformations in
the oligomethylene part makes the deliberate study of small
model compounds essential. With the knowledge of the single
crystal structures of such model compounds the correct polymer
structure is found with much greater reliability.

Examples of this approach in the field of biopolymers
are numerous. The structural analysis of proteins, polynucleotides
and polysaccharides has been immensely aided by this approach(4,5,6).
The model compound, especially the use of crystalline oligomers,
has been less prevalent in the area of the crystallography
of synthetic polymers even though the scope for synthesis of
suitable models is very great. An example of the model compound approach with vinyl polymers is found in the work of
Allegra et al (7) on the crystal structure of 2,2,4,4, tetra-
methyl adipic acid as a preamble to a study of the conformation
of polyisobutylene (8). This important crystallographic model
study showed that the valence angle along the vinyl chain was
a function of the substituents and it was deduced that in polyisobutylene the valence angles along the backbone were alternately
$110°$ and $124°$.

The polycondensation of terephthalic acid: $HOOC$-⟨○⟩-$COOH$, with a linear diol: $HO-(CH_2)_x-OH$, yields the aromatic polyesters of the type: $-O-(CH_2)_x-O-CO$-⟨○⟩-$CO-_n$. The first member of the series, poly(ethyleneterephthalate). $(x = 2$, abbreviated as PET or 2GT*) was discovered in 1946 by Whinfield & Dickson (9). This polymer, with a very high melting point ($265°C$) and a good resistance to hydrolysis, has since been widely studied because of its commercial applications as a fiber (10)**. Recently, another polymer of the series, poly(tetramethylene terephthalate) = poly (butylene terephthalate) = 4GT, has been used as a thermoplastic resin (11)

When this work was undertaken, only the crystalline structure of 2GT had been reported. Daubeny, Bunn & Brown (12), showed that this polyester belongs to the triclinic space group P1 and that there is only one chemical repeat unit per unit cell. The chain is in a planar trans conformation although some distortions had to be introduced to account for the 0.15 Å shortening of the crystallographic fiber repeat as compared to the length of a fully extended chain. In the crystal, the terephthaloyl groups of adjacent chains are face to face (and side by side). Although these structural features were later confirmed (13-15), no explanation was offered as to the short value of the observed fiber repat.

Apart from the academic interest that the answer to such a question could have, this series of aromatic polyesters is well suited to the development of a new approach to the structure determination of a crystalline polymer. Some of the advantages of this series are listed below:

1- The poly(x-methylene terephthalates) prepared by melt-phase polymerization under reduced pressure (16) are readily available. These compounds are made up of deformable x-methylene segments alternated with rigid terephthaloyl residues. The only degrees of freedom are within the methylenic section of the polymer.

2- There are no hydrogen-bonds in these polyesters. Consequently, only van der Waals interactions keep the chains together.

*

xGT is the abbreviation used in this paper to describe poly (x-methylene terephthalates).

**

This polyester has many trade names, among which: Terylene, Dacron, Kodel, Fortrel, etc....

3- The fibers of these polyesters are well crystallized
   and show good orientation so that a large number of
   diffraction spots (40 to 100) can be recorded on an
   X-ray fiber diagram. Furthermore, small lamellar
   single crystals of 2GT have been produced (17,18)
   from dilute solution making it possible to investigate
   these polyesters by electron diffraction as well as
   by X-ray diffraction.

Principle of the Model Compound Approach

The polymeric chain is symbolically cut into a number of
fragments called model molecules or model compounds. To best
reproduce the polymer, one model molecule must have a number
of atoms overlapping a similar grouping in the following fragment.
In the case of the aromatic polyesters, the rigid aromatic
ring is the obvious choice for a common group of atoms. In
this manner dimethyl terephthalate is the model compound for
the terephthalic part of poly(ethylene terephthalate) and other
poly(x-methylene terephthalates), while oligomethylene glycol
dibenzoate molecules, ⬡-CO-O-(CH$_2$)$_x$-O-CO-⬡ , are model
compounds for the x-methylenic part of the polyesters. Figure 1
represents such partitioning applied to poly(ethylene terephthalate).

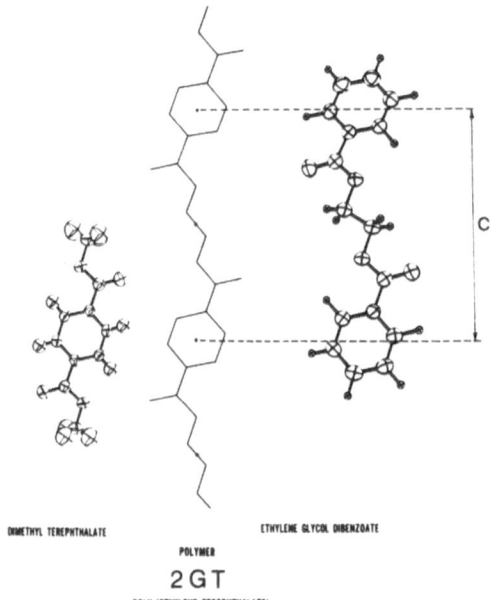

DIMETHYL TEREPHTHALATE   ETHYLENE GLYCOL DIBENZOATE

POLYMER

2 G T

POLY (ETHYLENE TEREPHTHALATE)

Figure 1. The two model compounds used to describe the chain
of poly(ethylene terephthalate).

Since it is common to all poly(x-methylene terephthalates), the structure of dimethyl terephthalate was determined first (19). The comparison of this molecule (Figure 2) with others which also contain a terephthaloyl segment reveals that there are oscillations, of small amplitude, between the carboxylic group and the aromatic ring.

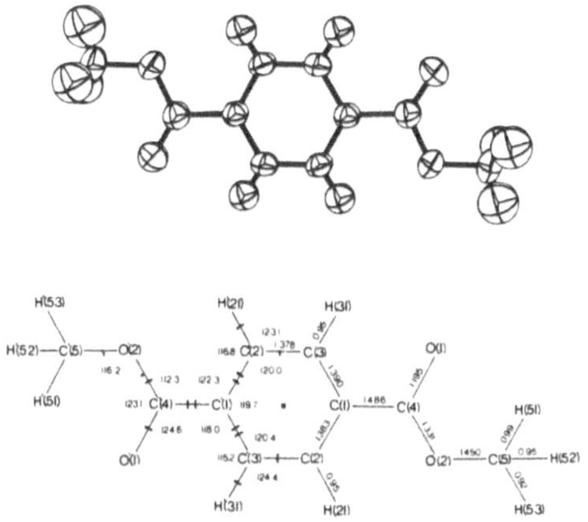

Figure 2.   The structure of dimethyl terephthalate.

Obviously it is in the study of the series of model compounds describing the methylene section of these polyesters that the important information relevant to the polymer structure will be found.

Poly(ethylene terephthalate) and Its Model Compounds

The crystal and molecular structure of ethylene glycol dibenzoate and ethylene glycol di-*para* chlorobenzoate have been determined (20,21) and are shown in Figure 3a and 3b respectively.

(a)   (b)

Figure 3.   (a) The structure of ethylene glycol dibenzoate
            (b) The structure of ethylene glycol di-*para* chloro-
                benzoate.

The interatomic bond distances and angles of these two molecules are shown in Figures 4a and 4b below.

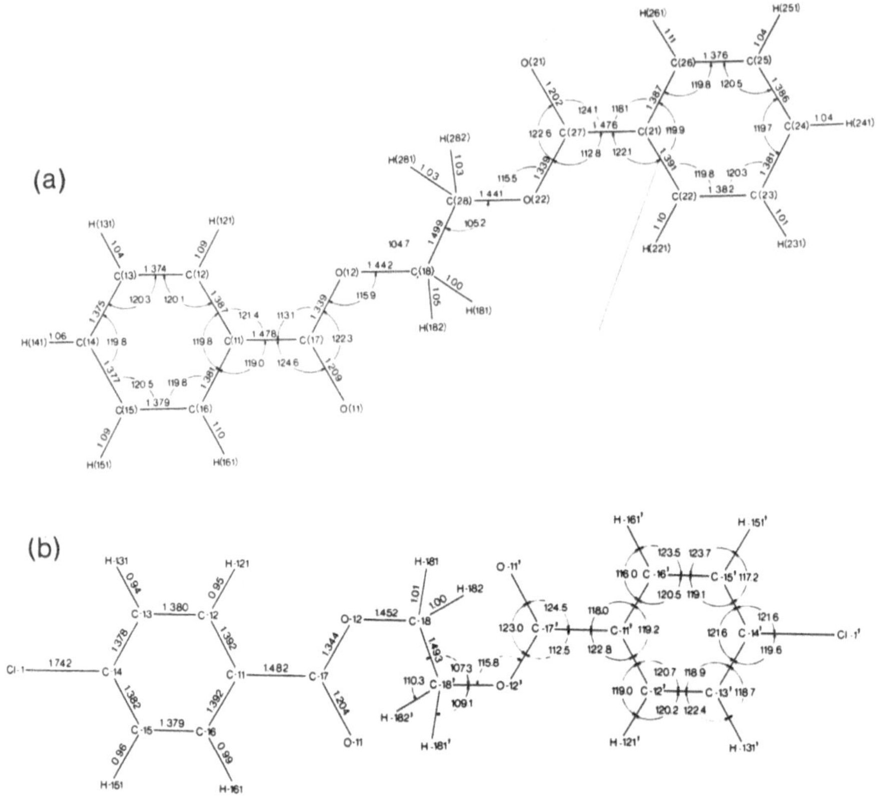

Figure 4. Bond distances and angles.
(a) Ethylene glycol dibenzoate
(b) Ethylene glycol di-*para* chlorobenzoate.

The geometry of the methylenic sequence ($-O-CH_2-CH_2-O-$) departs notably from the "standard" geometry. For instance, the CH_2-CH_2 distances (1.499(2) and 1.493(2)Å) are significantly shorter than the accepted value of 1.537Å for $Csp^3 - Csp^3$ type bonds (22), furthermore the $O-CH_2-CH_2$ bond angles are not as open, by about $5°$, as expected.

From the conformational point of view these two molecules behave in a similar way at the junction of the aromatic and carboxylic groups. However there are some drastic differences along the methylenic sequence of atoms. The conformation in the $CO-O-CH_2-CH_2-O-CO$ sequence is *trans-trans-trans* (175.6, -172.8, -176.1°) for ethylene glycol dibenzoate, while it is *trans-gauche-trans* (-175.0, 74.5, -175.0°) for the *para* substituted analog.

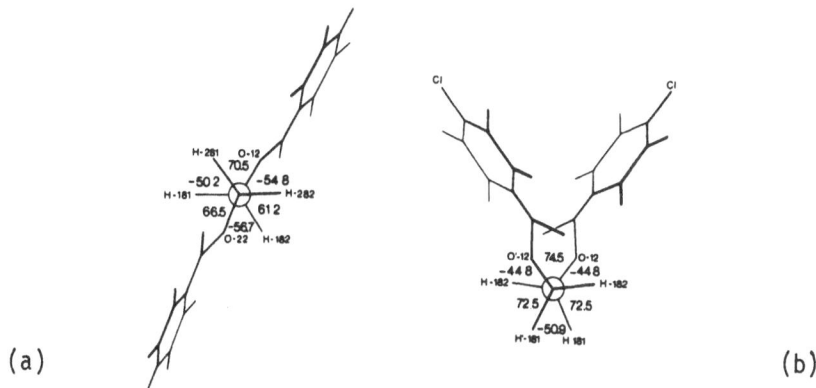

Figure 5.   Newman projection along the CH_2-CH_2 bond.
             (a) Ethylene glycol dibenzoate
             (b) Ethylene glycol di-*para* chlorobenzoate.

A conformational analysis has been undertaken in order to understand the surprising change in conformation of two molecules
differing only by the existence of a chlorine atom in *para* at
the terminal phenyl ring.

     The non-bonded interaction energy, the electrostatic energy
and the torsion energy were included in this analysis. Furthermore, a two-fold rotation axis was assumed in the middle of the
CH_2-CH_2 bond so that two torsion angles, ϕ_1 and ϕ_2, were sufficient
to describe the methylenic part of the molecule. The iso-energy
map (Figure 6) and Table 1 summarize this calculation.

Table 1

Minimum Energy Conformations of Ethylene Glycol Dibenzoate

Conformation #	ϕ_1	ϕ_2	$E(kcal\ mol^{-1})$
I	±100	∓ 60	0.0
II	180	180	0.5
III	±177.5	∓ 62.5	0.6
IV	± 80	± 55	0.7
V	± 80	±175	1.1

Observed Conformations of Model Compounds

	ϕ_1	ϕ_2
Ethylene glycol dibenzoate	-176.1 175.6	-172.8
Ethylene glycol di-*para* chlorobenzoate	-175.0 -175.0	74.5

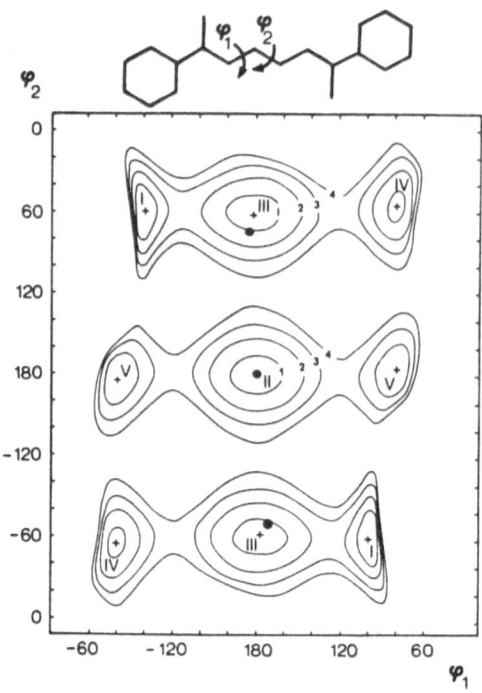

Figure 6.   Iso-energy contours for ethylene glycol dibenzoate
             as a function of ϕ_1 and ϕ_2. The crosses are
             minimum energy positions while the dots are
             experimental conformations from crystal structure
             analysis.

The two conformations observed in the solid state are among
the low energy conformations arrived at by conformational analysis
and there is a difference of only 0.1 kcal mol^{-1} between the
trans-gauche-trans and the *trans-trans-trans* conformations. This
explains the existence in two different conformations of molecules
chemically very similar. The calculation also reveals that the
differences between the observed (model compounds) and calculated
torsion angles are more important for ϕ_2, the torsion angle around
CH_2-CH_2.

    The *trans-trans-trans* conformation in ethylene glycol diben-
zoate corresponds to the planar zig-zag conformation established
for 2GT (**20**). The equivalent of the fiber repeat has been evaluated
using the geometry of the ethylene glycol dibenzoate molecule. This
is the distance between the centers of the phenyl rings of the
molecule. It is found that this distance, 10.75Å, is identical to
the actual fibre repeat c of 2GT (see Figure 1).

In conclusion, the ethylene glycol dibenzoate molecule is
a good model for 2GT since it not only reproduces the trans-trans-
trans conformation of the methylenic part, but it also has an equivalent fiber repeat which exactly matches the observed fiber repeat
of the 2GT polyester. This feature is arrived at with no distortion
of the molecule. The geometrical and conformational data can be
transferred as a whole to the polymer.

Model Compounds and the Structure of Poly(trimethylene terephthalate)

In the following paragraph, the model compound approach is
applied to the structure determination of an unknown polyester,
that of poly(trimethylene terephthalate) = 3GT.

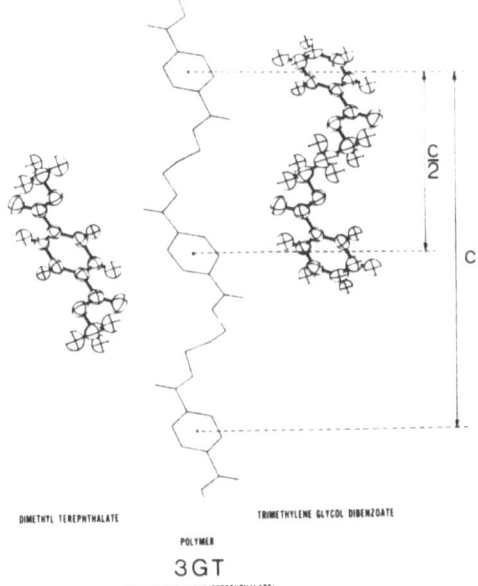

DIMETHYL TEREPHTHALATE   TRIMETHYLENE GLYCOL DIBENZOATE

POLYMER
3GT
POLY (TRIMETHYLENE TEREPHTHALATE)

Figure 7.   The 3GT chain and its model compounds. The latter are
            shown in their crystalline conformations from which
            the trial model of the chain in the crystal was proposed,
            as shown.

The crystal structure of trimethylene glycol dibenzoate and
trimethylene glycol di-para chlorobenzoate, both model compounds
for the trimethylene part of 3GT, have been determined by conventional X-ray technique (23,24). The two molecules are shown in
Figures 8a and b, while their bond distances and angles are presented in Figures 9a and b.

(a)   (b)

Figure 8.   (a) Trimethylene glycol dibenzoate
             (b) Trimethylene glycol di-*para* chlorobenzoate

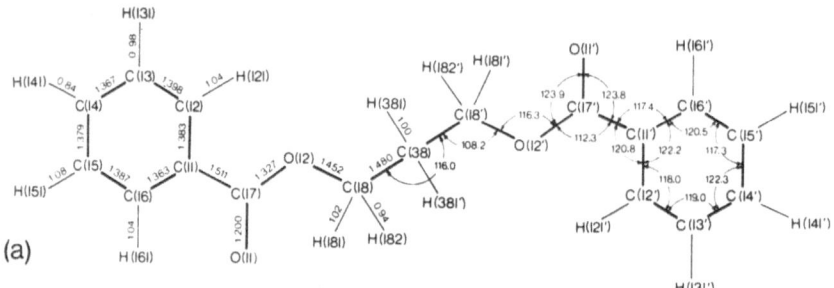

(a)

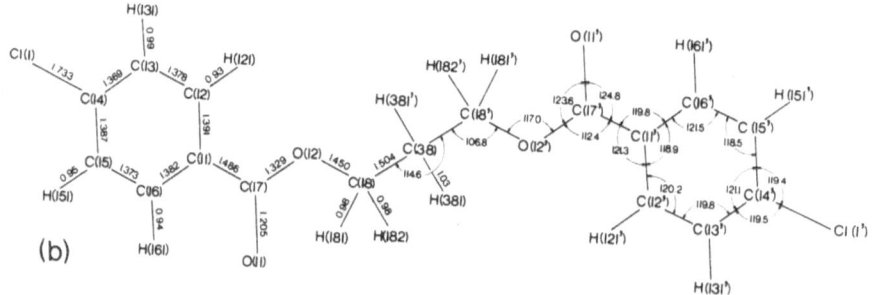

(b)

Figure 9.   Bond distances and angles
             (a) Trimethylene glycol dibenzoate
             (b) Trimethylene glycol di-*para* chlorobenzoate

The CH$_2$-CH$_2$ shortening observed in ethylene glycol dibenzoate
is also present here. The CH$_2$-CH$_2$ bond distances are 1.480(7) Å in
trimethylene glycol dibenzoate and 1.504(3) Å in the *para* chloro
substituted analog. The O-CH$_2$-CH$_2$ angles are also less open than
expected. However, there are no conformational differences
between the two molecules. The conformation of the trimethylene
CO-O-CH$_2$-CH$_2$-CH$_2$-O-CO sequence is *trans-gauche-gauche-trans*. The
average equivalent advance per monomer calculated from both

molecules in their conformation is 9.57 Å. This value is fairly close to the early reports of 9.1 Å (25,26).

X-ray diffracted intensities of 3GT were obtained from a polymeric fiber (Figure 10). Lamellar single crystals of 3GT could be prepared upon slow cooling from a nitrobenzene solution of the polyester. A view of a 3GT single crystal and its electron diffraction pattern are shown in Figure 11.

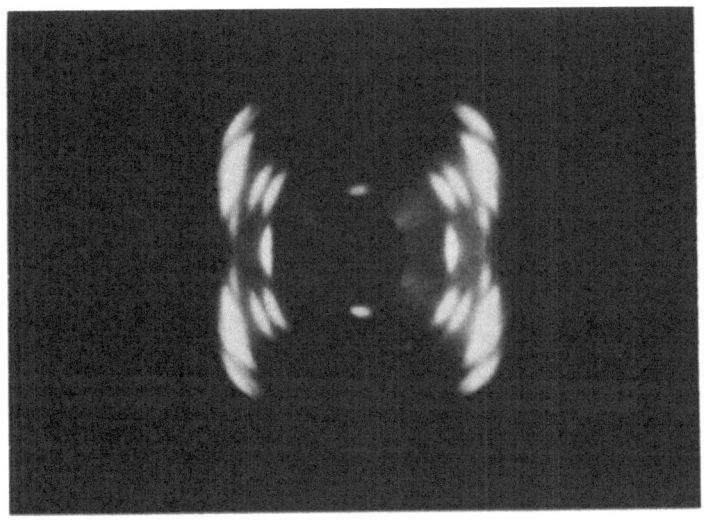

Figure 10.   X-ray fiber diagram of 3GT.

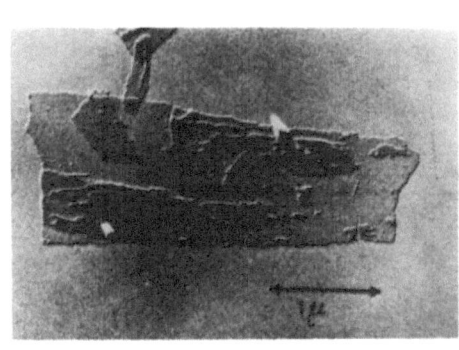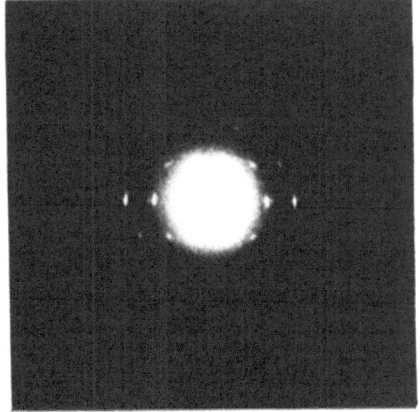

Figure 11.   Lamellar single crystal of 3GT (left) and its
             electron diffraction pattern (right).

3GT has a unit-cell of dimensions a = 4.637 b = 6.266
c = 18.64 Å (fiber repeat), α = 98.4 β = 93.0 γ = 111.1°
and the space group is PĪ. This, combined with the observed
density, reveals that the triclinic unit cell contains two
chemical repeat units which follow each other along c. (Figure 7).
As it happens the fiber repeat of 3GT, c = 18.64 Å, is quite
similar to twice the advance per monomer estimated from the model
compounds in their *trans-gauche-gauche-trans* conformation.

A conformational analysis calculation (Figure 12, Table 2)
finds the model compound conformation among the low energy ones.

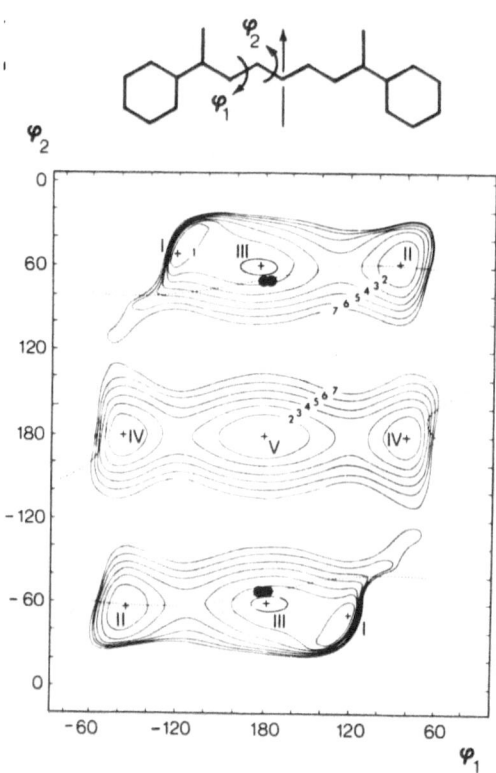

Figure 12.   Iso-energy contours for the trimethylene section in
             trimethylene glycol dibenzoate and 3GT. The crosses
             are minimum energy positions while the dots are
             experimental conformations from crystal structure
             analysis.

TABLE 2

Minimum Energy Conformations of Trimethylene Glycol Dibenzoate

Conformation #	ϕ_1	ϕ_2	$E(\text{kcal mol}^{-1})$
I	±125	∓ 50	0.0
II	± 80	± 55	0.9
III	±175	± 60	1.1
IV	± 80	180	1.5
V	180	180	1.7

Observed Conformations of Model Compounds

	ϕ_1	ϕ_2
Trimethylene glycol dibenzoate	±176.8	∓66.8
Ethylene glycol di-*para* chlorobenzoate	±176.7	±66.3

The remaining steps of the structure determination were done while deliberately keeping the macromolecular chain rigid. The chain, built from the model compounds data, was oriented within its unit cell using Williams' PACK5 program (27). For the best orientation, i.e., when the interactions between adjacent chains are minimal, the crystallographic R index = $\Sigma\Delta F/\Sigma Fo$ computed from all the observed X-ray data reached 0.255. As far as the electron diffraction data was concerned, among the 25 observed reflections, two were very strong and the remaining ones so weak that only a qualitative Fo-Fc agreement could be obtained.

The packing of 3GT is compared in Figure 13 with that of its model compounds. The packing of 3GT is similar to that of 2GT, in the sense that the terephthaloyl groups are facing each other and that there is lateral stabilisation by the H atoms from the aromatic rings. In the case of the model compounds, only in trimethylene glycol dibenzoate are the aromatic rings stacked, while the chlorine atoms play a dominant role in trimethylene glycol di-*para* chlorobenzoate.

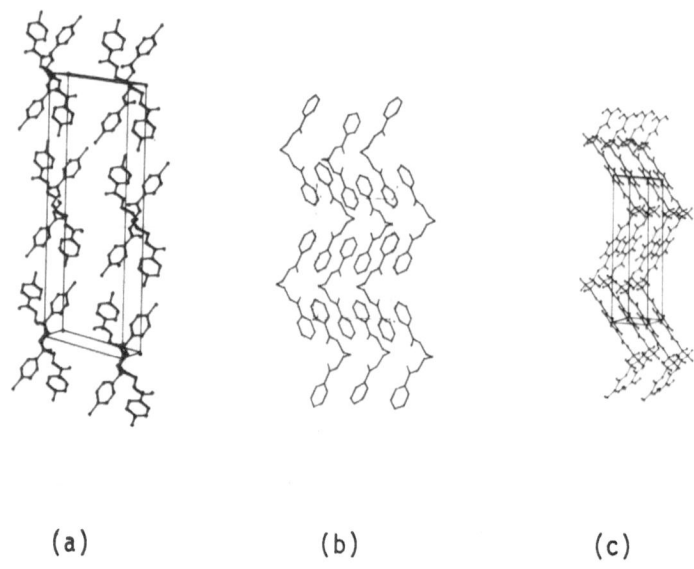

(a)   (b)   (c)

Figure 13.   Packing
             (a) Trimethylene glycol di-*para* chlorobenzoate
             (b) Trimethylene glycol dibenzoate
             (c) poly(trimethylene terephthalate)

CONCLUSIONS

It has been shown that, through the use of model compounds, it was possible to obtain precise geometrical data as well as the possible conformations of a macromolecular chain.

The crystal structure of hexamethylene glycol dibenzoate has been determined (28) in order to extend this methodology to poly (hexamethylene terephthalate) = 6GT. So far, it has been possible to confirm that the hexamethylene section of 6GT has the fully extended planar zig-zag conformation.

This approach is also being extended to the case of polymers with interchain hydrogen bonds. Work on poly(ethylene terephathal-amide) and its model compound, N,N'ethylenebisbenzamide, is now underway.

The use of polymer single crystal electron diffraction has been very helpful, in the determination of the polymer's unit cell dimensions and to confirm the correctness of the proposed model.

CONFORMATION AND CRYSTALLINE STRUCTURE OF POLYALKANOATES

The crystal structures of polyesters based on α-hydroxy and β-hydroxyacids have been extensively studied since the realization that a family of high melting polymers was achievable based on these structures (29,30). By suitable substitution of the monomers, optically active polymers can be prepared and the deliberate adjustment of the relative optical antipode content and distribution leads to steric copolymers which can present all the characteristics of isotactic, syndiotactic and stereoblock structures now familiar in the poly-α-olefin series.

Table 3 is a list of the important structures in the polar polyester family which have been extensively studied. The crystal structures display a variety of chain conformations although the planar zig-zag and 2_1 structures predominate.

ALPHA HYDROXY POLYALKANOATES

Two members of the series: poly (glycollide) and poly (lactide), (cf. Table 3) have proven to be biodegradable man-made polymers (31,32). Indeed, the long sought after goal of an absorbable synthetic suture has been achieved in a com-mercial (31) oriented fiber based on poly (glycollide). The strength retention and in vivo absorbability of this material are exactly matched to the needs of the surgeon. By analogy with the long used collagen sutures it might have been expected that the optically active poly (S-lactide) would be superior as an absorbable suture. The matter hinges on whether a hydrolytic or enzymatic process is involved in the in vivo absorption. This field, along with the whole domain of biodegradability of polyester copolymers involving poly (lactic acid), is still in a state of development (32).

One interesting α-hydroxy polyester whose crystal structure is still unknown is poly(tetramethyl-glycollide) (33,34). This polymer is reported to exist in two crystalline conformations (33) and melting points between 190-200°C have been reported (34). Using the computer model building approach (35) where angle ω is set at zero the other two angles are systematically varied leading to the isoenergy and superposed helical parameter map shown in Fig. 14. Because of the symmetry of the monomer the map is symmetric about the n = 2 locus (where n is the number of monomers per helix turn). Two stable conformations (each one either right or left handed) are predicted (35):

TABLE 3

Conformation of Polyesters Possessing Chain "Sense"

Polymer	Formula	Fiber Period	Conformation	Melting Pt. °C
Polyglycollide	${CH_2COO}_n$	7.02	zigzag (2/1)	227
Poly-S-lactide	${CH-COO}_n$ CH_3	27.8	10/3	160
Poly-β-propriolactone	${CH_2CH_2COO}_n$	4.82 7.02	zigzag 2/1 helical	122
Poly-R-β-hydroxybutyrate	${*CH-CH_2-COO}_n$ CH_3	5.96	2/1 helical	184
Poly-pivalolactone	${CH_2-C-COO}_n$ CH_3 CH_3	4.74 5.97	zigzag 2/1 helical	238
Poly-ε-caprolactone	${(CH_2)_5-COO}_n$	17.30	zigzag (2/1)	58
Poly-ω-hydroxydecanoate	${(CH_2)_9-COO}_n$	27.1	zigzag	80

-        one$_o$helix is 7/2 (i.e. n = 3.5) with an advance per monomer,
h, of 2.5Å i.e. a fiber repeat of 17.5Å,

-        the other is also a 7/2 helix but h = 1.5Å   i.e. a fiber repeat
of 10.5Å.   The two possible chain conformations are shown in Fig. 15.
The former is fairly close to that proposed for poly(S-lactide).
The latter is energetically more favorable and has helical
parameters which are very similar to that of the polypeptide
alpha helix. As can be judged from Fig. 15 the two models have
a distinctly different carbonyl orientation: either perpendicular or parallel to the helix axis. Unlike the poly(S-lactide)
where only one helix handedness is stable, both right and left-
handed helices are equally stable in this calculation. This
is evident from the symmetry in Fig. 14.

BETA HYDROXY ALKANOATES

In the β-hydroxy series of polyalkanoates, poly-β-propiolactone first revealed the "alpha" to "beta" reversible transition
in the polyalkanoates. Two distinct melting points corresponding
to crystalline polymorphs with a helical (alpha) and an extended
(beta) conformation were found (36). This discovery was followed
by a report (37) that a family of symmetric α,α' substituted
β-hydroxy polyesters existed whose melting points were in the
vicinity of 240°C. Polypivalolactone is the simplest example
of this family and is also found in several crystalline forms:
helical and one planar zig-zag (38,39).

When α,α' substitution is unsymmetric and the catalyst
produces an optically neutral polymer, a drastic drop in melting
point is observed as a function of the ratio of the number
of carbons in the substituents (37). However since the materials
studied must have been nearly atactic the true crystalline
melting point of these materials must be considered unknown.
Their crystalline conformation is like that of polypivalolactone
as regards the backbone atoms in the 2_1 helical conformation.
A syndiotactic structure i.e. optical antipode alternation
along the chain is reported for at least one case (40).

When the substituent is in the β position one encounters
a family of naturally occuring optically active polyesters
of which poly-β-hydroxybutyrate (PHB) is the most common.
This material is a reserve substance in bacteria and in the
natural state is a 0.5 μm diameter crystalline granule. The
polyester biosynthesis is coupled with the dynamics of the
"log phase" growth of bacterial systems (41). With the onset
of sporulation, a rapid depolymization to monomer and turnover
occurs. A full understanding of the nascent morphology of
native granules PHB has been achieved (42).

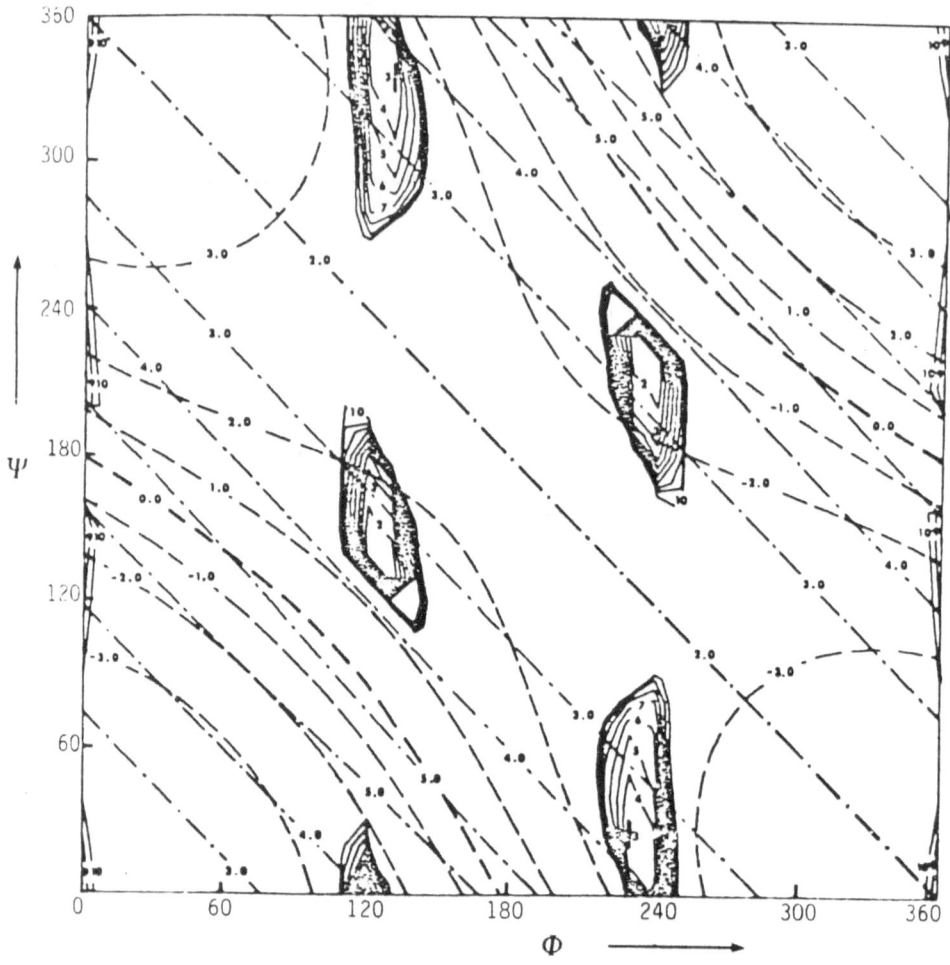

Fig. 14.   Isoenergy contour map for poly(tetramethyl-glycollide) with
            ω = 0°. The plot shows conformational angles Ψ and Φ with
            energy contours (full lines) in kcal/mole. The helical
            parameters: n (number of monomers per turn) -.- and h (advance
            per monomer along the helix axis) ---, are superimposed.

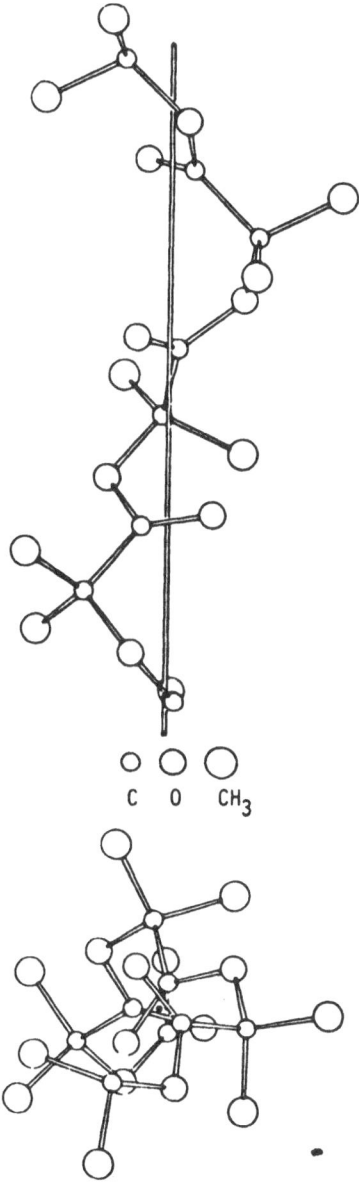

C   O   CH₃

Fig. 15A. Proposed 7/2 crystalline conformation for poly(tetra-methyl-glycollide) with fiber repeat of 17.5A.

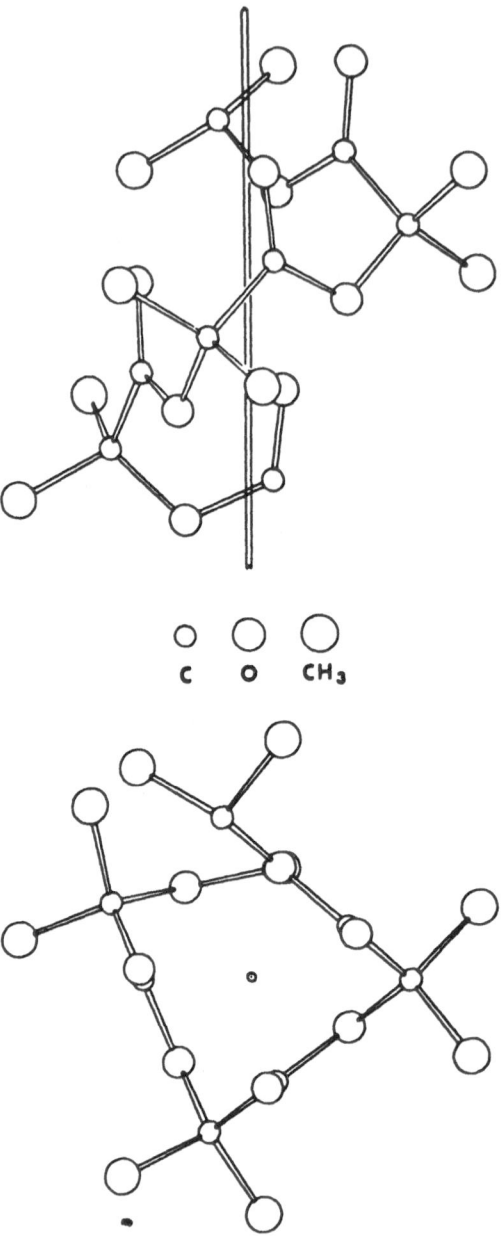

Fig. 15B.   Proposed 7/2 crystalline conformation for
            poly(tetramethyl-glycollide) with fiber
            repeat of 10.5Å.

A second β-substituted polyester: poly-β-hydroxy pentanoate (PHP) has been extracted from sewage (43). The polyester is probably similar to PHB with respect to bacterial physiology. However the polymer has its own characteristic melting point and crystalline structure (44).

Both PHB and PHP have been prepared synthetically (44,45), the former in optically active form. The x-ray crystal structure of the native PHB and that of the synthetic non-optically active material are reported to be the same (46). This result is difficult to explain except in terms of a stereoblock structure in the synthetic polymer and a segregation of crystallites involving right and left handed helices.

The polyalkanoates based on α,α' disubstituted and simple β-substituted poly (β-propiolactone) form a family of isohelical structures with a fiber repeat of about 6A. These polyesters all display the same twofold screw symmetry because of similar polar and non-bonded intramolecular interactions which dominate the structure independently of the substituents, this phenomenon has been termed conformational isomorphism (47). The transition from helical (alpha) to extended (beta) structures under the influence of tension is a general feature of the α,α' disubstituted systems but not of the β substituted ones (38, 48).

CONCLUSIONS

Because of the polar nature of the chains of polyalkanoates they pack in the crystal with antiparallel chain orientation. This favors crystallization of the folded chain lamellar type (49). Furthermore the extended chain-helical transformation (alpha → beta) offers a unique energy absorbing mechanism which can lead to tough thermoplastic materials. The possibility of grafting reactions on elastomeric backbones has shown how these polyesters can transform ordinary elastomers into thermoplastic elastomers (50, 51).

Finally the biodegradability of these polyesters is far from having been exploited. For example block-copolymers incorporating the glycollide or lactide monomers offer the possibility of easily synthesized and processed materials. The mechanical properties of such systems is only beginning to be examined. The biodegradation or chemical hydrolysis of poly(β-hydroxybutyrate) can provide a large-scale source of optically active monomer (52) for studies of this kind.

REFERENCES

1. W. H. Carothers (1940) High Polymers vol. 1, ed. by H. Mark and G. S. Whitby, Interscience Publishers, New York.

2. T. Ooi, R. A. Scott, G. Vanderkoor and H. A. Scheraga (1967), J. Chem. Phys. 46, 4410.

3. J. Cornibert, N. V. Hien, F. Brisse and R. H. Marchessault (1974) Can. J. Chem., 52, 3742.

4. D. A. Brant, A. E. Tonelli and P. J. Flory (1969) Macromolecules, 2, 228.

5. S. Perez and F. Brisse (1978) Biopolymers, 17, 2083.

6. S. Perez and R. H. Marchessault (1978) Carbo. Res. 65, 114-120.

7. E. Benedetti, C. Pedrone and G. Allegra (1970) Macromolecules 3, 727.

8. Go Wasai, Takeo Saegusa and Junji Furukawa (1965) Makromol. Chem. 86, 1-8.

9. J. R. Whinfield and J. T. Dickson (1946) British Patent No 578079.

10. A. F. Brown and K. A. Reinhart (1971) Science, 173, 287-293.

11. M. P. Van Der Wielen (1975) Polymer Eng. Sci. 15, 102-106.

12. R. de P. Daubeny, C. W. Bunn and C. J. Brown (1954) Proc. Roy. Soc., A226, 531-542.

13. S. Arnott and A. J. Wonacott (1966) Polymer 7, 157-166.

14. Y. Y. Tomashpol'skii and G. S. Markova (1964) Polym. Sci. USSR., 6, 316-324.

15. J. S. Tse and T. C. W. Mak (1975) J. Cryst. Mol. Struct., 5, 75-80.

16. J. G. Smith, C. J. Kibler and B. J. Sublett (1966) J. Polym. Sci., A1, 1851-1859.

17. Y. Yamashita (1965) J. Polym. Sci., A3, 81-92.

18. M. Hachiboshi, T. Fukuda and S. Kobayashi (1969) J. Macromol. Sci., B3, 525-555.

19. F. Brisse and S. Perez (1976) Acta Cryst., B32, 2110-2115.

20. S. Perez and F. Brisse (1976) Acta Cryst., B32, 470-474.

21. S. Perez and F. Brisse (1975) Canad. J. Chem., 53, 3551-3556.

22. L. E. Sutton (1965) Tables of Interatomic Distances and Configuration in Molecules and Ions. Supplement 1956-1959. London: The Chemical Society.

23. S. Perez and F. Brisse (1977) Acta Cryst., B33, 3259-3262.

24. S. Perez and F. Brisse (1976) Acta Cryst., B32, 1518-1521.

25. J. Goodman (1962) Angew. Chem., 74, 606-612.

26. R. Jakeways, I. M. Ward, M. A. Wilding, I. H. Hall, I. J. Desborough and M. B. Pass (1975) J. Polym. Sci., (Polym. Physics Ed.), 13, 799-813.

27. D. E. Williams (1969) Acta Cryst., A25, 464-470.

28. S. Perez and F. Brisse (1977) Acta Cryst., B33, 1673-1677.

29. R. Thiebont, N. Fischer, Y. Etienne and J. Coste (1962) Industries des Plastiques Modernes 14, 1

30. Y. Chatani, K. Suchiro, Y. Okita, H. Tadokoro and K. Chujo (1968) Makrom. Chem., 113, 215.

31. E. Schmitt, U. S. Patent 3,736,646

32. Richard D. Sinclair, U. S. Patent 4,057,537

33. D. G. H. Ballard and B. J. Tighe (1967) J. Chem. Soc. (B), 702.

34. T. Alderson, U. S. Patent 2,811,511

35. J. Cornibert, PhD Thesis, Chemistry Dept., University of Montreal (1972).

36. G. Wasai, T. Saegusa and J. Furukawa (1964) J. Ind. Chem. Soc. Japan, 67, 601.

37. Y. Etienne and R. Soulas (1963) J. Polymer Sci. (C), 4, 1061.

38. F. W. Knobloch and W. O. Statton, U. S. Patent 3,299,171.

39. R. Prud'homme and R. H. Marchessault (1964) Makrom. Chem. 175, 2705.

40. R. H. Marchessault, J. St. Pierre, M. Duval and S. Perez, (1978) Macromolecules 11, 1281.

41. R. Alper, D. G. Lundgren, R. H. Marchessault and W. A. Cote (1963) Biopolymers, 1, 545.

42. D. Ellar, D. G. Lundgren, K. Okamura and R. H. Marchessault (1968) J. Mol. Biol., 35, 489.

43. Lowell Wallen (1974) Environmental Sci. and Tech. 8, 576-579.

44. M. Yokouchi, Y. Chatani, H. Tadokoro and H. Tani (1974) Polymer Journal 6, 248.

45. J. R. Shelton, J. B. Lando and D. E. Agostini (1971) J. Polymer Sci. (A1), 9, 2789.

46. D. E. Agostini, J. B. Lando and J. R. Shelton (1971) J. Polymer Sci., 9, 2775.

47. J. Cornibert and R. H. Marchessault (1975) Macromolecules 8, 296.

48. J. Cornibert, R. H. Marchessault, A. E. Allegrezza and R. W. Lenz (1973) Macromolecules, 6, 676.

49. R. Prud'homme and R. H. Marchessault (1974) Macromolecules, 7, 541.

50. S. A. Sundet, R. C. Thamm, J. M. Meyer, W. N. Buck, S. W. Cuzwood (1976) Macromolecules, 9, 371.

51. R. H. Marchessault, P. Noe, S. Perez, J. St. Pierre, R. Jorgensen J. Polymer Sci., Physics (in press).

52. S. Coulombe, P. Schauwecker, R. H. Marchessault and B. Hauttecoeur (1978) Macromolecules 11, 279.

PRESENT UNDERSTANDING OF THE VISCOELASTIC AND MECHANICAL PROPER-
TIES OF STEREOREGULAR POLYMERS

W. J. MacKnight

Polymer Science and Engineering Department
Materials Research Laboratory
University of Massachusetts
Amherst, Massachusetts 01003

INTRODUCTION

In the previous chapter a description of the behavior of
the glass transition temperature (Tg) of vinyl and vinylidene
polymers was presented and a theory (1) was put forward rationalizing the dependence of Tg on configuration in the case of
vinylidene polymers. Here we review thermal, dynamic mechanical, and dielectric relaxation data for a series of poly(alkyl
α chloroacrylates) of varying configurations (2,3,4). These
data are further interpreted in the light of the theory mentioned above and the results compared with the corresponding
poly(alkylmethacrylates).

In order to determine the effect of tacticity on properties it is necessary to prepare polymers of varied stereochemical structures ranging from the highly syndiotactic to the
highly isotactic. Further, it is necessary to characterize the
various stereoisomers both with regard to tacticity and to
stereosequence distribution. These requirements have never been
completely met. However, the poly(alkyl α chloroacrylates)
studied in these laboratories probably represent as close an
approximation to the requirements as has yet been attained. It
was possible to synthesize stereoregular poly(α chloro methyl
acrylates), poly(α chloro ethyl acrylates), and poly(α chloro
isopropyl acrylates) (3), characterize them by NMR (5), and
study their transition and relaxation behavior (4). Early work
aimed at the synthesis of isotactic poly(alkyl α chloroacrylates) by the use of homogeneous anionic systems or Grignard reagent initiators invariably resulted in products with $\bar{M}_n$ <
10,000 (6). In most instances these polymers would not form

431

R. W. Lenz and F. Ciardelli (eds.), Preparation and Properties of Stereoregular Polymers, 431–447.
Copyright © 1979 by D. Reidel Publishing Company.

films suitable for mechanical properties studies, and their Tg's
were low, necessitating a long extrapolation to obtain limiting
values. It was only possible to obtain reasonably high molecular weight products by the use of a heterogeneous catalyst
formed by the 1,4 addition reaction of a Grignard reagent to an
unsaturated ketone (7).

EXPERIMENTAL

The preparation of the stereoregular poly(alkyl α chloroacrylates) used in this study has been described in detail previously (3). In summary, the highly syndiotactic polymers were
obtained by a free radical, UV initiated polymerization at -50°
or lower, the "conventional" polymers were obtained by free radical polymerization initiated by benzoyl peroxide at -70°, and
the highly isotactic polymers were obtained by use of the Grignard reagent complex as initiator. The polymers were characterized primarily by the use of 300 MHz proton NMR (5). Relevant
data for the samples used for property studies are collected in
Table I.

Table I
Characterization Data For Stereoregular Poly(α Chloroacrylates)

Polymer	Fraction m Dyads(a)	$\overline{M}_n$(b)(g/mole)
Poly(methyl α chloroacrylate)	0.70	80,000
" "	0.29	270,000
" "	0.25	400,000
Poly(ethyl α chloroacrylates)	0.73	309,000
" "	0.29	362,000
" "	0.20	181,000
Poly(isopropyl α chloroacrylates)	0.95	50,000
" "	0.64	50,000
" "	0.36	77,300

(a) From 300 MHZ NMR (5)
(b) From membrane osmometry (3)

Films were prepared for mechanical and dielectric testing
by casting 4% - 12% (w/v) solutions of the polymers in $CHCl_3$ onto clear plate glass and allowing the solvent to evaporate slowly.
This procedure produced films of 4-8 mils thickness which were
then dried in a vacuum oven at 60°C. for one to two days. After
this treatment, it was found that the residual solvent, estimated to be less than 0.5% by weight by pyrolysis gas chromatography caused a decrease in Tg of up to 50°C. in the case of
the syndiotactic poly(methyl α chloroacrylates) and about 25° to
30° C. in the case of the syndiotactic poly(ethyl α chloroacrylates) and poly(isopropyl α chloroacrylates). This residual solvent could not be removed by heating in a vacuum oven, and it

was found necessary to extract the films with methanol at ambient temperature for several days followed by a similar extraction with water. It is postulated that the large effect on Tg was due to some sort of complex formation between the polymer and the $CHCl_3$. Further studies are necessary to elucidate the nature of this complex.

The mechanical measurements were carried out on a Rheovibron Dynamic Viscoelastometer, Model DDVII. The frequencies employed were 3.5 H_z, 11 H_z, and 110 H_z and the temperature range was from -100°C. to +170°C. depending on the Tg of the polymer under observation. Tan δ, E' and E" were obtained at the quoted frequencies as functions of temperature by standard techniques.

Dielectric measurements were carried out with a General Radio capacitance bridge, type 1620-A, in conjunction with a Balabaugh Lab. three terminal cell, type LD-3. The dielectric loss tangent and the real and imaginary parts of the dielectric constant, ε' and $\varepsilon"$ were obtained as functions of temperature at 6 frequencies: 200 H_z, 500 H_z, 1 kH_z, 2 kH_z, 5 kH_z, and 10 kH_z. The temperature range investigated was from room temperature to +190°C.

Differential Scanning Calorimetry (DSC) measurements were carried out using a Perkin Elmer instrument, model DSC-1B. The scanning rate was 20°C./minute in all cases and calibrations for determining heats of fusion were accomplished using naphthalene, adipic acid, or benzoic acid for the appropriate temperature ranges.

RESULTS AND DISCUSSION

Figure 1 presents the dyad tacticity determined from 300 MH_z NMR (5) versus the Tg measured by DSC for the poly(alkyl α chloroacrylates) studied. All values are corrected to "infinite" molecular weight as follows. The molecular weight dependence of Tg for a linear polymer is given by the Flory-Fox relationship

$$Tg = Tg\infty - K/\overline{M}_n \qquad\qquad (1)$$

where Tg is the observed value for a polymer of a given molecular weight and Tg∞ is the limiting value. The constant, K, is typically in the range of 10^5 for poly(styrene), poly(methyl methacrylate), and many other non-polar polymers. Thus for $\overline{M}_n >$ 10^5, the Tg's of such polymers are essentially independent of molecular weight. In the case of stereoregular poly(methyl methacrylates), it has been stated that K depends on tacticity, varying from about 1.0 x 10^5 for 100% isotactic dyads, to 4.0 x 10^5 for 100% syndiotactic dyads (8). The results of the

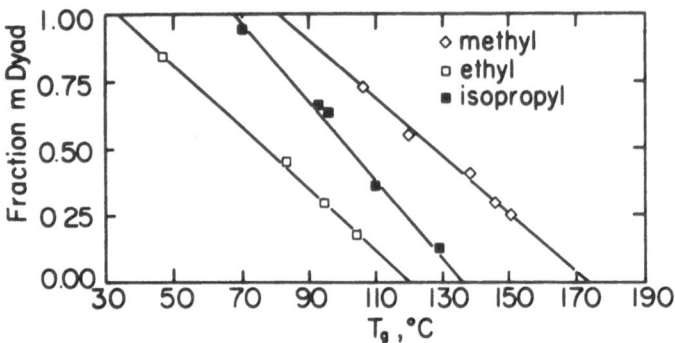

Dependence of Glass Transition
Temperatures (DSC) of Poly (Alkyl
α-Chloroacrylates) on Tacticity
(300 mHz NMR).

Figure 1

simultaneous solution of eq. (1) for various molecular weight
stereoregular poly(alkyl α chloroacrylates) gave K = 1.0 x 10^5
for the highly isotactic polymers and K = 3.0 x 10^5 for the
highly syndiotactic polymers (3). These values were then used
to determine the Tg_∞'s which are plotted in Figure 1. The observed dyad tacticity dependences of Tg are good straight lines
in all cases making it possible to extrapolate the observed values to the 100% syndiotactic case on the one hand and to the
100% isotactic case on the other. Despite the apparently
straightforward extrapolations displayed in Figure 1, the procedure may be criticized on several grounds. First, the highly
isotactic poly(alkyl α chloroacrylates) prepared by the use of
the modified Grignard reagent catalyst are extremely blocky,
while the "conventional" and highly syndiotactic polymers prepared by free radical initiators are all more or less random (5).
It is well known that sequence distributions can have a large
effect on Tg. Thus the type of relationships implied by Figure
1 are really only valid for a series of polymers in which the
configuration varies but the sequence distribution remains the
same. Second, Bywater and Toporowski (9) have shown that in the
case of the stereoregular poly(methylmethacrylates), Tg approaches
an asymptotically limiting value of 120° at high degrees of syndiotacticity. Previously, the assumption of a straight line relationship between Tg and configuration in poly(methylmethacrylate) resulted in an extrapolated value for the 100% syndiotactic
polymer of 160° (8). It is therefore apparent that the Tg values
for 100% isotactic and syndiotactic poly(alkyl α chloroacrylates)
obtained from Figure 1 must be only provisionally accepted pend-
ing further elucidation of stereosequence effects and behavior at

high degrees of syndiotacticity.

Tm's are also observable by DSC for the most highly isotac-
tactic poly(alkyl α chloroacrylates), and an intermediate transition labelled T_s is observable for isotactic poly(methyl α chloroacrylates) at 159°C. These data are summarized in Table II.

As discussed in detail in the previous chapter, the Tg difference between any pair of stereoisomers is given by

$$Tg(syndiotactic) - Tg(isotactic) = 0.59 \ \Delta\epsilon/k \qquad (2)$$

where Tg(syndiotactic) is the glass transition temperature of the
syndiotactic isomer, Tg(isotactic) is the glass transition temperature of the isotactic isomer, $\Delta\epsilon$ is the difference in the
flex energy between the syndiotactic isomer and the isotactic,
and k is Boltzsmann's constant. It was originally deduced that,
for the methacrylates,

$$Tg(syndiotactic) - Tg(isotactic) = 112° \qquad (3)$$

Subsequently the relationship was applied to stereoregular poly
(ethyl α chloroacrylates) (2) and was found to be

$$Tg(syndiotactic) - Tg(isotactic) = 90° \qquad (4)$$

In the present work, the values obtained are based on more complete NMR analyses of tacticity at higher frequencies and a much
greater range of stereoregular isomers. These results are for
the methyl ester

$$Tg(syndiotactic) - Tg(isotactic) = 92° \qquad (5)$$

for the ethyl ester

$$Tg(syndiotactic) - Tg(isotactic) = 80° \qquad (6)$$

for the isopropyl ester

$$Tg(syndiotactic) - Tg(isotactic) = 68° \qquad (7)$$

The Tg differences decrease with increasing ester side chain
length or bulkiness in contradiction to the simple theory. Also,
the values are all considerably smaller than that originally
quoted for the stereoregular poly(methylmethacrylates). The
value quoted in eq. (3) was based on an extrapolated Tg for the
100% syndiotactic poly(methylmethacrylate) of 160°. If Bywater's
value of 120°, (discussed above), is accepted we obtain an
amended figure of 80° for the Tg difference between the stereo-
isomers of poly(methyl methacrylate). In view of the uncertain-

Table II

Thermal Transitions of Stereoregular Poly(alkyl α chloroacrylates)

Polymer	Tacticity (m dyads)(a)	T_g(b)(°C.)	T_s(°C.)	T_m(°C.)
Poly(methyl α chloroacrylate)	0.70	90	152	186
"	0.29	152	--	--
Poly(ethyl α chloroacrylate)	0.73	52	--	109
"	0.20	104	--	--
Poly(isopropyl α chloroacrylate)	0.95	70	--	191
"	0.36	110	--	--

(a) Derived from 300 MHZ NMR (5)
(b) Corrected to "infinite" molecular weight (3)

ties in the extrapolation procedure, the difference between the methyl esters, eq. (5) and the ethyl esters, eq. (6) are probably not significant. However, the value for the isopropyl esters, eq. (7), is certainly lower. The theory embodied in eq. (2) rests on the validity of the Gibbs-DiMarzio approach to the glass transition. In addition it is assumed that Tg is an iso-free volume state and that inter- and intra-molecular effects on Tg are strictly separable. The experimental disagreement with the theory for the case of the isopropyl esters is probably mainly due to the breakdown of this latter assumption. That is, the isopropyl ester side group has a significant effect on the energetics of chain conformations as well as affecting the free volume. In spite of these considerations, it is clear that the simple theory does remarkably well in rationalizing configurational effects on the Tg's of both poly(alkyl methacrylates) and poly(alkyl α chloroacrylates). Its predictive value is certainly limited inasmuch as Δε cannot at present be calculated from a knowledge of molecular structure.

Turning to the crystallization behavior, it is perhaps surprising that any crystallization occurs at all in the isotactic poly(methyl α chloroacrylates) and poly(ethyl α chloroacrylates) which have only about 0.7 fractions of m dyads. The effect, however, is traceable to the blocky nature of those polymers prepared using the modified Grignard reagent catalyst and to the fact that there is considerable stereo inhomogeneity in these polymers as demonstrated by fractionation which results in portions of greatly enhanced stereoregularity (3). It should be pointed out that the observed enthalpies of fusion are small, of the order of 2 cal/g for both the methyl and ethyl polymers and of the order of 8 cal/g for the more highly isotactic isopropyl polymer. The enthalpies of fusion for the 100% crystalline poly (alkyl α chloroacrylates) are unknown, but if the reported value for isotactic poly(methylmethacrylate) of 22 cal/g (10) is used as a rough basis of comparison, it is apparent that the degrees of crystallinity are low.

A quantitative assessment of the weight fraction crystallinity is not possible in the absence of information about the theoretical enthalpies of fusion and the size of the crystals. The appropriate relationships are:

$$W_c = \frac{\Delta H \text{ observed}}{\Delta H \text{ theoretical } (\infty)} \tag{8}$$

where W_c is the weight fraction crystallinity, ΔH observed is the experimental enthalpy of fusion and ΔH theoretical (∞) is the enthalpy of fusion for the hypothetical 100% crystalline polymer. ΔH theoretical is obtained on the basis of several well established extrapolation procedures (11). Eq. (8) applies to the case where the crystal is "infinitely" thick. For these crystals, ΔH theoretical must be modified as follows

$$\Delta H_{theoretical} = \Delta H_{theoretical(\infty)} - 2\Delta H_e/\ell \qquad (9)$$

where $\Delta H_{theoretical(\infty)}$ refers to the "infinitely" thick perfect crystal, ΔH_e is the excess enthalpy associated with forming the basal surfaces of the lamellar crystal and ℓ is the crystal thickness. Even if the $\Delta H_{theoretical(\infty)}$ were known for the poly (alkyl α chloroacrylates), it is probable that the correction factor, $\Delta H_e/\ell$, would be of considerable importance, since the rather low degrees of stereoregularity would favor the formation of small crystals. Wide angle x-ray diffraction patterns of the unoriented materials reveal a very low degree of crystalline order, although it has been possible to obtain lattice parameters from such measurements on specimens oriented above their Tg's (3). A frequently quoted empirical observation is that the ratio (Tg/Tm) lies in the range of 0.50 to 0.75 for many polymers. If we evaluate this ratio for the isotactic poly(alkyl α chloroacrylates) studied we find values of 0.79, 0.85 and 0.74 for the methyl, ethyl, and isopropyl polymers respectively. Using literature values for isotactic poly(methylmethacrylate), a value of 0.74 is obtained (12). Thus, only poly(isopropyl α chloroacrylate) behaves in a "normal" fashion. It is well known that many exceptions to the empirical generalization exist (13). In the present case, the probable presence of small crystals would indicate depressed melting points as well as low enthalpies of fusion and this might account for the high ratios for the methyl and ethyl polymers.

The presence of a third transition between the Tg and Tm in the case of the isotactic poly(methyl α chloroacrylate) has been commented on previously and evidence has been presented for the assignment of this phenomenon to the glass transition of a stereocomplex of isotactic and syndiotactic polymer molecules (3). It is not possible to confirm or reject this hypothesis on the basis of available evidence.

The results of the mechanical and dielectric relaxation studies are embodied in Table III and Figs. (2), (3), and (4). An inspection of Figs. 2-4 reveals that all of the polymers show a prominent relaxation associated with the glass transition and labelled α in Table II. As expected, the temperatures and activation energies of these relaxations are strong functions of the steric structures of the polymers. The activation energies listed in Table III are obtained from the slopes of plots of the logarithm of the maximum frequency of the dielectric loss

TABLE III

Activation Energies of Relaxation Processes in Poly(alkyl α chloroacrylates)

Polymer	m Dyads	Temp. Range For Dielectric (D) & Mechanical (M) Loss Maxima (°C.)	Activation Energy[a] k cal/mole	Relaxation Process
Poly(methyl α chloroacrylate)	0.70	117-127 (D)	120 $\pm$5	α
"	=	103-111 (M)		
"	=	~180 (D)		α$_{sc}$
"	=	~160 (M)		
"	0.29	169-174 (D)	150 $\pm$9	α
"	=	154-161 (M)		
"	=	120-150 (D)		β
"	=	100-110 (M)		
"	0.25	176-185 (D)	154 $\pm$16	α
"		161-166 (M)		
"		125-160 (D)	38 $\pm$4	β
"		100-110 (M)		
Poly(ethyl α chloroacrylate)	0.73	92-119 (D)	44 $\pm$1	α
"		68-74 (M)		
"	0.29	126-142 (D)	85 $\pm$5	α
"		101-111 (M)		
"	0.29	~70 (M)		β

(cont'd.)

TABLE III (cont'd.)

Polymer	m Dyads	Temp. Range For Dielectric(D) & Mechanical(M) Loss Maxima (°C.)	Activation Energy[a] k cal/mole	Relaxation Process
Poly(ethyl α chloroacrylate)	0.20	132-146 (D)	99 +5	α
		109-117 (M)		
		~70 (M)		β
Poly(isopropyl α chloroacrylate)	0.36	128-148 (D)	63 +3	α
		106-117 (M)		

(a) From Dielectric Data only.

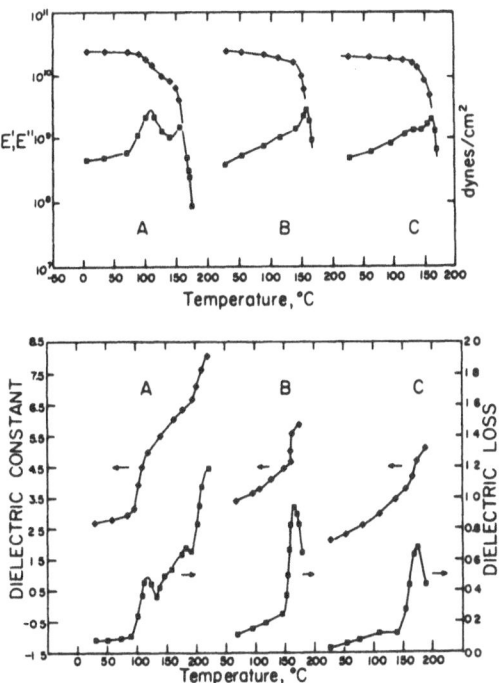

Figure 2: Upper temperature dependence of E' and E", at 200 Hz for (A) isotactic poly(methyl α-chloroacrylate), (B) atactic poly(methyl α-chloroacrylate), and (C) syndiotactic poly(methyl α-chloroacrylate). Lower: temperature dependence of the dielectric constant and loss at 200 Hz for (A) isotactic poly(methyl α-chloroacrylate), (B) atactic poly(methyl α-chloroacrylate), and (C) syndiotactic poly(methyl α-chloroacrylate).

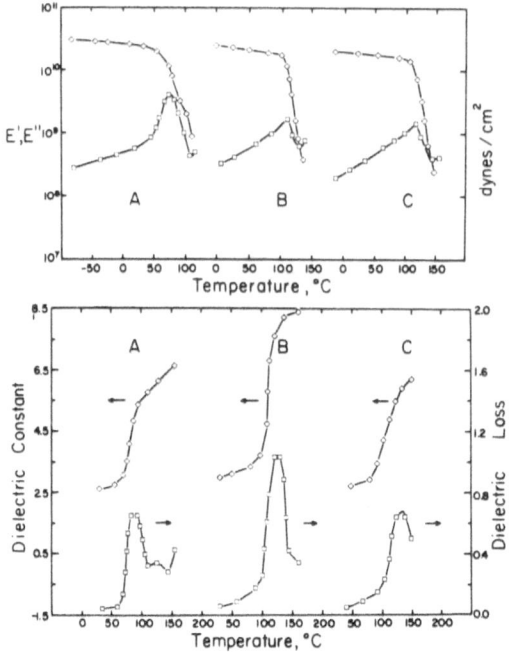

Figure 3:   Upper temperature dependence of E' and E" at 200 Hz
for (A) isotactic poly(ethyl α-chloroacrylate), (B) atactic
poly(ethyl α-chloroacrylate), and (C) syndiotactic poly(ethyl
α-chloroacrylate).   Lower: temperature dependence of the dielectric constant and loss at 200 Hz for (A) isotactic poly(ethyl
α-chloroacrylate), (B) atactic poly(ethyl α-chloroacrylate),
and (C) syndiotactic poly(ethyl α-chloroacrylate).

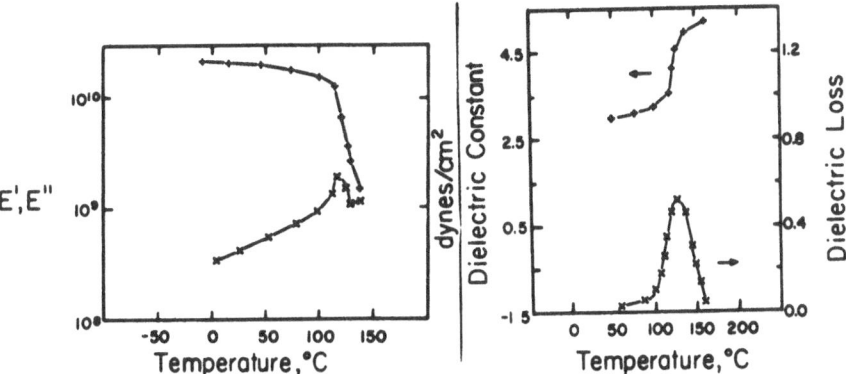

<u>Figure 4</u>: Left: temperature dependence of E' and E" at 200 Hz for syndiotactic poly(isopropyl α-chloroacrylate). Right: temperature dependence of the dielectric constant and loss at 200 Hz for syndiotactic poly(isopropyl α-chloroacrylate).

constant (log f_{max}) versus $1/T$. Inasmuch as only 3 frequencies are available mechanically (3.5, 11 and 110 H_z), it is difficult to derive reliable values from the mechanical results. It can be said, however, that the dielectric and mechanical α relaxations correlate well with one another and are consequences of the same molecular motions (i.e., microbrownian segmental motions accompanying Tg).

Activation energies for relaxation phenomena associated with the glass transition are commonly interpreted according to the WLF formalism (14)

$$\log a_T = \frac{C_1(T - Tg)}{C_2 + T - Tg} \tag{10}$$

where a_T is the "shift factor", and C_1 and C_2 are "universal" constants for all amorphous polymers.

a_T may be written as the ratio of the relaxation time at temperature T to that at Tg. If we further interpret the frequency of maximum loss in either the mechanical or dielectric relaxation curves as being equal to the reciprocal of the relaxation time a plot of the logarithm of the maximum loss frequency versus reciprocal temperature will yield a temperature dependent activation energy according to eq. (10) as follows

$$E_a = (2.303 \, C_1 C_2 \, RT^2)/(C_2 + T - Tg)^2 \tag{11}$$

At Tg, eq. (10) reduces to

$$E_a = 2.303(C_1/C_2)R \; Tg^2 \tag{12}$$

In eqn's. (11) and (12) E_a is the activation energy and R is the gas constant. The preceding analysis is only approximate because it is based on a number of assumptions which are more or less unrealistic. It is beyond the scope of this treatment to discuss these points in detail, but standard treatises such as that of Ferry (15) may be consulted. Although the actual plots of log f_{max} vs $1/T$ from which the data in Table III are derived are not shown, they were all straight lines over the frequency and temperature range studied. This is because of the limited ranges available. Presumably, if a greater frequency and temperature range had been examined the plots would have exhibited curvature in at least qualitative accord with eq. (11). It may be further noted, that according to eq. (12), the activation energies for the α relaxation process at Tg should be in the ratio

$$E_a(syndio)/E_a(iso) = [Tg(syndio)/Tg(iso)]^2 \tag{13}$$

In eq. (12) E_a(syndio) is the activation energy for the α process in the syndiotactic isomer and E_a(iso) is the corresponding activation energy for the isotactic isomer. The Tg vales refer to the calorimetric Tg's for the appropriate stereoisomers. A comparison of the observed activation energy ratio with those calculated from eq. (13) for the various esters yields, for the methyl esters

$$(E_a \; syndio/E_a \; iso) \; observed \; = \; 1.3$$
$$(E_a \; syndio/E_a \; iso) \; calculated = 1.2 \tag{14}$$

For the ethyl esters

$$(E_a \; syndio/E_a \; iso) \; observed = 2.3$$
$$(E_a \; syndio/E_a \; iso) \; calculated = 1.4 \tag{15}$$

In view of the experimental uncertainties and the approximations inherent in the analysis, it is apparent that the simple treatment embodied in eqn's. (10)-(13) accounts for the observed activation energy trends in the stereoisomers studied.

A secondary relaxation occurring in the glassy state had previously been reported for "conventional" poly(methyl α chloroacrylates) prepared by free radical techniques (16). This β relaxation occurs also in the poly(alkylmethacrylates) and is assigned to motions of the ester side group about the carbon-carbon bond joining it to the main chain. It is discernible only in the highly syndiotactic methyl and ethyl polymers and is absent in the syndiotactic isopropyl polymers. It is a minor feature throughout and is not observable dielectrically for the syndiotactic ethyl polymer. The dielectric behavior of the β relaxation in the poly(alkyl α chloroacrylates) is in marked contrast to its behavior in the poly(alkylmethacrylates). In the latter case, "conventional" or syndiotactic poly(methylmethacrylate) has a dielectric β relaxation considerably greater in magnitude than the α relaxation in the same polymer, while the situation is reversed in isotactic poly(methylmethacrylate) (17). This observation is interpretable on the basis that the side chain relaxation occurs independently of the main chain relaxation in the syndiotactic poly(methylmethacrylates) but is correlated with the main chain relaxation in the isotactic poly(methylmethacrylates). The greater dipole moment of the ester side chain compared to the α methyl group then accounts for the observed dielectric relaxation magnitudes. In the poly(alkyl α chloroacrylates), the polar α Cl group introduces a main chain dipole moment of much greater magnitude than that of the ester side chain, and so the α relaxation is the predominant dielectric feature regardless of the tacticity of the polymer. The absence of the β relaxation in the isotactic poly(alkyl α chloroacrylates) does suggest that the ester side chain motion may merge with that of the main chain in these polymers in a similar manner to what is postulated to occur in the isotactic poly(methylmethacrylates).

The relaxation labelled α_{sc} for the isotactic poly(methyl α chloroacrylate) in Table II is thought to correlate with the transition observed by DSC at 150° and is tentatively assigned to motions accompanying the glass transition of the "stereocomplex" postulated to occur in these polymers, composed of isotactic and syndiotactic chains.

The only previous report of activation energies for poly (methyl α chloroacrylates) is for the "conventional" free radical polymer which can be assumed to be reasonably syndiotactic (16). The values quoted are 130 kcal/mole for the α relaxation and 26 kcal/mole for the β relaxation. The present results (Table II) are in qualitative agreement with these values. In general the activation energies for the α relaxations decrease with increasing ester side chain length or bulkiness and the isotactic isomers have α relaxation activation energies about 35-50 kcal/mole lower than the comparable syndiotactic isomers. This effect, as already discussed, is a consequence of the Tg difference between the isomers.

CONCLUSIONS

(1) The Tg's of stereoregular poly(alkyl α chloroacrylates) are strong functions of tacticity and can be rationalized, to a first approximation, on the basis of the previously proposed theory.

(2) Evidence exists for the presence of a stereocomplex composed of a combination of isotactic and syndiotactic poly (methyl α chloroacrylate).

(3) Crystallinity can be observed for the highly isotactic poly(alkyl α chloroacrylates).

(4) The relaxation behavior accompanying the glass transitions (α relaxation) in stereoregular poly(alkyl α chloroacrylates) is generally in accord with previous results on atactic polymers. The activation energies for the α relaxation process in the various stereoisomers can be accounted for on the basis of the WLF treatment.

(5) The secondary or β relaxation, assigned to motions of the ester side group is a minor feature dielectrically in contrast to the behavior of the poly(alkylmethacrylates).

REFERENCES

1. F.E. Karasz and W.J. MacKnight, Macromolecules, 1, 537 (1968).
2. B. Wesslen, R.W. Lenz, W.J. MacKnight and F.E. Karasz, ibid., 4, 24 (1971).
3. G. Dever, R.W. Lenz, F.E. Karasz and W.J. MacKnight, J. Polym. Sci., Chem. Ed., 13, 2151 (1975).
4. G. Dever, F.E. Karasz, W.J. MacKnight and R.W. Lenz, Macromolecules, 8, 439 (1975).
5. G. Dever, F.E. Karasz, W.J. MacKnight and R.W. Lenz, J. Polym. Sci., Chem. Ed., 13, 1803 (1975).
6. B. Wesslen and R.W. Lenz, Macromolecules, 4, 20 (1971).
7. D.S. Breslow and A. Kutner, J. Polym. Sci., B9, 129 (1971).
8. E.V. Thompson, J. Polym. Sci., A-2, 4, 199 (1966).
9. S. Bywater and P.M. Toporowski, Polymer, 13, 94 (1972).
10. J.M. O'Reilly, F.E. Karasz and H.E. Bair, Bull. American Phys. Soc., 9, 285 (1964).
11. B. Wunderlich and K.M. Cormier, J. Polym. Sci., A2 5, 987 (1967).
12. O.G. Lewis, "Physical Constants of Linear Homopolymers", Springer Verlag, New York, 1968, p. 106.
13. W.A. Lee, Brit. Polym. J., 5, 3762 (1971).

14. M.L. Williams, R.F. Landel and J.D. Ferry, J. Am. Chem. Soc., 77, 3701 (1955).
15. "Viscoelastic Properties of Polymers, 2nd. Edition", by John D. Ferry, Wiley, New York, 1970, Chapter 11.
16. K. Deutsch, E.A.W. Hoff and W. Reddish, J. Polym. Sci., 13, 565 (1954).
17. H. Shindo, I. Murakami and H. Yamamura, J. Polym. Sci., A-1, 1, 297 (1969).

EFFECT OF STEREOREGULARITY ON BULK PROPERTIES OF POLYMERS

Frank E. Karasz

Materials Research Laboratory
Polymer Science and Engineering Department
University of Massachusetts
Amherst, Massachusetts 01003

I.   INTRODUCTION

Information on the effect of stereoregularity on transitions and thermodynamic properties of bulk macromolecular systems has been developed, largely on an empirical basis, as methods for synthesizing and characterizing polymers of differing
configurations became available. In this report we shall consider (a) generalizations that are available in this context,
and, in particular, (b) the effect of stereoregularity on glass
transition phenomena together with a discussion of specific systems.

II.   TRANSITIONS: General Comments

The most obvious effect of stereoregularity in polymers con-
cerns their crystallizability and hence the potential existence
of a melting transition with a concomitant heat of fusion. It
is generally accepted that with one or two possible exceptions
(eg. polyvinyl alcohol), substantial stereoregularity is required to permit crystallization in vinyl polymers and, indeed,
the appearance of a melting transition is sometimes adopted as a
crude assay of synthetic procedures. It should be noted, of
course, that the absence of a fusion transition cannot in itself
be taken as evidence for atacticity, since thermal and solvent
exposure history strongly influence the extent of crystallization
in isotactic and syndiotactic systems. For example, isotactic
polystyrene ($T_m = 240^\circ C$) can be bulk annealed to yield material
with a degree of crystallinity (x_c) between 0.4 and 0.5, and

449

R. W. Lenz and F. Ciardelli (eds.), Preparation and Properties of Stereoregular Polymers, 449–457.
Copyright © 1979 by D. Reidel Publishing Company.

yet can be readily quenched to produce totally amorphous polymers with bulk properties indistinguishable in most respects
from those of the amorphous atactic polymer.

Only scattered comparisons of the melting transition properties of a syndiotactic vis-à-vis an isotactic polymer exist.
In polymethyl methacrylate (PMMA) for example, bulk forms can be
crystallized (T_m for iso-PMMA is $160^{\circ}C$) but neither the extents
of crystallization nor the ΔH_f's, the heats of fusion of the respective crystals, is known. A similar situation exists with respect to the corresponding volume parameters. Systematic studies of ΔH_f as a function of tacticity would be of value as an indication of the capability of the lattice to accommodate configurational impurities.

In the case of the glass transition, more complete experimental data are available, at least with respect to vinyl polymers. As will be discussed below, for certain types of structure a substantial effect of stereoregularity on the glass transition temperature, T_g, is found, and in addition a theoretical
treatment is available which accounts for these effects. However, very little data is available with respect to the effect
of stereoregularity on, for example, the configurational entropy
S_c at T_g, on ΔC_p, the size of the heat capacity increment at T_g,
or on other key thermodynamic parameters of the glassy state.

In the PMMA's and to some extent in the polystyrenes there
is some evidence for small density variations (other than those
arising from crystallinity) with tacticity (1). These may extend to the molten state.

Relaxations below the glass transition have received wide-
spread study, using principally dynamical mechanical or dielectric relaxation techniques (2). Some effects of tacticity have
been noted but will not be discussed here. The effect of configuration on the properties of the quasi-isolated macromolecule,
i.e. in dilute solution under θ-condition, is also of great interest but is outside the scope of this report.

III. TRANSITIONS: Specific Examples

Detailed heat capacity studies of atactic and isotactic
polystyrene (PS) can be used to illustrate the above discussion
(3). In crystalline iso-PS such studies (Fig. 1) reveal two
well defined transitions - a melting transition ($T_m = 240^{\circ}C$,
heat of fusion, $\Delta Q_f = 34$ joules/g) and a glass transition ($T_g =
95^{\circ}C$, $\Delta C_p \sim 0.30$ j/g). After quenching, the transition prop-
erties of the amorphous i-PS (T_g, ΔC_p at T_g) are virtually identical to that of a-PS. In addition, the heat capacities (C_p)

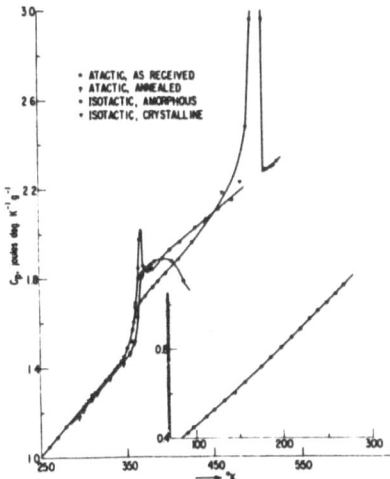

Figure 1: Heat capacity of polystyrene.

themselves seem independent of tacticity both below and above T_g.

These facts suggest that the configuration entropy of the two stereo-isomers in the melt may be equal. This assumption permits calculation of S_c of the glass from the experimental data at any temperature below T_m using

$$S_c(T) = \{S_i(T_m) - S_i(T)\} - \{S_a(T_m) - S_a(T)\}$$

where S_i and S_a refer to the experimentally determined entropies of the isotactic and atactic isomers (referred to an arbitrary zero). Fig. 2 shows $S_c(T)$ vs. T below T_m. The dashed line is a calculated curve based on extrapolated C_p^m values for the amorphous and semicrystalline polymers below T_g. The intercept, representing the point at which the residual entropy of the glass and crystal would be equal, has some significance in glass transition theory. The Gibbs-DiMarzio treatment, for example, predicts the existence of such a temperature (the so-called T_2) at about 50°C below the experimental T_g. The PS data, in which $T_g - T_2 \sim 80°C$, is in fair agreement (4).

The calculation of $S_c(T)$, above, depends, ultimately, on the correctness of the assumption regarding the equality of the absolute entropies of the atactic and isotactic forms in the melt. An absolute test of this hypothesis is not possible; however, consideration of the configurational enthalpy, $H_c(T)$ offers further possibilities through heat of combustion measurements. The latter provides a common reference state and for PS a direct comparison is possible in that amorphous forms of both isomers can be obtained. The results for PS indicate that the H_c's at ambient

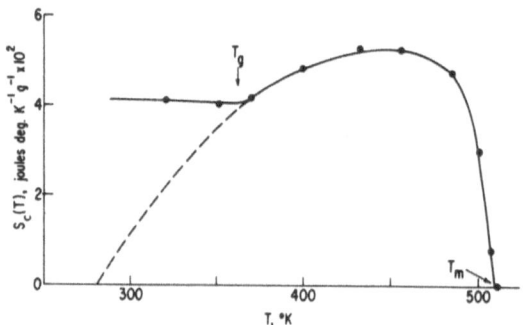

Figure 2:   Configurational entropy of polystyrene.

temperatures for atactic and isotactic polymers are indeed identical with experimental error (5).

In polymethyl methacrylate (PMMA), heat capacity studies
confirm that T_g is a substantial function of tacticity (T_g = 45°,
100°, 115°C for the isotactic, atactic and syndiotactic forms respectively) and that ΔC_p at T_g and C_p itself are different for
the stereo-isomers, (Fig. 3,4) (6).

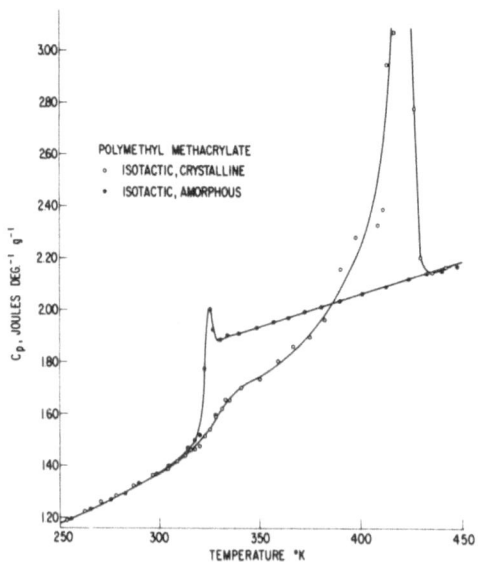

Figure 3:   Heat capacity of isotactic polymethyl methacrylate.

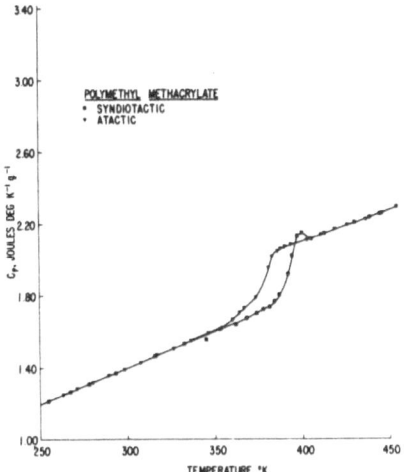

Figure 4:   Heat capacity of syndiotactic and atactic polymethyl methacrylate.

The densities of the stereo-isomers are also known to differ slightly around ambient temperatures. In view of these results, it seems unlikely that configurational entropy in this polymer could be independent of tacticity (2).

The effect of tacticity has not been specifically investigated in the case of poly(4-methyl-pentene-1). A peculiar feature of this polymer is that the reported specific volumes of the amorphous and semicrystalline forms are virtually equal at the T_g (~25°C) (7). An explanation for this was provided indirectly from C_p studies of the crystalline polymer (8). The configurational entropy at T_g, calculated on the assumption that the entropies of the crystal above T_m and that of the non-crystallizable polymers at the same temperature are identical, yields a value which is much lower than that normally found, and which is inconsistent with the predictions of theory (Fig. 5). The conclusion is that one cannot in this case equate the melt entropies of the atactic (i.e. amorphous) and the isotactic (crystalline) stereo-isomers. This is borne out, to some extent, by the finding that the melt volumes of the amorphous and crystalline fractions also differ by about 1% (7). It seems therefore that the peculiar room temperature volumetric behavior is merely a reflection of the difference in stereo-isomeric properties.

We have provided examples of three extreme cases of the effect of stereo-regularity on thermodynamic properties. In PS, tacticity seemingly has no effect on T_g nor on any other property of the amorphous phases of the respective isomers. In contrast, the T_g's as well as many of the other thermodynamic

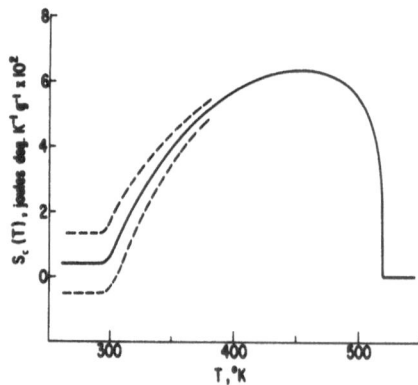

Figure 5: Configurational entropy of poly(4-methyl-1-pentene).
The dashed lines represent the effect a ± 10% error
in the assumed ΔC_p for the amorphous phase at T_g.

parameters of the stereo-isomers of PMMA display substantial
differences. Finally, in P4MP, the T_g appears to be independent
of tacticity, but configurational entropy and specific volume at
least, appear to be dependent on this structural parameter.

IV. EFFECT OF STEREOREGULARITY ON GLASS TRANSITIONS IN VINYL POLYMERS

Certain regularities have been observed in the T_g's of substituted vinyl polymers of the type $(CH_2CXY)_n$ (9). Thus, if X=H
(i.e. monosubstituted vinyl polymers) there is no effect of tacticity on the T_g's of the respective polymers. Examples are
polystyrene (Y = -ϕ), polypropylene (-CH_3), alkyl acrylates
(-COOR) and in the longer branched α-olefins (-$(CH_2)_nCH_3$). In
contrast, in unsymmetrically disubstituted vinyl polymers there
is a large effect of tacticity on T_g. The best documented examples are in the methacrylate polymers (X = -CH_3; Y = -COOR)
and in the α-methyl styrenes (X - -CH_3; Y = -ϕ).[3] In the latter
polymers, the T_g of the syndiotactic is invariably substantially
above that of the isotactic isomer. This effect must ultimately
be related to the conformational properties of the macromolecular chains and as a first approximation may be regarded as an
intra-molecular effect.

A second generalization may be noted with respect to the
effect of side group chain length of T_g's. Here, the effect of
an increasingly long side chain is always to lower the T_g's
within a given polymer series, irrespective of tacticity.[9] The
best documented examples of this again are to be found in the

alkyl acrylate and methacrylate series; in the latter case, data
is available for both the atactic and isotactic cases. These
effects are quantitatively shown in Table 1. Since the effect
is independent of configuration, it must be primarily an intermolecular one.

Table 1
Effect of Tacticity and of Side-Chain Length on T_g's
($^\circ$C) in the Polymethacrylate and Polyacrylate Series

	Methacrylates		Acrylates	
R	Conven-tional[a]	Iso-tactic[c]	Conven-tional[c]	Iso-tactic
Methyl	105	43	8	10
Propyl	35		-44[b]	
Isopropyl	81[c]	27	-6	-11
Butyl	20	-24	-49[b]	
Isobutyl	53[c]	8	-24	
sec-Butyl	60[b]		-22	-23
Cyclohexyl	104[b]	51	19	12

(a) S.S. Rogers and L. Mandelkern, J. Phys. Chem., 61, 985 (1957)
unless otherwise noted.
(b) O.G. Lewis, "Physical Constants of Linear Homopolymers",
Springer-Verlag, New York, N.Y., 1968.
(c) J.A. Shetter, J. Polym. Sci., Part B, 1, 209 (1963).

There appears to be no well founded exceptions to these empirical rules. The effect of tacticity on the T_g of polyvinyl
chloride has been a matter of long standing controversy, but
demonstrations of any differences (which would compromise the
first rule expounded above) appear to be complicated by the concomitant effects of branching and crosslinking related to the
differing synthetic procedures. The increases that have been observed in T_g for vinyl polymers with very long side groups reflect the onset of independent crystallization of these groups.

A basis for the effects described above can be developed in
terms of the Gibbs-DiMarzio theory of the glass transition, using
the following two postulates (4).

1. The tacticity effect arises from the fact that, when
neither substituent is hydrogen, the energy difference between
the two predominant rotational isomers is greater for the syndiotactic configuration than for the isotactic configuration. However, when one of the substituents is hydrogen, the energy difference between the rotamers of the two configurations is the same.

2. The effect of increasing the ester side chain length in the poly(alkyl methacrylates) and poly(alkyl acrylates) is to modify intermolecular interactions, and to leave the intramolecular interactions unchanged.

The T_g's for the "conventional" methylacrylates shown in Table 1 of course are lower than those of the pure syndiotactic isomers which in most cases have not been prepared. Various extrapolating schemes suggest that pure syndiotactic PMMA has a T_g in the vicinity of 160°C.

From the Gibbs-DiMarzio theory it can be shown that

$$T_g \Delta\alpha = \frac{V_0 \ln[(1+V_0)^2/4V_0]}{[2(1-V_0)/(1+V_0)]-[4V_0/(1+V_0)] \ln[(1+V_0)^2/4V_0]-[1-V_0]} \tag{1}$$

Where V_0 is the fractional unoccupied volume at T_g, (or actually at T_2) and $\Delta\alpha$ is the change in expansion coefficient at T_g. $T_g\Delta\alpha$ is seen to be a function of V_0 only, and since, according to quite well established observations of Simha and Boyer this product is a constant for all polymers (10), eqn. (1) implies an iso-free volume glass transition state. In fact, using the empirical Simha-Boyer constant of 0.113 for $T_g\Delta\alpha$, it is found that $V_0 = 0.025$ at T_g, identical with the free volume estimated from the WLF equation (11). The hole formation energy, E_0, in the Gibbs-DiMarzio treatment is a function of T_g alone and is directly proportional to the latter, as is seen in eqn. (2).

$$E_0 = \frac{kT_g}{2} [(1 + V_0)/(1 - V_0)]^2 \ln [(1 + V_0)^2/4V_0] \tag{2}$$

The flex energy, ε, is similarly dependent also on V_0

$$-\ln [(1 + V_0)/2] + [V_0/(V_0 - 1)]\ln[1 + V_0)^2/4V_0] =$$

$$\ln [1 + 2 \exp(-\varepsilon/Kt_g)] + \frac{(2\varepsilon/kT_g)\exp(-\varepsilon/kT_g)}{1 + 2 \exp(-\varepsilon/kT_g)} \tag{3}$$

and using the above numerical results, it can be shown that $\varepsilon/T_g = 1.70$ for any polymer obeying the "universal" relations.

If this result is applied to PMMA we find that the difference in flex energies, $\Delta\varepsilon$, between the syndiotactic and isotactic isomers of any polyalkyl methacrylate configurational pair is 191R or about 380 cals/mole, and, it is predicted the T_g difference for any member of the series should be 112°. This is based on the assumption, it must be emphasized, that the configurational effect is an intramolecular effect.

The above analysis and correlation has now been applied to other disubstituted acrylate polymers, as is discussed by W.J. MacKnight elsewhere in this volume and largely corroborated (12).

References

1. T.G. Fox et al., J. Am. Chem. Soc., 80, 1768 (1958).
2. N.G. McCrum, B.E. Read and G. Williams, Anelastic and Dielectric Effects in Polymeric Solids, Wiley, London, 1967.
3. F.E. Karasz, H.E. Bair and J.M. O'Reilly, J. Phys. Chem., 69 2657 (1965).
4. J.H. Gibbs and E.A. DiMarzio, J. Chem. Phys., 28, 373 (1958).
5. R.M. Joshi, B.J. Zwolinski, J.M. O'Reilly and F.E. Karasz, J. Polym. Sci., A-2, 5, 705 (1967).
6. J.M. O'Reilly, F.E. Karasz and H.E. Bair, Bull. Am. Phys. Soc., 9, 285 (1964).
7. B.G. Ranby, K.S. Chan and H. Brumberger, J. Polym. Sci., 58, 545 (1962).
8. F.E. Karasz, H.E. Bair and J.M. O'Reilly, Polymer, 8, 547 (1967).
9. F.E. Karasz and W.J. MacKnight, Macromolecules, 1, 537 (1968).
10. R. Simha and R.F. Boyer, J. Chem. Phys., 37, 1003 (1962).
11. M.L. Williams, R.F. Landel and J.D. Ferry, J. Am. Chem. Soc., 77, 3701 (1955).
12. B. Wesslen, R.W. Lenz, W.J. MacKnight and F.E. Karasz, Macromolecules, 4, 24 (1971).

INDEX OF SUBJECTS